T0181451

Mobilität – Innovation – Transformation

Reihe herausgegeben von

Wolfgang H. Schulz, Zeppelin Universität, Friedrichshafen, Deutschland

Oliver Franck, Zeppelin Universität, Friedrichshafen, Deutschland

Stanley Smolka, Zeppelin Universität, Friedrichshafen, Deutschland

Mobilität ist eine wesentliche Voraussetzung für die internationale Arbeitsteilung, das Wachstum von Volkswirtschaften, der Möglichkeit Einkommen zu erzielen sowie der Teilhabe am sozialen und kulturellen Leben. Dabei sind die Anforderungen an die Mobilität von Nachhaltigkeit über Klimaneutralität vielfältig. Um diese Herausforderungen zu bewältigen sind Innovationen und Transformation wesentliche Treiber, damit die Mobilitätsindustrie sich schnell an neue Bedingungen anpassen kann. Diese Schriftenreihe soll mit herausragenden Wissensbeiträgen dieses Forschungsdreieck aus Mobilität, Innovation und Transformation stärken. Herausgeberwerke sollen aktuelle Diskussionen vorantreiben, Doktorarbeiten neueste Forschungsergebnisse präsentieren und Lehrbücher die gewonnenen Erkenntnisse in die Bildung integrieren.

Leander Kauschke

The Transition to Smart Mobility

Acceptance and Roles in Future Transportation

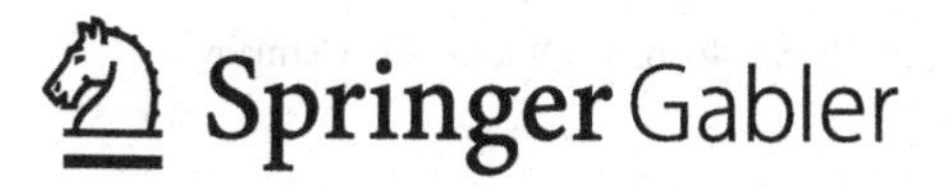

Leander Kauschke
Burg (bei Magdeburg), Germany

This work was written as a dissertation at Zeppelin University, Friedrichshafen, Germany. It was supervised by Wolfgang H. Schulz and Horst Wieker and disputed on 2023-04-18.

Mobilität – Innovation – Transformation
ISBN 978-3-658-43000-9 ISBN 978-3-658-43001-6 (eBook)
https://doi.org/10.1007/978-3-658-43001-6

This Springer Gabler imprint is published by the registered company Springer Fachmedien Wiesbaden GmbH, part of Springer Nature.
The registered company address is: Abraham-Lincoln-Str. 46, 65189 Wiesbaden, Germany

Paper in this product is recyclable.

- for father -

Preface

I was fortunate to start my Ph.D. within a unique setup between an applied and an academic university, a situation all parties benefited from in the end. It is generally very rewarding to work in multiple environments with so many different skilled people and ideas.

Starting with project-related IT-engineering work in a full-time job in 2016, socio-economic models, statistics, and survey evaluation became a kind of passion for me. As mobility technology progressed incredibly fast and the research was interdisciplinary, I invested a lot of energy in the thesis to be able to publish it quickly. Consequently, the empirical part went comparably smoothly, although my two children were born in 2019 and 2021, which probably made me lose some good sleep.

I want to thank everyone who supported me on the way, especially my partner Laura, who helped me not to lose track during the Corona crisis from 2020 to 2022. Many thanks also to all reviewers, without whom the work would have been much more cumbersome.

Now I am curious to hear from you, the reader, whether you find this dissertation and its findings insightful as well as enjoyable to read. I personally hope I will be able to continue working on a smarter and more sustainable mobility in the future.

Leander Kauschke

Abstract

Most contemporary notions of the fairly recent concept 'smart mobility' portray an imminent transition of similar socio-economic consequences as the shift from horses to automobiles did 100 years ago. In this context, the present dissertation offers an in-depth look at the variables involved in the equation of smart mobility acceptance (1) and envisaged institutional change (2). Both views are embedded in the transition framework of the multi-level perspective (MLP).

Study 1 is user-orientated. It develops an UTAUT2 from a comprehensive literature review. Three scenarios validate the structural equation model (SEM) in SmartPLS. Use cases cover eBikes (N = 537), Mobility-as-a-Service (N = 531), and fully automated vehicles (N = 558). Based on the results, acceptance primarily relies on five factors: performance expectancy, facilitating conditions, social influence, habit, and hedonic motivation. Further findings and implications are discussed, both for theory and practice.

Study 2 applies the Institutional Role Model (IRM) to mobility transitions, prior research having highlighted the urgency to better understand system dynamics. Thus, this innovative approach enables the smart mobility ecosystem to be mapped in a structured manner. Nine institutions, four technical roles, and four economic roles are identified. Results of interviews with experts (N = 8) validate and specify the model. The IRM's significance as a vision provider for the mobility regime is elicited.

In the end, these two perspectives amalgamate. Following philosopher Juergen Habermas this opens up a more sophisticated space for public debate about the commencing transition to smart mobility.

Keywords: Smart mobility · Acceptance · Transition · SmartPLS · MLP · UTAUT2 · IRM

Contents

Abbreviations

AIC	Akaike information criterion
ANOVA	Analysis of variance
ANT	Actor network theory
ARTS	Automated road transport shuttle
AV	Automated vehicle
AVE	Average variance extracted
B2B	Business-to-business
BEV	Battery electric vehicles
BI (Intent)	Behavioral intention to use
CB	Covariance-based
CCA	Confirmatory composite analysis
CE (CollEff)	Collective efficacy
CFA	Confirmative factor analysis
C-ITS	Cooperative intelligent transport systems
CMV	Common method variance
DV	Dependent variable
EB (eBike)	Electric bicycle
EE (Effort)	Effort expectancy
EFA	Exploratory factor analysis
EoT	Economy-of-things
EPM	Ecosystem pie model
EV	Electric vehicle
EXP	Experience
FAV	Fully automated vehicle
FC (FaCon)	Facilitating conditions
FCEV	Fuel cell electric vehicles

GoF	Goodness-of-fit
HAM	Habitual momentum
HE (Habit)	Habit (expectancy)
HM (Hedo)	Hedonic motivation
HMI	Human-machine-interface
HTMT	Heterotrait-monotrait ratio of correlations
ICT	Information and communication technology
IoS	Internet-of-services
IoT	Internet-of-things
IPMA	Importance-performance map analysis
IRM	Institutional role model
IS	Information systems
IT	Information technology
ITS	Intelligent transport systems
IV	Independent variable
LSD	Least significant difference
M	Mean
MA (MaaS)	Mobility-as-a-Service
MANOVA	Multivariate analysis of variance
MDN	Median
MGA	Multi-group analyses
MICOM	Measurement invariance of composite models
MIS	Management information systems
MLP	Multi-level perspective
MMTI	Multimodal traveler information
MP	Motile pleasure
NAM	Norm activation model
OLS	Ordinary least square
PE (Perform)	Performance expectancy
PEOU	Perceived ease of use
PLS	Partial least squares
PR (Risk)	Perceived risk
PU	Perceived usefulness
PV (Price)	Price value
RMSEA	Root mean square error of approximation
SD	Standard deviation
SEM	Structural equation modeling
SI (Social)	Social influence
SM	Smart mobility

SMC	Sociomaterial consumerability
SME	Small and medium enterprises
SO	Sociality
SRMR	Standardized root mean square residual
TAM	Technology acceptance model
TLI	Tucker-Lewis index
TPB	Theory of planned behavior
TRA	Theory of reasoned action
TTF	Task-technology-fit model
UB	Use behavior
UTAUT	Unified theory for the acceptance and use of technology
UX	User experience
VIF	Variance inflation factor

List of Figures

List of Tables

Introduction 1

At this moment in time, the world of transportation is experiencing a variety of transitional movements, ranging from rapidly improving technology to an overall shift from transportation products to mobility services. A century of automobility is coming to an end (Docherty et al., 2018).

A major factor driving this transformation is climate change. In the face of environmental exploitation and dwindling resources, economies are struggling to implement activities to combat global warming while simultaneously sustaining efficient, socially equitable, and economically competitive (National People's Congress, 2014; The White House, 2016; European Commission, 2019). Many hopes are thus pinned upon new information technologies (IT) to master these challenges (Seele and Lock, 2017). Whereas in some areas of human existence, corresponding conceptions are already in place (e.g., smart home) or are being developed in an integrated manner (e.g., smart grid), progress in the transport sector is often scattered and solutions remain proprietary (Wieker et al., 2014a; Wieker, 2016). Nonetheless, the techno-economic advances in mobility and logistics are numerous and significant. Recent discussions have revealed four key trend areas: electrification, automation, sharing, and ubiquitous connection through IT. That said, the successes of these developments in terms of innovation diffusion have often fallen short of initial expectations due to certain systemic hinderances and societal resistance (e.g. König and Neumayr, 2017; Kanger et al., 2019). Moreover, there is hardly any effect on CO_2 reduction at all (European Parliament, 2019). Over the past few years, numerous research disciplines have improved our understanding of challenges in mobility transitions. Five prominent schools of thought to cause a change in the mobility system are stylized as follows:

L. Kauschke, *The Transition to Smart Mobility*, Mobilität – Innovation – Transformation, https://doi.org/10.1007/978-3-658-43001-6_1

(1) *Engineers* focus on technology. They relate environmental problems to a lack of efficiency and are certain that research and development will be able to solve problems (Lyons, 2011). They commit themselves to future technology such as clean drive trains, material efficiency, and fully closed material loops.
(2) *Social ecologists* advocate a cultural shift toward approaches that embody green values. They criticize modernism, hypermobility, and anthropocentrism (Devall, 1991). This translates into a desire for locality, deceleration, and a conscious reduction of mobility.
(3) *Business scientists* regard alternative forms of transport as challenging for business model development (van den Heuvel et al., 2020). They pursue customer groups and the potential to create value for investors. Regarding mobility change, business science provides hands-on evidence of best practices in future transport solutions.
(4) *Neo-classical economists* understand environmental externalities as market failure. They hence endorse the idea of marginal social cost pricing. This concept entails internalizing external costs elicited by transport through measures such as missions trading or CO_2 taxation. Such policies will potentially encourage companies to find cost-effective solutions and thus limit the adverse effects of transportation to a minimum (Maibach et al., 2008).
(5) *Psychologists* promote behavior change by targeting specific beliefs (Fishbein and Ajzen, 2005). They try to understand how humans make their decisions and derive recommendations from this knowledge on how concrete factors of human perception can be shifted away from detrimental behavior. Interventions may work on a personal or collective level. In sum of all human decisions, a societal change is assumed.

Other disciplines discussing the issue include cultural geography (Urry, 2004; Barr, 2018) and political sciences (Shaw and Docherty, 2014; Nikolaeva et al., 2019). They provide evidence on the efficacy of policy interventions and recommend different pathways of transport governance.

Ultimately, all these framings consistently deliver well thought out solutions for the road to future mobility. However, what restrains our understanding is that each of these disciplines relies on a limited set of dimensions and that relating these different steams of research to one another is overcomplicated.

1.1 The Multi-Level Perspective

The need for an interdisciplinary approach is the reason why system innovationist Frank Geels developed the Multi-Level Perspective (MLP), which thinks in terms of co-evolution of different socio-technical systems and accepts interactions between technology, market, policy, culture, and society. The MLP originates from sociology, institutional theory, and innovation studies. It enables the analysis of shifts from one socio-technical system to another (Geels, 2004). It achieves this by conceptualizing three levels of structuration. According to Giddens (1984), higher structuration is associated with a tendency to move beyond the control of the individual. As depicted in Figure 1.1, the model frames 'the big picture'. Its premise is that transitions are non-linear processes interplaying at three analytical levels:

(1) The first level is *niche-innovations.* This is where novelties emerge in protected domains such as R&D laboratories, demonstration projects, or real-world field trials. Actors of niche technology form smaller networks and are open to radical innovation. Niche technology can diffuse to the next highest level once an idea gains momentum and stabilizes in a market-dominating design (Geels, 2004). Niches are still essentially free of market mechanisms.
(2) From the start, the dynamics in these niches are not solely shaped by general societal developments but also by existing regimes in their field of application, i.e. by the given socio-economic, technological, and institutional structures, market constellations, and societal problems to be solved in a specific sector, for example. This second sphere is thus referred to as the *socio-technical regime.* Well-rehearsed regime structures can provide stability and certainty of expectation over long periods. However, they might also induce path dependencies and resistance to necessary change.
(3) The third level of the MLP is the concept of *socio-technical landscape.* Its structuration is so strong that actors cannot change it at one's discretion. As well as including norms, rules, and shared cultural beliefs and values, it also involves humankind's infrastructural framing: cities, motorways, electricity, and data structures (Stryker, 1989). Furthermore, it includes. Changes in the landscape level are likely to pressure the regime to adapt or become more resilient.

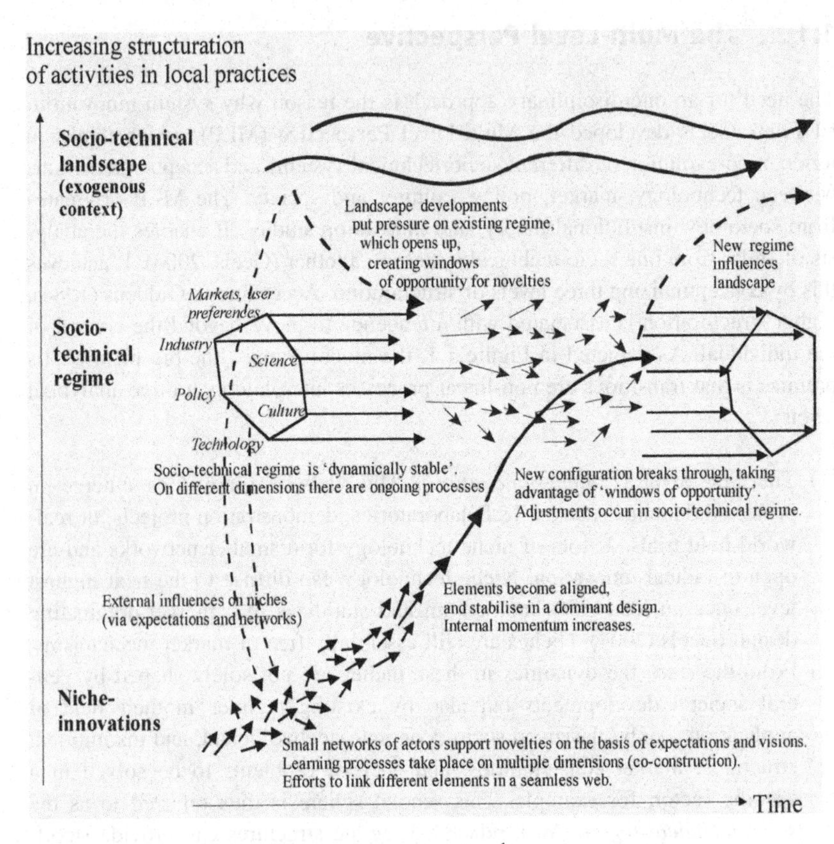

Figure 1.1 The multi-level perspective of transitions[1]

As evident in Figure 1.1, the MLP aims to trace the broad lines of socio-technical change and in this respect to provide a universal ordering grid for the intertwined evolutionary dynamics at different levels of society. Different transformation paths are possible depending on the state of the regime, the niches, and the landscapes. The existence of several possible paths indicates that it is possible to help shape which path will ultimately unfold. In subsequent work,

[1] Reprinted from Research Policy, Vol.3, Geels, Frank and Schot, Johan, *Typology of socio-technical transition pathways,* pp. 399–417, 2007, with permission from Elsevier. License No.: 4945831036714.

Geels and Schot (2007) discovered and categorized historical patterns of technology transitions. Furthermore, some researchers even applied the MLP to the transport market (van Bree et al., 2010; Geels, 2012; Kanger et al., 2019). The authors elaborate that the regime of mobility is currently caught between stabilizing and destabilizing forces. For instance, while on the one hand landscape pressures from climate change are destabilizing the regime, other forces such as automotive culture or growing user requirements are equally working against change. As a result, niche technologies are seldomly able to gain a foothold in the current system. The cracks in the regime are showing, however. Having said that, rather than disruptive change, these seem to lead to a sequence of substitution and reconfiguration processes. Before going into detail about how this work believes in helping understand these developments, the next section will elicit the object of study: smart mobility.

1.2 Smart Mobility

While the current regime is often referred to as that of automobility (Urry, 2004; Geels, 2012; Hoffmann et al., 2017), there is an active ongoing debate about where the various socio-technical dynamics of mobility will lead us to next. Experts agree on the one hand that emerging technology such as automated vehicles (AVs), electric drive trains, and Internet-of-Things (IoT) ecosystems will revolutionize individual and collective mobility (Docherty et al., 2018). On the other hand, new socio-economic practices like peer-to-peer sharing or teleworking arguably shift how mobility needs to be organized and operated (Whittle et al., 2019). Overall, four vital technological shifts constitute the transition from automobility to the next level: automation, electrification, connected mobility, and collaborative-shared mobility. Several studies (e.g. Sprei, 2018; Parkhurst and Seedhouse, 2019; Holden et al., 2020) analyzed these trends and concluded that not one of them has enough potential alone to transition the regime due to either market barriers (e.g., connected mobility), substitutional character (e.g., electrification) or ambiguous outcome (e.g., automation). An immense transitional power hence lies in connecting the developments. Since all these trends are enabled by information and communication technology (ICT) (Benevolo et al., 2016), they can be combined as smart mobility, analogously to smart homes or smart cities.

The need to adopt such a systemic perspective is also reflected in the issue's visibility in social and political debates. Since the 1990 s, it has been observed that new technologies are constantly in the spotlight in debates and funding programs but then quickly disappear (e.g., fuel cells or biofuels). Not least of all because, in the last ten years, smart mobility has been gaining interest, especially in progressive regions such as the Netherlands, the United Kingdom, and Israel[2]. It is possible that it is becoming increasingly recognized that the actual value of mobility digitization lies in making changes measurable and thus being able to manage them efficiently (Wintermann, 2021). From a systems theory perspective (e.g. Pepper, 1926), it is hoped that interaction and spontaneous self-organization lead to the whole, smart mobility here, being more than the sum of its parts.

Smart mobility is a relatively young concept. As such, terminology and understandings vary. Public hypes and occasional misconceptions about new technologies are emerging. Fashionable terms such as 'autonomous driving' contrast with technical terms such as 'automated driving'. Ridesharing versus carsharing or connected cars versus smart cars are other poorly defined terms. In their study, Cookson and Pishue (2017) found that innovative systems are the most important feature when buying a new car, yet few people even know what a connected car is. Schwanen (2015) observed a similar phenomenon: people are often unclear about the functionality of apps and mobile applications while simultaneously being keen to use them.

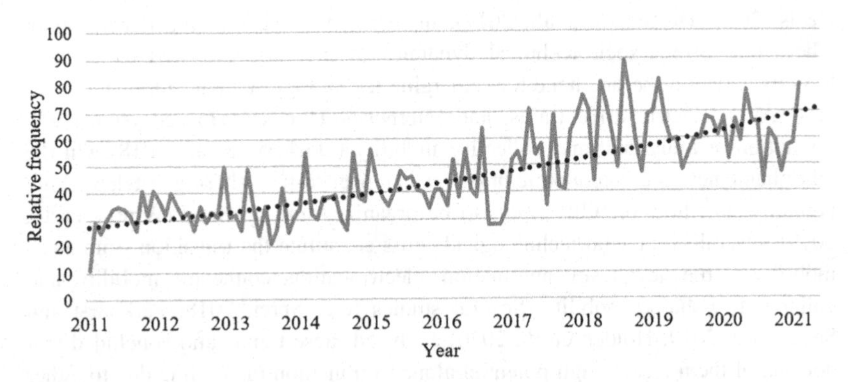

Figure 1.2 Worldwide search interest in 'smart mobility'. (Google Trends, 2022)

Findings of semi-structured interviews with smart mobility experts from different fields (N = 3) conducted as part of this work also reflect this. Results in Table 1.1 show that smart mobility is an expression that can be deconstructed into technical, social, or economic terms. Different research needs are highlighted depending on the subject area in which the experts classify smart mobility. It is

[2] Please see Figure 1.2, Section 1.2. *Worldwide search interest in 'smart mobility'.*

noticeable that social-psychological phenomena such as acceptance are mentioned several times. Overall, there is disagreement about when smart mobility will become established, as smart mobility is too unspecific. However, there is consensus on the assessment that transformation will occur. This change represents both an opportunity and a risk, according to the experts.

Table 1.1 Selected paraphrases from preliminary interviews. (Method adapted from Meuser and Nagel (2009))

Question	Politics expert	Economy expert	Research expert
1. What is smart mobility?	Fair and individual mobility	Mobility automation technology	100% flexible travel
2. Where is a need for research?	Change of eco-awareness, older people's acceptance	Technical safety, technology reliability	Lack of market readiness
3. When will it be common?	2025	2040?	Already exists
4. How will it be accepted?	About the price of new mobility	'Keep it simple and stupid'	Launch visible projects
5. How can technology breakthrough?	Will, project partners, and money	Somewhat minor, but more demand-oriented solutions	Cooperation with major companies

Smart mobility currently finds growing interest in interdisciplinary researchers and scholars. However, it is versatile and dynamic. Finding a suitable definition for smart mobility is therefore not trivial. A few definitions shall be highlighted. In consulting, for example, people like to discuss revolution and elaborate further (Geotab, 2018): *'Smart mobility refers to using modes of transportation alongside or even instead of owning a gas-powered vehicle.'* While this definition remains somewhat vague, Ahmed et al. (2020) explain the smartness of things in more detail. In their view, it depicts systems' intelligence and self-learning capabilities. Thus, in the first place, smart mobility provides efficient and convenient travel with a minimum of '*human interference*'. Further approaches see smart mobility primarily as a control entity of smart cities. Accordingly, Mueller-Seitz et al. (2016) write: *'Smart mobility includes as a performance dimension factors such as sustainable, innovative and safe transportation systems.'* Flügge (2016) describes the numerous target dimensions of smart mobility and thus imbues the term with more substance. She states: *'Smart mobility is defined as a proposition that enables energy-efficient, low-emission, safe, comfortable, and cost-effective mobility, and that is accessed intelligently by transport users.'* Other authors discuss the term by distinguishing it from related concepts such as sustainable mobility. Thus, Jeekel (2017) notes: *'Smart mobility is user-oriented, technology-oriented, IT-oriented, developed-world-oriented. And, quite important, smart mobility is action–oriented'*. Lyons (2018) notes it conjures up a sense of new opportunity and progress. He continues to explain that smart mobility needs to broaden its horizons and include sustainable mobility as well as livable mobility spaces. Manders and Klaassen (2018)'s text-mining approach adds that *'smart mobility seems to be about optimization, rather than challenging the status quo.'* These findings make it imperative not to neglect the social and sustainability-related goals of smart mobility but to include them in a definition. After all, studies expect reductions in emissions of up to (cumulatively) 50% through smart mobility technologies (Barth et al., 2015; Jochem et al., 2015; Pribyl et al., 2020). Overall, reviewing the definitions reveals the scope of meaning that exists. Therefore, I would like to propose a more precise definition of smart mobility:

> *Smart mobility advocates a smooth future of ubiquitous and connected transport, i.e., an effective system optimized for sustainability, attractiveness, and affordability with the accompanying introduction of automated and electrified vehicular systems.*

By means of this definition, smart mobility embraces technologies as well as environmental challenges to improve quality of life. In addition, the definition includes electrification, automation, as well as the primary goal of digitization: process optimization. This conception thus aims to expand the field of use cases included in smart mobility meaningfully.

1.3 Challenging Automobility

We must contemplate the status quo to understand how to transition to smart mobility. Within the transport domain, the hegemonic system is the regime of automobility (Geels, 2012). This term refers to the utilization of automobiles as the major means of transport. Historically, the mass spread of automobiles was one of the greatest socio-economic shifts of the 20th century. Automobility fundamentally changed the economy and society in favor of freedom, prosperity, and quality of life. In contrast, negative impacts of transport led to health problems, global warming, and inefficient resource management (Urry, 2004, 2008).

The regime

Today, automobility is deeply embedded in the world and its lifestyles. In spite of growing public criticism, it is still thriving and expanding around the globe. A variety of systems have adapted to this; the technical aspects of the transportation system (vehicles, infrastructures, etc.), the organizational models (ownership, services, etc.), the legal framework (emission thresholds, trade, etc.), user habits, etc., are all evolving together. This interactive dynamic has created and continues to create path dependencies that make it difficult to change the overall direction of market and societal development. It is further maintained through ongoing investment in roadway projects and advertising lifestyle norms that sustain car use. It follows that the regime has proven stable against changes in the first two decades of the new millennium[3] (Rotmans et al., 2001; Geels, 2012).

The landscape

At the meso level of automobility, we find stabilizing and destabilizing forces. The first category includes economic considerations, automotive culture, and user preferences. Likewise, a shock event such as the Corona crisis is part of the landscape. Corona stabilized the existing system in this specific case since people have retreated into known protective spaces (Gkiotsalitis and Cats, 2021). In stark contrast, there are destabilizing forces such as digitization and sustainability concerns. Both challenge existing practices and business models. They build up more pressure by growing public awareness and influencing policy debates. A destabilizing shock such as Dieselgate accelerated the rethink of mobility process (Hoffmann et al., 2017). Overall however, the two types of forces seem to be balancing each other out at the moment.

The niches

In this context, technical niche innovations that may challenge the regime's assumptions and norms are born. Typically, niche innovations are local initiatives on a city scale. Thus, Helsinki and Singapore, for example, are testing comprehensive Mobility-as-a-Service (MaaS) concepts (ITS International, 2019; CCV, 2021). In other trend fields of smart mobility, innovation tends to occur in the laboratory or within an internationally connected IT community. In networking, for example, new forms of data ecosystems are currently being developed that will enable the connection of everything with everything else in the 'Internet

[3] As of 2022, the latest uptake of electric vehicles in western nations can be seen as a first severe crack in the regime. Nevertheless, it remains to be determined where this development will lead to (see the grand narratives in Section 2.4.5.3).

of Things'. This creates a basis for many other systems. Furthermore, decentralized digital identities are being tested in pilot projects to ensure the secure identification of road users. In vehicle-related technologies, alternative forms of powertrains are competing against each other, with battery-electric drive currently leading the way. This development is also thanks to new players such as Tesla, who thus play a crucial role in defining a vision for new mobility. For the field of automation, a similar picture can be drawn with internet companies such as Google entering mobility. However, automation technology is still in its infancy and thus almost exclusively happening in protected spaces or showcase projects. Hoffmann et al. (2017) emphasize the support provided by public authorities in this context. In summary, electromobility is the only smart mobility technology to have made it to large-scale implementation so far.

Conclusion
For smart mobility to become the epitome of new mobility, three strategic approaches from MLP theory can be utilized (Geels and Schot, 2007): (1) strengthen landscape pressures, (2) weaken regime resistance, or (3) strengthen niche development. However, this is still relatively unspecific since the MLP is typically a global model that illustrates the entire process. By its nature, it does not yet provide a helpful starting point for investigating the concrete barriers and drivers of transition. Overall, the MLP tends to pay less attention to actors (Geels, 2019). For the present work, I hence aim to enrich the MLP with two actor-oriented perspectives enabling deeper considerations.

1.3.1 The Importance of a User Perspective

'The worm must taste right to the fish, not to the fisherman'. Unfortunately, this marketing truism has not yet been sufficiently incorporated into the design of new mobility, as recent literature highlights (e.g. Whittle et al., 2019). Thus, the first perspective to be added to the MLP should be one of the potential users. These fulfill vital socio-economic functions in innovation processes.

User perspective is essential for smart mobility development; public behavior, in particular that of car drivers, needs to change in order for them to be more likely to switch to combining different modes of transport (Mulley and Kronsell, 2018). They need to have confidence in the service so that they are less dependent on their cars. There is a plethora of components necessary to make smart mobility more attractive. For instance, smart services must be flexible and available at short notice. The price must also be able to compete with automobility.

Furthermore, the transport must be as fast as using a car, and the service should be offered via easy access brokers (Jittrapirom et al., 2018). Besides this, establishing services is related to using the mobility service providers' offers. Suppose customers booked the services individually and directly with the providers. In that case, the concept of smart mobility would be useless because the service would not be used, and thus there would be no demand. By increasing smart mobility, a broker or intermediary can offer individual services more favorably to providers since it can negotiate the offers more economically (Giesecke et al., 2016). Thus, from the user's perspective, there are significant research areas on the development of smart mobility. If the benefits increase for consumers, they are more likely to demand smart mobility services and the service providers are more willing to disclose mobility data or provide access to their apps (MaaS Global, 2017; Uber, 2017). Overall, the next mobility system must be developed to appear attractive to users (Lyons et al., 2019).

A review of Bögel and Upham (2018) on psychology in sociotechnical transitions highlights that while user motivations are often implicitly referred to, they are rarely theorized explicitly in psychological models. Thus, bringing acceptance modeling[4] to the MLP can offer much value for research and practice. Therein, behavioral economists and social psychologists adopt the perspective of potential users. They analyze attitudes and perceptions and their effect on desired behaviors. From this, they derive recommendations on promoting acceptance in a transdisciplinary manner. Such user insights may help anticipate innovations' adoption, usage, and, in the sense of the MLP, the impacts on the regime (Axsen and Sovacool, 2019).

1.3.2 The Importance of an Institutional Perspective

The second perspective on smart mobility transition is the examination of institutions that constitute the regime. Like the user perspective, this view must be seen as pivotal, since without institutionalizing business models, no scale-up of technologies can take place (Nelson, 1994).

As new technologies bring specific physical or intangible properties, institutions must be adapted or even reinvented to resolve specific conflicts. In the case of smart mobility, for example, this could involve new digital value chains, ownership rights, or liability issues. However, transition research suggests that in the absence of deep regulatory and market reforms, no system-wide change can be

[4] See Chapter 2 for further information on theory and application.

induced, even by powerful niche innovations. An example is the introduction of Uber in Germany, which, due to solid conflicts with the cab industry, only allows minimal operation (Nietsch and Schott, 2021). In the same context, but with the example of energy markets, Wolsink (2018) argues that an institutional lock-in often leads to regime change failing. Hence, to avoid getting into such a situation, the co-evolutionary processes of smart mobility require an analysis of the institutional context and potential fits and misfits.

In socio-technical studies, institutions are often referred to as either regulative (laws and policies), normative (ethics and responsibility), or cultural-cognitive (shared beliefs and values) (Geels, 2004). However, this view excludes corporations and businesses, which since the early 20th century have become increasingly important social actors and can thus become institutions in their own right. Sociologists and institutionally oriented economists recognized this and postulated that the decision about a new system could only be determined within a network of circumstantially relevant institutions and pending roles (Coase, 1998). The overall focus must be on minimizing transaction costs for system transformations. Hence, in the sense of this stream of research, a new system will be successful once the institutions involved can agree on an economic setup that provides enough benefit for each participant.[5] Transferring this idea to the MLP, it is imperative to define the design of an institutional target setup at an early stage where institutions have to develop too. The Institutional Role Model (IRM)[6] represents a suitable tool for this purpose (Schulz et al., 2021b).

1.3.3 The Enriched MLP of Smart Mobility

Figure 1.3 integrates the institutional (domain of the IRM) and the user perspective (domain of acceptance) into the MLP. It presents how the consideration of acceptance can lead to increasing landscape pressure as user requirements can be better managed. Understanding acceptance can moreover strengthen technology if designed in a user-orientated way. Finally, the vision of future mobility tasks and institutions can amplify the development in niches and landscapes by providing guidance for decision-makers and entrepreneurs.

[5] In reality, the so-called new economy established its systems with hardly any institutional cooperation. However, this led to numerous social and legal problems that remain unresolved to this day. In order to better distribute the prosperity resulting from the new service world, Herzenberg et al. (2018) persuasively call for establishing decentralized, supra-enterprise institutions that can and should regulate the new economy.

[6] See Chapter 3 for further information on theory and application.

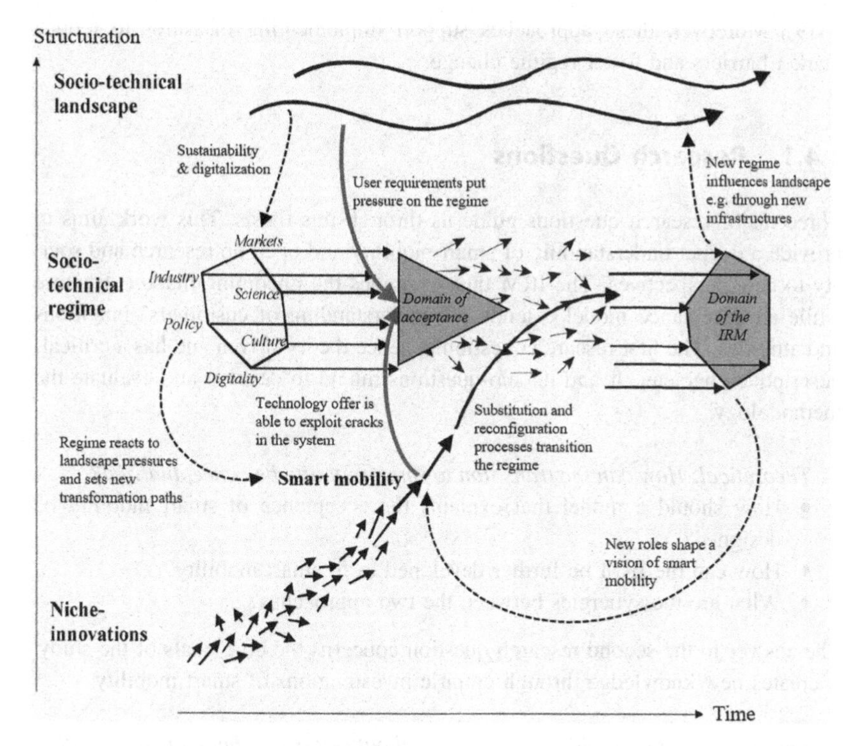

Figure 1.3 Two views on smart mobility regime change within the MLP

1.4 Scope and Design

Despite such enormous potential, the diffusion of a variety of smart mobility technologies in Germany and other countries is still significantly falling short of expectations. Thus, the transition to smart mobility, which is likely based on electrification, automation, and smart connection, is stagnating in many areas (Kagermann, 2017). This leaves researchers wondering what the reasons might be.

The value of this dissertation is to enrich our view on change in mobility by investigating often-disregarded transition factors. For the first time, perspectives of cross-technology acceptance and a visionary role model are examining smart mobility. Both views are particularly important since research has demonstrated that they can accelerate transitions (e.g. Schlüter and Weyer, 2019; Schulz et al.,

2019). Moreover, these approaches support implementing measures to reduce market barriers and foster regime change.

1.4.1 Research Questions

Three major research questions guide us through this thesis. This work aims to provide a deeper understanding of smart mobility and open up research and society to new perspectives. The IRM thus examines the upcoming market structure while an acceptance model extends our understanding of customers' intentions and attitudes. The first research question is hence theory-driven and has a critical-descriptive character. It and its sub-questions intend to develop and evaluate the methodology.

1. *Theoretical: How can the transition to smart mobility be conceptualized?*
   - How should a model that explains the acceptance of smart mobility be designed?
   - How can the IRM be further developed to fit smart mobility?
   - What are the synergies between the two approaches?

The answer to the second research question concerns the core goals of the study. It creates new knowledge through empiric investigations of smart mobility.

2. *Empirical: What can the models tell us about status and functionalities within the transition?*
   - Which factors determine the acceptance of smart mobility, how, and to what extent?
   - What role do contextual influences such as experience or demographic characteristics play?
   - Which technical and economic roles have to be covered?

Finally, the third research question is practical in nature. It aims to conclude practical consequences from the findings and deliver a utopian message to decision-makers and practitioners.

3. *Practical: How can the transition to smart mobility be successfully designed?*
   - What can decision-makers learn from acceptance results?
   - What is the practical relevance of IRM utilization in the context of the MLP?
   - What recommendations for action result from this?

1.4.2 Dissertation Structure

The study's composition derives from the research questions and is based on the classic structure of empirical research. The work has two main parts: a larger acceptance study and a smaller spotlight on institutional roles. Figure 1.4 illustrates the structure of the study based on the main research content of the individual chapters. The central themes are captured on the left.

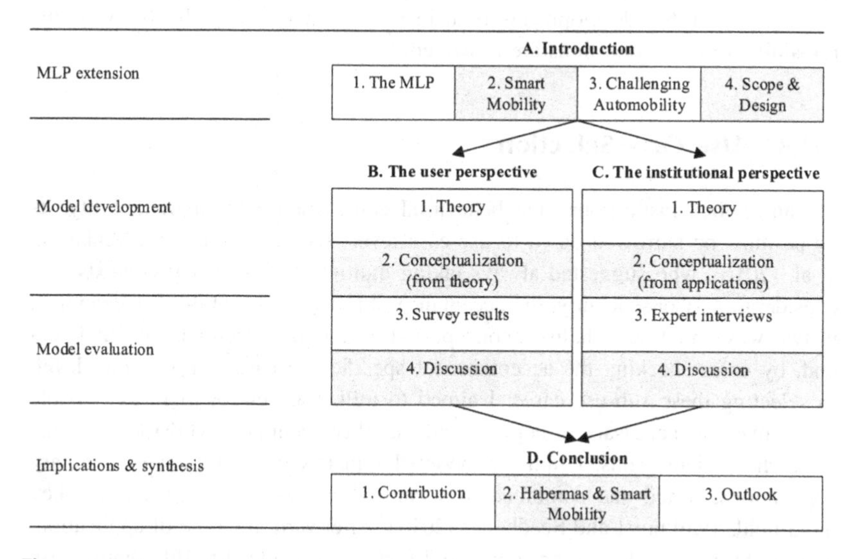

Figure 1.4 Research layout

The introduction in Chapter 1 revealed how the two perspectives of the study embed themselves in an overarching framework. These paragraphs thus already form part of research question 1's theory development. This is subsequently pursued in Chapters 2 and 3 by reviewing the background literature.

Chapter 2 provides a comprehensive empirical acceptance analysis of smart mobility. It constitutes the main focus of the work, with modeling and methodology deriving from a broad spectrum of literature. Quantitative survey data are used as input to validate the model. Large parts of the second and third research questions can be answered by capitalizing on these results and the subsequent discussion.

Chapter 3 also begins with a presentation of the theory, in this case that of the institutional role model. The research model is derived from prior and present applications. Next, the model is submitted to experts and discussed qualitatively. Given that the focus is on the development of a vision of the institutions of smart mobility, a quantitative application of IRM is not the subject of this research.

Finally, Chapter 4 summarizes the results of each perspective and attempts to develop synergies between them. The thoughts of the sociologist Habermas take us through an excursus on public communication theory. Finally, an outlook is given on further developments both in practice and theory. In this way, the remaining research questions are answered.

1.4.3 Use Case Selection

For an applied institutional and behavioral consideration of smart mobility, its scope must be narrowed down to use-case level. This is in line with Venkatesh et al. (2016), who suggested always taking multiple technological contexts into consideration. Consequently, the acceptance object 'smart mobility' is examined in two ways: first, as a holistic concept and a system of technologies, and second, by cross-checking the acceptance of specific use cases on a feature level. In selecting these sub-use cases, I aimed to utilize as few of them as possible yet as many as necessary to depict smart mobility accurately whilst keeping the research efficient. Thus, I initially modeled a matrix of well-known exemplary applications, sorted and combined by means of transport (x-axis), and the key trend fields (Parkhurst and Seedhouse, 2019). Afterward, the type of application, be it a physical product or an immaterial service, was added to differentiate use cases even more successfully (y-axis).

The matrix in Figure 1.5 was applied as a selection tool. The first step was the exclusion of previously examined use cases such as cooperative intelligent transport systems (C-ITS) and electric vehicles (EVs) (e.g. Geis et al., 2016; Kauschke and Schulz, 2017). Next, I integrated the MaaS scenario to cover almost all service-related smart mobility use cases. The main focus for users of MaaS includes apps, public transport, and the integrative use of sharing services. It integrates both public and private operators. The relation to cars and bikes in MaaS is rather weak and for the most part not product-related. Hence, I selected the eBike (EB) and the fully automated vehicle (FAV) SAE Level 5, which also helped yield the trends of automation and electrification into this smart mobility representation. I did not include other automation technology, such as automated

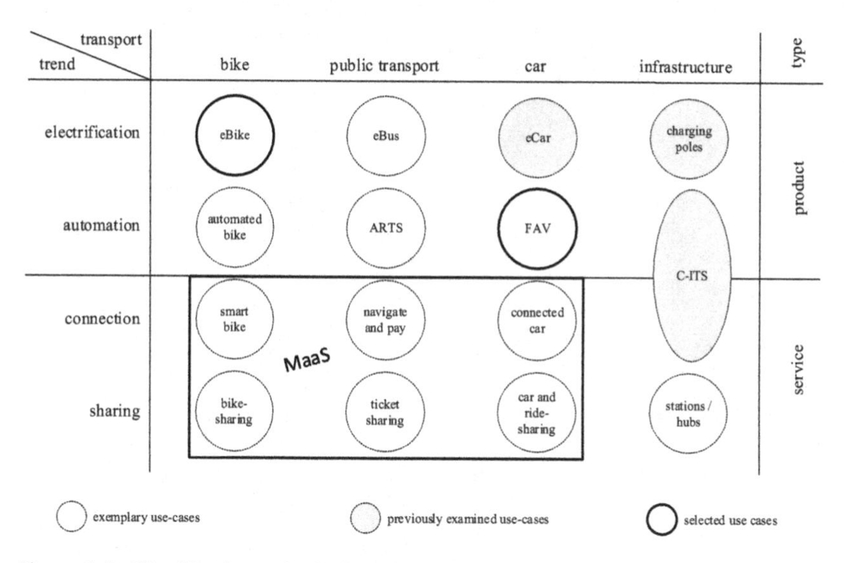

Figure 1.5 Classification and selection of smart mobility use cases

road transport shuttles (ARTS), as they had also been studied quite extensively in prior research (e.g. Madigan et al., 2016; Nordhoff et al., 2018b).

A charming side effect hereby is that eBikes and FAVs form two very different technological contexts. One technology is already widespread and can be experienced while the other is more in its visionary stage. MaaS, however, is very different. It can furthermore be assumed to be easier for users to evaluate MaaS alone than individual sub-services (e.g., ridesharing versus car sharing) that might be difficult to differentiate for the average consumer. Finally, I chose three sub-use cases besides the main use case of smart mobility to proceed with acceptance and institutional analysis.[7]

[7] The selection procedures were supported and approved by the doctoral supervisors.

2 The User Perspective

2.1 Theoretical Framework

This section provides an overview of acceptance theory, from the basic concept of acceptance to the latest revelations in research. Generally, the results of the below theory recap indicate that research on acceptance of new technology is a heterogeneous field depending on whether the focus is conceptual, methodological or technological. I conclude that for the analysis of technology systems such as smart mobility, comprehensive behavioral modeling is the most promising approach. This creates the basis for state-of-the-art acceptance research for smart mobility.

2.1.1 The Concept of Individual Acceptance

In spite of public excitement, the diffusion of new technologies often fails due to a lack of user acceptance. Acceptance research, therefore, attempts to grasp and systematize the relevant influencing factors, as well as to describe their mode of operation and present them in models. It is a positivistic discipline (Straub and Gefen, 2004). Present work understands acceptance as the result of a chain of social, cognitive, and economic processes. It is hence a pivotal measure for forecasting the success of technological innovations (Kollmann, 2013). Regarding the transition to smart mobility, acceptance can be the missing link in understanding human decision-making.

Acceptance research can be distinguished from adoption and diffusion research in that acceptance research mostly occurs mostly at micro-level, i.e., the level of the individual (Rogers, 1995). Only a limited prognostic objective is pursued.

L. Kauschke, *The Transition to Smart Mobility*, Mobilität – Innovation – Transformation, https://doi.org/10.1007/978-3-658-43001-6_2

What is more important is the understanding of structures and the derivation of enhancement strategies. With this in mind, acceptance research focuses on users' acceptance or rejection of innovations at a given time. Diffusion research, on the other hand, considers the speed at which a target group accepts an innovation as a step-by-step process. Thus, both theoretical streams are not to be seen in opposition but in connection via time and experience components.[1]

Despite the growing importance of acceptance in transportation research (Angelis et al., 2017; Geis and Schulz, 2017; Kapser and Abdelrahman, 2020), the definitions of acceptance often remain vague. Demarcations to other terms, i.e., adoption, diffusion, or the concept of acceptability, are also necessary. This non-standardization of acceptance currently limits our understanding and overall outcome comparability (McAndrews et al., 2018). Thus, this section offers some clarification and adds reflections on smart mobility acceptance in the context of the present study. Based on Adell et al. (2014), I identify four categories to frame the basic concept of acceptance (Figure 2.1). Elements of these categories are frequently used to explain user technology acceptance but they almost always lack coherent perspective and standardized wording (Huijts et al., 2012).

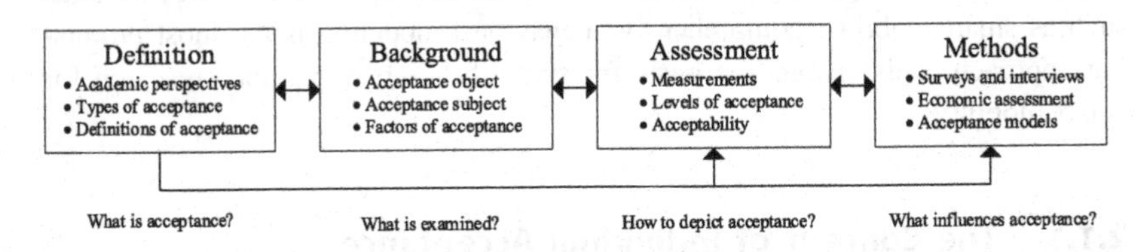

Figure 2.1 The basic concept of acceptance

2.1.1.1 Definition of Acceptance

In order to understand its conception, acceptance must first be defined. This is vitally important as it forms the basis for depicting and evaluating acceptance. Literature thereby reveals definitions that originate from different contexts and various academic understandings of the same term. Economics, psychology, and social sciences are the main disciplines that deal with acceptance.[2] These academic perspectives led to various types of acceptance including *emotional* and *rational* acceptance (Davis, 1989; Van der Heijden, 2004), *practical* and *public* acceptance (Dillon and Morris, 1996; Vlassenroot et al., 2010) or *behavioral* and

[1] See Figure 2.2, Section 2.1.1.3 for a modeling attempt.

[2] See Section 2.1.2 for details on the different approaches.

habitual acceptance (Ajzen, 1991; Chen and Chao, 2011). Something these types all have in common is centering the user as an agent of technology and regarding them in the context of socio-technological systems.

On the other hand, they strongly differ about interpretations of acceptance. In their review of definitions, Adell et al. (2014) conclude with a progression of definitions ranging from user attitudes (*'Acceptance is basically the question of whether the system is good enough to satisfy all the needs and requirements of the users and other potential stakeholders'* (Nielsen, 1994)) over willingness to use (*'Acceptance is the will to adopt an innovation'* (Chismar and Wiley-Patton, 2003)) toward the actual use (*'Acceptance is the demonstrable willingness within a user group to employ information technology for the task it is designed to support'* (Dillon and Morris, 1996)). Finally, incorporating the multitude of different views, Adell et al. (2014) decide on a definition that contextually also qualifies to assess smart mobility acceptance:

> *'Acceptance is the degree to which an individual incorporates the system in his/her behavior, or, if the system is not available, intends to use it.'* (Adell et al., 2014)

This suitability has four reasons. Firstly, the definition emphasizes acceptance as a continuous measure that is seldom either fully present or absent. As the shift to smart mobility is an incremental societal process, so is its acceptance. Secondly, the definition focuses on the individual with its beliefs and attitudes that subjectively form behavior. This conception is in line with the predominant ideas of acceptance in the context of the most important research (Davis, 1989; Venkatesh et al., 2003). It can therefore also be assumed for smart mobility. I want to emphasize that individual acceptance must not be confused with public acceptance, often measured in representative large-sample surveys (Wolsink, 2018).

Thirdly, the presented definition highlights system acceptance instead of pure technology acceptance. Thus, it includes processes and stakeholders surrounding a technology. Smart mobility is more of a system and hence yields this interpretation. Fourthly, behavior can be understood as an intention to use, actual usage, or both. This depends on the availability of the system. Smart mobility is not yet available to everyone; consequently, ideas about availability will vary, as will the evaluation of acceptance via intention or usage. Eventually, both will have to be deliberated.

2.1.1.2 Background of Acceptance: Subject, Object, and Context

Definitions of acceptance can be sharpened considering the socio-economic *context* (Saad, 2006). Science additionally distinguishes between the *subject* and the *object* of acceptance (Lucke, 1995). Factors of acceptance are assigned to one of these three elements.

The starting point of acceptance is the acceptance *subject*. This individual develops an attitude related to the acceptance object, including a behavioral tendency that can be activated on a recurring basis, which may result in observable utilization. In smart mobility-related acceptance research (Huijts et al., 2014; Panagiotopoulos and Dimitrakopoulos, 2018; Lyons et al., 2019), the acceptance subject is the individual (future) user. Acceptance factors usually applied to this user are (Breckler, 1984):

- Personal attitudes (e.g., habits, concerns, or extraversion)
- Norms (e.g., personal standards or group beliefs)
- Emotional affects (e.g., general level of reaction intensity toward a topic)
- Sociodemographic groups (e.g., age, gender, or job)

The second element of acceptance is the *object* of acceptance. In the present examination, this is the system of smart mobility and related technology. This system is defined by attributes, which can be categorized as follows. Depending on the user (and its context), the same attributes can cause different assessments (Schäfer and Keppler, 2013):

- Cost-benefit evaluations (e.g., economic, ecological, or social)
- Rational benefit evaluations (e.g., ease or satisfaction)
- Emotional evaluations (e.g., aesthetics, enjoyment, or risk evaluation)

Thirdly, the *context* of acceptance includes factors that neither directly assign to the *object* nor the *subject* but influence the overall acceptance process. Relevant constructs, for instance, are (Venkatesh et al., 2003):

- Social influence factors (e.g., expectations or conflicts)
- Control factors (e.g., self-efficacy or perceived limitations of personal action possibilities toward the system)
- Supporting frameworks (e.g., technology, law, culture)
- Other circumstances and use contexts (e.g., weather, time, location)

2.1.1.3 Assessment: Acceptance or Acceptability

The conception of individual acceptance is also different, depending on how acceptance is asserted. While some approaches use freely designed qualitative and quantitative data (Kyriakidis et al., 2015; Polydoropoulou et al., 2018), economic measures (Lin and Tan, 2017; Zhou et al., 2020), or economically-measured usage data (Straub et al., 1995; Dudenhöffer, 2013), the dominant way to acknowledge acceptance scientifically is through the formation of theoretical constructs such as norms, attitudes, or emotions. Unfortunately, these constructs cannot be captured directly. Thus, they are operationalized for indirect measurement using so-called latent variables.[3]

The key to correctly designing these measures is the time at which acceptance is measured. Levels of experience define this point. Schade and Schlag (2003) describe *acceptance* as the sum of behavior and attitudes after having personally been in touch with new technology and *acceptability* as a 'prospective' judgment of a future introduction. Sadly, other definitions contradict this. Grey areas between acceptability and acceptance are revealed. For example, Pianelli et al. (2007) differentiate between *a-priori* and *a-posteriori acceptability.* They thereby find that *a-posteriori acceptability* after a first experience does not necessarily include a behavioral reaction. *A-priori acceptance,* on the other hand, is a kind of acceptance based mainly on personal attitudes. This is grounded by Nielsen (1994), who introduces *social* and *practical acceptability. Social acceptability* refers to the process of accepting a system of considering individual social surroundings, hence adding the dimension of norms and social space. *Practical acceptability* combines these dimensions of costs and overall daily compatibility. As a final step, real-world acceptance is achieved when behavior becomes *habitual* (Chen and Chao, 2011). In trying to transfer these thoughts to a five-level taxonomy of acceptance assessment, one can hypothesize that acceptance is a continuous process with varying relevance of acceptance factors, depending on the stage of cognitive effort and experience with a system. The assumed levels of experience in Figure 2.2 are inspired by Rogers (1995)' steps toward innovation adoption. In gaining experience, one's perception and acceptance shifts. The cognitive effort to render a new system tangible to an individual is higher the less experience one has, which is then again influenced by the level of integration into the mobility system (Lyons et al., 2019). Not least because of the known positive effects of experience on user acceptance, one has to assume that acceptance generally increases within this process. However, acceptance always

[3] See operationalization of variables in Section 2.2.

remains a relative parameter of the relationship effects between the influencing factors.

Figure 2.2 summarizes these deliberations as a taxonomy proposed by Lyons et al. (2019). Parallelly, influencing factors for the respective levels were added based on the considerations of Nordhoff et al. (2019)'s multi-level model and research on initial trust (Zhang et al., 2019) and continuing acceptance (Chen, 2016). Furthermore, Venkatesh et al. (2003) stated, inter alia, that, in voluntary contexts, *social influence* (SI) only impacts behavioral intentions after a period of use. These thoughts formed the basis for a set of hypotheses to be tested in this work.

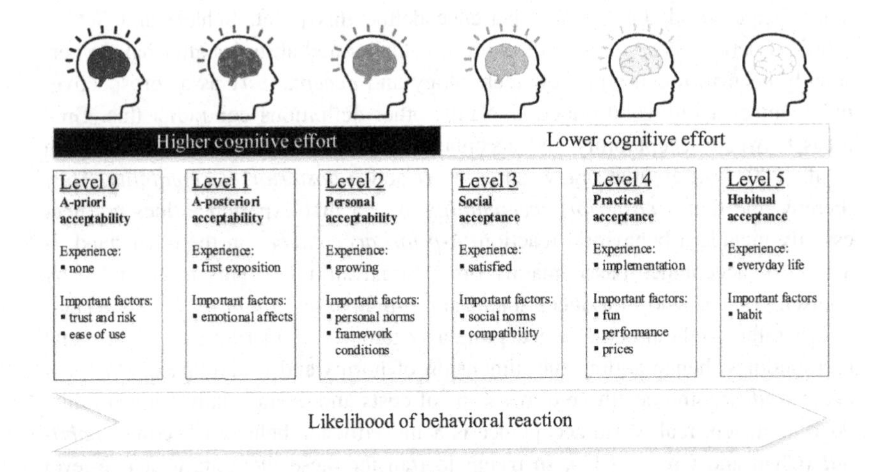

Figure 2.2 A taxonomy of individual acceptance

As sketched previously, the meanings put into the term *acceptability* are manifold, which is also based on the fact that defining experience in acceptance research is 'problematic' overall (Adell et al., 2014). In the end, borders between *acceptance* and *acceptability* are blurring.

To clarify, I define *acceptability* in line with Huijts et al. (2012) as a-priori perceived use, respectively the individual judgment about a technological system including attitudes, intentions and first experiences before any behavioral reaction toward using it has been caused. This reaction can occur at any level, but in gaining experience, the likelihood of such behavioral reactions increases. Ideally, they become more likely between Level 2 and Level 3 in Rogers (1995)'s process of innovation adoption, as experience and information gathering needs are about to be satisfied within this period, and an adoption decision takes place. As a result, for my conception of acceptance, I will consider the first three levels *acceptability* and all the above levels of *acceptance*. In contexts other than in-depth acceptance analysis, for example, the term acceptance will be preferably used in the theoretical surrounding and practical implications.

2.1.1.4 Methods of Acceptance Analysis

The fourth dimension of individual system acceptance relates to the methods field applied to analyze acceptance. This is closely related to the measurement of acceptance, as most methods have more or less standardized or common data collection procedures. Yet still, it is different because methods link data to the bigger picture and thereby open up our understanding to diverse points of view on the same issue. As previously mentioned, methodological access to acceptance can be found in social sciences, economics, and psychology. Historically, psychology-based acceptance research has received significant attention because acceptance has become the key term within social and scientific discussions about measuring and forecasting the success of technological innovations (Kollmann, 2013). This trend is further encouraged by accelerated living environments, globalization, and digitalization. Due to the overall importance of acceptance modeling in related research, I systematize methods by non-model and model-based approaches in Table 2.1 (inspired by Schmidt and Donsbach (2016)).

Table 2.1 Key methodologies of acceptance research

Type	Approach	Methods examples
Non-model-based (descriptive)	General attitudinal research	Surveys, focus groups, field trials
	Forecasting	Scenario technique, expert interviews
	Evaluation and selection	Contingent evaluation, stated choice experiments, willingness-to-pay
	Other	Cluster analysis, media analysis
Model-based (explicative)	Innovation diffusion models	Delphi technique, survey panels, econometrics
	Risk and trust models	Neurological and game-theoretical modeling methods
	Causal models	Controlled experiments
	Behavioral models	Structural equation models, linear regression models

The frequency of a method's application varies according to acceptance background and discipline. In the end, it has dominantly been behavioral models, which have thus been taken up and further developed by other disciplines such as engineering, informatics, or marketing. Finally, a relatively independent stream of transdisciplinary research with different modeling streams resulted.[4] Authors later had cross-disciplinary fertilization in their further theoretical developments to render proper disciplinary distinctions imprecise. After all, acceptance research can today most likely be considered a part of behavioral economics or social psychology. Depending on the research object, it can also be part of management studies, engineering research, as well as education, transport, or environmental sciences.

2.1.2 Behavioral Acceptance Modeling

As one can see, there is no standard way to assess the acceptance of technology. Yet, behavioral models have become the method of choice in explicative acceptance research since they can look under the surface of acceptance (Schäfer and Keppler, 2013). These representations of reality help us to structure the influencing factors and mechanisms gathered in the context of initial empirical studies or derived from existing theoretical approaches. They merge them into the context of interrelationships. In addition to a verbal description, they also include a visual representation[5], making them easy to comprehend relative to their enormous explanatory potential. The hundreds of models, extensions, and elaborations that exist are often transdisciplinary, and consider social, psychological along with economic factors. This variety yields a deeper and more complex understanding. Venkatesh et al. (2003) provide the general underlying concept in Figure 2.3. Hence, acceptance in acceptance modeling is regarded as a process of individual reactions, intentions, and actual use. The relations to be detected in between are what matter to researchers.

[4] Commonly referred to as information system (IS) research.

[5] See Figures 2.4–2.6, Sections 2.1.2.2 ff. for examples.

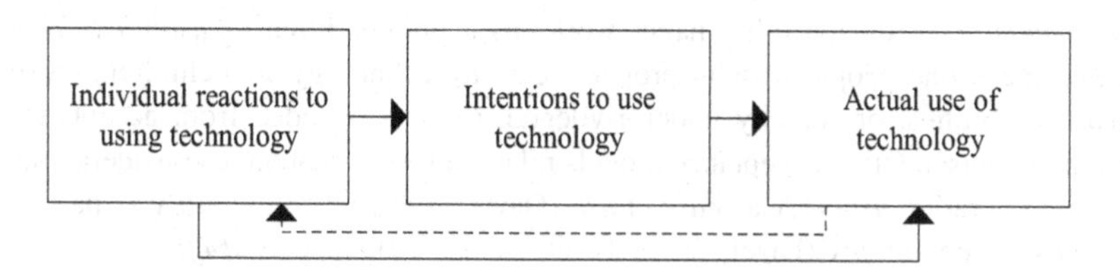

Figure 2.3 The basic concept for acceptance modeling (Venkatesh et al., 2003)

Mentioned models can be aggregated under the topic of behavioral economics. This is the consideration and description of social, cognitive, and emotional factors in the human decision-making process or situations of uncertainty (Tversky and Kahneman, 1974). Many streams of research can be recognized within this wide scope of inquiry.

In this section, I want to give an overview of technology acceptance modeling streams and stress the importance of comprehensive modeling, as introduced in the Universal Theory of Acceptance and Use of Technology (UTAUT). Moreover, the suitability and importance of acceptance modeling for smart mobility are outlined. Alexandre et al. (2018)'s examination of the acceptability criteria of new technology provides a review. They historically identify five approaches to acceptance modeling from different academic scholars that influenced one another. The most significant commonality of the approaches is that, historically, they have been used to explain acceptance phenomena associated with new (information) systems. Present work advances a sixth area to Alexandre et al. (2018)'s observations that is aiming toward utilizing a universal model.

2.1.2.1 Ergonomics Approach

The first area evolves from psycho-ergonomics and represents an engineering perspective. In this technology-centered approach, it is assumed that manifest functional and ergonomic aspects will bring a technology into use and acceptance. The primary constructs in modeling are *usability* and *accessibility*. *Usability* has been decomposed into sub-criteria such as effectiveness, satisfaction, ease, and reliability (Shackel, 1981). Nielsen (1994) outlines five usability metrics in his model: efficiency, learnability, memorability, satisfaction, and error. The ergonomics perspective has, inter alia, led to the development of ISO 9241-11, which contains guidance and modeling on ergonomics and usability for office workplaces with visual displays (ISO, 1998). These fundamental models are also often employed in transportation, such as in analyzing public transportation information provision (Hussain and Mkpoguijogu, 2017).

Accessibility, on the other hand, works as a pre-condition of *usability.* This could mean that people without proper access to technology, like children, those with disabilities, or socially underprivileged, can be excluded from acceptance from the outset. Other acceptance models relate to these ergonomic considerations by incorporating *usability* as *ease of use* (Davis, 1989) or *accessibility* as part of *framework conditions* (Fazel, 2013; Huijts et al., 2014)), or *self-efficacy* as the *perception of external control* (Ajzen, 1985).

The strength of the ergonomic perspective is the in-depth analysis of the human-technology interaction. However, it discards many other aspects of acceptance, such as motivations, attitudes, and norms. As a result, models are best applied in a prototype validation phase, where technology can already be fully experienced in an experimental setting, but little is known about their acceptance.

2.1.2.2 Social-psychologic Approach

The starting point for acceptance research in social psychology might be the work of Fishbein and Ajzen (1975), in which they formulated the Theory of Reasoned Action (TRA). As a general behavior model in the field of acceptance research, the TRA forms one of the most influential theoretical explanations for human social behavior and is the basis for further models such as the technology acceptance model (TAM). The idea of TRA is that behavior is a rational decision based on salient beliefs of a social agent: the user. The agent's beliefs are subjectively perceived norms and attitudes. These then form the *behavioral intention* (BI), which is the key concept to influencing behavior in this stream of research. TRA has been expanded to the Theory of Planned Behavior (TPB) by Ajzen (1991). Thereby, the construct *perceived behavioral control* was added as a determinant of intention. *Perceived behavioral control* is an individual's evaluation of necessary personal skills, availability of resources, and social accessibility to a particular technology. It originates from Bandura (1977)'s seminal theory on self-efficacy being the critical determinant of *use behavior* (UB). It also helps to predict actual behavior. A rationale for this can be the example of two people with the same *intention to use* an innovative public transport system. Most likely, they will not perform the same behavior if one has limited skills to use smart devices or if no substitutive service is offered on the desired route.

As depicted in Figure 2.4, the three independent antecedents of *intention* in TPB are influenced by each other. Furthermore, *perceived behavioral control* directly shapes intentions and behavior. Control norms are parts of subjective norms, and vice versa influence *perceived behavioral control* (Ajzen and Madden, 1986; Ajzen, 2002). Since attitudes develop from beliefs (Fishbein and Ajzen, 1975) another correlation is assumed between *subjective norm* and *perceived behavioral control.* These hypotheses are the foundation for the development of utilitarian (Venkatesh and Davis, 2000; Venkatesh et al., 2003) or hedonic models (Davis et al., 1992; Van der Heijden, 2004).

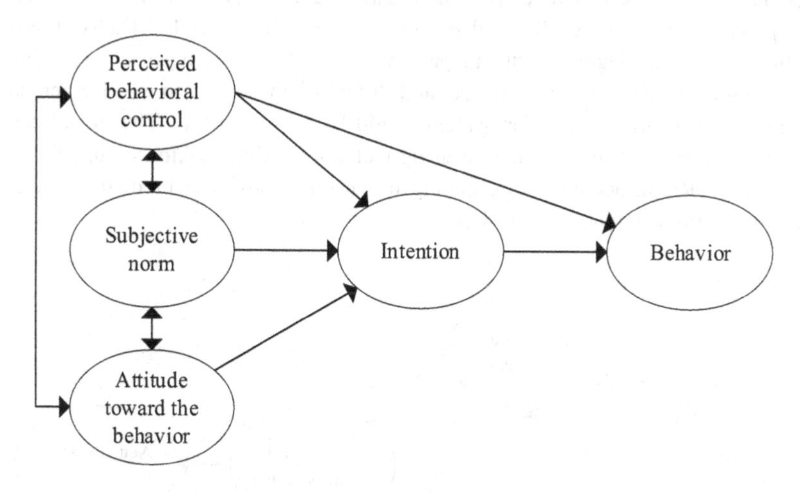

Figure 2.4 The theory of planned behavior (Ajzen, 1991)

TRA and TPB have been widely applied and validated across disciplines and research objects due to their simplicity and robust explanatory power. Yet still, critics argue that the model does not capture many vital determinants of acceptance like moral norms, habits, or self-identity (Conner and Armitage, 1998). Besides Bandura, Fishbein, and Ajzen, Rogers (1995)'s work concerning the diffusion of innovations, deserves particular attention when it comes to acceptance in consumer research. The author thereby integrates social-psychological and sociological knowledge on adoption into a five-step process. This knowledge about the social process dimension of acceptance has been seminal to research and has thus been integrated into the following studies and research streams. Overall, the social-psychological way of acceptance modeling can be considered the most influential in modern academic acceptance analysis.

2.1.2.3 Utilitarian Approach

With the rise of personal computers and increasing software utilization, different user-centered hypotheses about information system usage behavior evolved in the 1980s. They were based on the assumptions of rational human decision-making to maximize personal productivity. The technology acceptance model

TAM (Davis, 1989) of all the theories became the most valued approach. It adapts the spirit of innovation diffusion and combines it with ergonomics' usability and social psychology's perspective on behavior and attitudes. It does this in a simple yet powerfully robust manner (Lee and Larsen, 2003). The two key constructs are *perceived usefulness* (PU) and *perceived ease* of use (PEOU). Davis (1989) defines PU as 'the degree to which a person believes that using a particular system would enhance their job performance' and PEOU as 'the degree to which a person believes that using a particular system would be free of effort'. The model can account for 40% of the variance in using technology, thus making it superior to TRA and TRB in accurately explaining usage intentions. The predictive power therein mainly originates from PU.

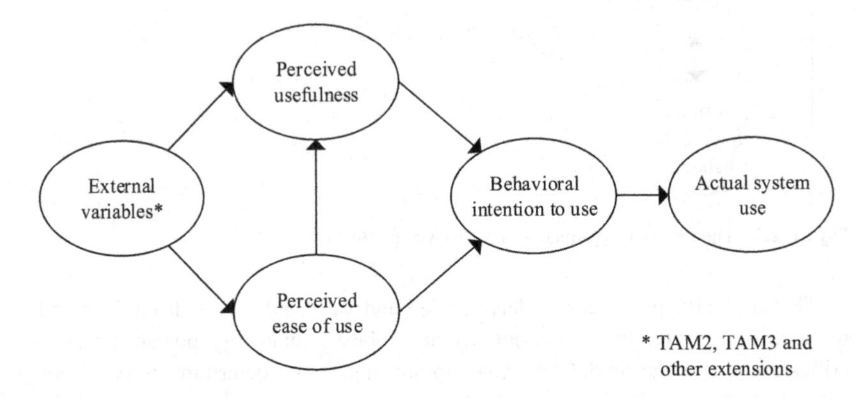

Figure 2.5 The technology acceptance model (Davis, 1989)

The TAM in Figure 2.5 has been extended by its developers toward TAM2 (Venkatesh and Davis, 2000) and TAM3 (Venkatesh and Bala, 2008) to explain acceptance more thoroughly. Constructs of social influence (*subjective norm, image,* and *voluntarism*) and instrumental processes (e.g., *task relevance, quality of output,* or *result demonstrability*) were added. Also, *experience* was first considered, while the *attitude toward using* dropped more during the validation process. This helped carry the model to an average explanatory power of 60%, which presents a substantial improvement.

Many more extensions and elaborations of the original model exist. Alone Lee and Larsen (2003) present 101 TAM-based studies incorporating other models or adding new external variables. They summarize that TAM is '*richly confirmed*' and yet still provides many exciting directions for future discoveries. Researchers value the simplicity and adaptability of the approach. On the other hand, the malleable use of the TAM has led to severe issues like contradicting findings and hypothesis inconsistencies. The undertone of the exhaustive review of Legris et al. (2003) is that researchers agree on the core model but cannot achieve unanimity concerning the importance of each criterion. They argue that TAM cannot well be applied to systems outside the world of information systems, as its utilitarian nature excludes intrinsic motivational components. Chuttur (2009) adds that TAM generally needs more relevant and rigorous applications. He finds that the TAM might be parsimonious in explaining deeper psychological processes and generally driven by innovation optimism. More comprehensive modeling efforts must try to capture a systemic perspective to balance more or less standardized criteria against each other.

2.1.2.4 Hedonic Approach

To overcome some of these issues, other approaches from the 90s do not regard the user as a rational decision-maker but as a subjectively judging and emotionally controlled being. The basic idea of this hedonic perspective is that objects inherit a joy of use that cause an emotional reaction. The more fun an individual experiences in utilizing a certain system or technology, the more likely they will continue to use it. This value is captured as *perceived enjoyment* (Davis et al., 1992), *satisfaction of expectations* (DeLone and McLean, 1992), or *intrinsic motivation* (Venkatesh, 2000).

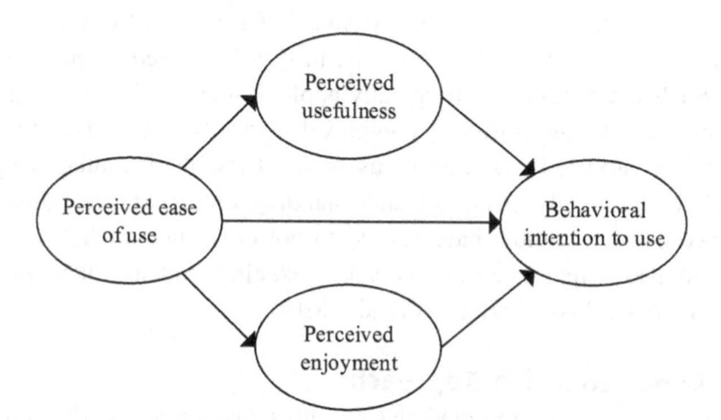

Figure 2.6 The acceptance model of hedonic information systems (Van der Heijden, 2004)

Van der Heijden (2004) eventually points out that researchers might generally have to distinguish between hedonic and utilitarian systems. Hedonic systems e.g., provide self-fulfilling value (e.g., home or leisure technology) as opposed to productive value (e.g., work or personal efficiency). The acceptance structure in Figure 2.6 is thus different, based much more on fun and ease than *perceived usefulness.* In the real world, clear distinctions between the two types are meanwhile often impossible. The natures of many technical systems, personal vehicles in transport or mobile communications apps for instance, are ambiguous. Nonetheless, adding hedonic features such as gamification to a system remains a sound and smart way to achieve user acceptance. Hence, following acceptance research (Teo and Noyes, 2011; Madigan et al., 2017; Alalwan et al., 2018) used the hedonic perspective and integrated it into their models.

2.1.2.5 User-experience Approach

Some other literature regards the determination of acceptance as a 'holistic unit' (Alexandre et al., 2018) of situations, contexts, and experiences. This stream of research thus considers the dynamics of user and system interactions evolutionarily based on *user experience* (UX). It includes and extends aspects of previous acceptance frameworks according to the examination goals. A famous example is the UX acceptance model of Mahlke and Thüring (2007). In this model, users evaluate a system based on a characteristic setting. This setting is defined by individual *context*, *user characteristics,* and *system properties.* Within this experience setting, users then individually appraise a system based on utilitarian (e.g., *controllability* or *effectiveness of use*) and non-utilitarian qualities (e.g., *aesthetics* or *personal identification*). This process is, in turn, mediated through *emotional reactions.* Such UX models are frequently applied in evaluating existing human-computer interactions and clarifying complex design issues. They help researchers understand that varying the setting of users and technologies alters acceptance. UX models consequently shape an understanding of acceptance influenced by circumstance. Hence, authors have started to not only build models around UX situations but also to integrate *experience* into existing models (Chen et al., 2011; Punel and Stathopoulos, 2017; Tsou et al., 2019).

2.1.2.6 Comprehensive Approach

Each of the approaches presented has its advantages and disadvantages with respect to the specifics of the elements it refers to (e.g., subject, object, and context of acceptance) and examination goals (e.g., the exploration of relationships, finding the best explanation of acceptance, or group comparisons). In order to level this heterogeneity and overcome conflicting evidence in the literature, researchers have proceeded in two different ways.

Initially, many studies conceptualized new models by more or less thoroughly combining two theories. Researchers achieved this by identifying models that have different, though overlapping perspectives on system utilization behavior. In validating sets of new hypotheses, precisely tailored research tools for specific application areas emerged that can explain acceptance better and more profoundly than each original model alone. Dishaw and Strong (1999), for instance, combined the TAM with the task-technology-fit (TTF) (Goodhue and Thompson, 1995) and thus increased the explained variance in the actual use of computer software from 40 to 50%. They incidentally discovered that system functionalities and system experience strongly influence *perceived ease of use*. Park and Ha (2014) extended TPB with another seminal behavior theory, the norm activation model (NAM) (Schwartz, 1977). NAM primarily concerns itself with sacrificing a person's own interest for the well-being of others via the construct of *personal norms*. In the combined model, *personal norms* function as a second significant determinant of behavior besides TPB's BI. Other interesting examples of model combinations can be found in Koufaris (2002) (TAM and flow theory), Morgan-Thomas and Veloutsou (2013) (TAM and online UX) or Richards and Johnson (2014) (TRA and parallel process model). Albeit the fact that a lot of new knowledge could be generated, the mere combination of theories has disadvantages, too: results are hard to validate and challenging to compare.

Thus, a second way to overcome proprietary methods has received attention. In this stream of research, the aim is to integrate and elaborate different models intelligently to arrive at comprehensive and transferable frameworks that coherently capture all acceptance aspects. This is vital for developing systemic and unified perspectives on acceptance. These are urgently needed, especially in contemporary ICT, as system complexity and innovation speed are large and still growing through the increasing functional connections between, inter alia, mobility, energy, and social systems (Gimpel et al., 2020). Kormos and Gifford (2014) conducted a meta-analysis regarding how people think about transitioning toward sustainability and found a large amount of unexplained variance (79%) between the reported and the objective behavior. Hence, improved explanatory frameworks are called for to accommodate confusing data.

Following this path, a recent stream of more adaptive modeling can be identified (Park et al., 2017b; Baptista and Oliveira, 2019; Gimpel et al., 2020). It is interesting to note that many models analyze acceptance beyond the original workplace environment within this stream. The field of new mobility technology and ICT is an excellent example of this (Vlassenroot et al., 2010; Huijts et al.,

2012; Sovacool, 2017). However, transport systems cannot be regarded as isolated from accompanying technologies, societal trends, and eco-political frameworks. Hence, the analysis must not rely on a limited set of factors. It must instead be able to relate factors to one another. All this is to establish a systemic and yet application-specific view in acceptance research.

2.1.2.7 The UTAUT

The origin of the above comprehensive acceptance modeling is probably the Universal Theory of Use and Acceptance of Technology (UTAUT). For the first time, Venkatesh et al. (2003) sketched an underlying structure in acceptance modeling based on previously discovered similarities (Moore and Benbasat, 1991). They then developed their theory from eight differently approaching models at that time to explain the acceptance of new technology, namely:

(1) The theory of reasoned action (Fishbein and Ajzen, 1975)
(2) The theory of planned behavior (Ajzen, 1991)
(3) The technology acceptance model (Davis, 1989)
(4) The combined TAM and TPB (Taylor and Todd, 1995b)
(5) The innovation diffusion theory (Rogers, 1995)
(6) The theory of human behavior (Triandis, 1977)
(7) The motivational model (Davis et al., 1992)
(8) The social cognitive theory (Bandura, 1986b)

As a result of stepwise integration, the model proposes that four key constructs determine technology use via BI. The authors argue that these elements account for 70% of the variance in BI ($R^2 = 0.7$) and about 50% of variance in *use behavior* ($R^2 = 0.5$). Relationships are thereby moderated[6] by *age*, *gender*, and *experience*, which paths the way to understanding acceptance as a means to explain adoption processes stepwise in longitudinal studies (Hu et al., 2003; Klöckner, 2014). Compared to TAM, the dimensions of *social influence* and *facilitating conditions* (FC) were added (Table 2.2).

[6] Moderation refers to a linear effect of a third variable on the relation between two variables. Moderation may be so substantial that an increase in the third variable significantly strengthens or weakens the effect between the other two.

Table 2.2 Key elements of the UTAUT model

Construct	Definition from Venkatesh et al. (2003)	Derived from
Performance expectancy	'The degree to which the user expects that using the system will help him or her attain gains in (job) performance.'	(2), (4), (7), (8)
Effort expectancy	'The degree of ease associated with consumers' use of technology.'	(3), (4), (5), (6)
Social influence	'The degree to which an individual perceives that important other (e.g., family or friends) believe that [they] should use a new system.'	(1), (2), (5), (6), (8)
Facilitating conditions	'The degree to which an individual believes that an organizational and technical infrastructure exists to support the use of the system.'	(2), (4), (5), (6)

The UTAUT has eventually been developed toward UTAUT2 by its original theorists (Venkatesh et al., 2012). This extension helps the model to move beyond the application in working and computer environments with three more predictors of intention, namely: *hedonic motivation*[7] (HM), *price value*[8] (PV), and *habit*[9] (HE). As depicted in Figure 2.7, all seven elements positively affect BI, while *habit* and *facilitating conditions* are at the same time predictors of actual *use behavior*. *Age*, *gender* and *experience* (EXP) moderate all of these hypotheses, sometimes as single moderators and sometimes as multiple additive moderators (Montoya, 2019). This, inter alia, means that the older, the more male or more experienced a user is, the more likely *habit*, for instance, will influence their intentions and behaviors.

Due to its explanatory power, simplicity, and transferability, 11000 articles had cited UTAUT in 2011 (Williams et al., 2011). The influence of the model on academic scholarship was hence considerable in the fields of banking (Martins et al., 2014; Baptista and Oliveira, 2015; Alalwan et al., 2018), education (Šumak and Šorgo, 2016; Almaiah et al., 2019; Chao, 2019), and online services (Casey and Wilson-Evered, 2012; Escobar-Rodríguez and Carvajal-Trujillo, 2014; Herrero et al., 2017). Moreover, further meta-analyses found its instruments highly reliable (Dwivedi et al., 2011), and its explanatory strength confirmed (Khechine et al., 2016) in even more applications and extensions. A further advantage of

[7] A key construct from consumer psychology.

[8] A key construct from economics.

[9] A key construct from sociology.

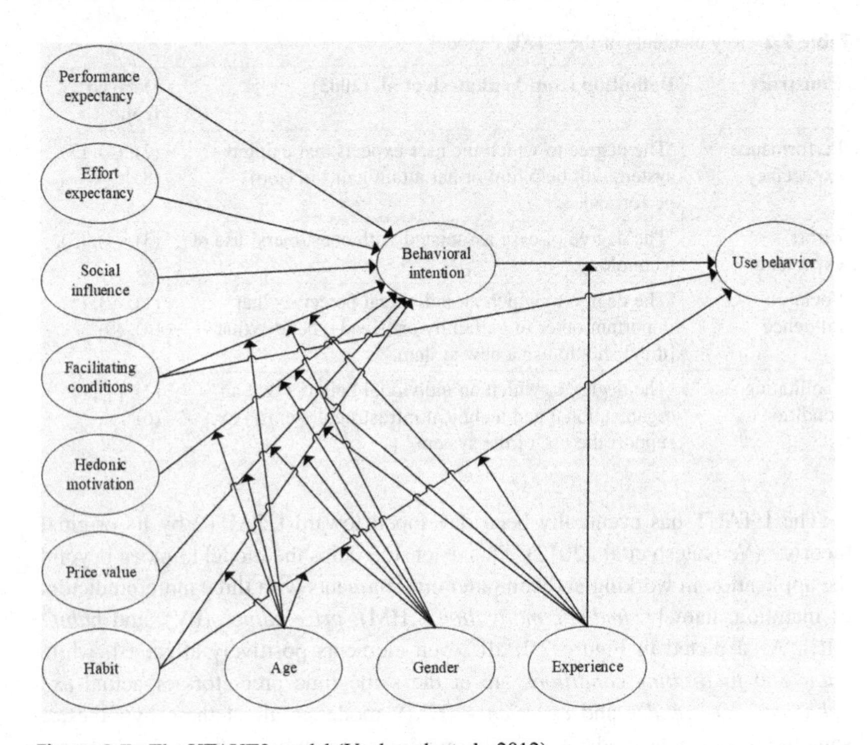

Figure 2.7 The UTAUT2 model (Venkatesh et al., 2012)

UTAUT is that it does not focus on one of the aspects of previous approaches (e.g., utility, social differences, hedonic value, experience) but considers all of them in a balanced way.

Critics, on the other hand, argue that the integration of many differently grounded models causes '*chaos*' for the theory of acceptance research (Bagozzi, 2007) and that high R^2 values are '*artificial*' when they are solely achieved in unpractical multiple additive moderations (van Raaij and Schepers, 2008). The second might be a valid argument since only a few following UTAUT studies even applied or validated the proposed moderators *age*, *gender*, and *experience* (Khechine et al., 2016; Tamilmani et al., 2017). Additional inconsistencies relate to the significances of the proposed predictors of *use behavior* and BI. Especially the assumed roles, effect sizes, and relations of *effort expectancy* (EE), *hedonic motivation*, *habit*, and *price value* are subject to academic dispute, as is the relation between *facilitating conditions* and BI. Other limitations of UTAUT studies are: (Dwivedi et al., 2011; Tamilmani et al., 2018a; Tamilmani et al., 2019b; Tamilmani et al., 2019a):

- Problems with measurement of usage (self-reported usage ≠ actual usage)
- Short-term measurement (measured at a single point of time)
- Single background (one subject, object, and context lead to non-transferable results)
- Predominantly student sampling (non-generalizable data)
- Often too small sample sizes (potential bias and lower explanatory power)

It can be concluded that UTAUT2 is the most comprehensive access to acceptance, but still, its findings are not necessarily generalizable for different application contexts. Venkatesh et al. (2016) confirm this in an extensive literature review and prepare a framework for future research, which I will build upon in Section 2.2 The baseline idea is that UTAUT and UTAUT2 should be regarded as flexible models that must always be explicitly extended to the topic. Furthermore, context such as time, system technology, or user types should be added to refine and understand results better. Generally, a systematic investigation is needed to clarify the formulation of the UTAUT2 and to harmonize conflicting evidence (Herrero et al., 2017).

2.1.3 Literature Review

The present collection of articles is to summarize what is already known about the acceptance of smart mobility and to uncover those issues not yet understood. This review hence describes and analyses the logical context of this thesis. Additionally, the aim is to reveal recurring positive and negative technology acceptance factors. From these findings, a robust model shall emerge.

The search for appropriate literature was conducted in Scopus, Web of Science, local library catalogues, and Google Scholar. I used both the terms 'acceptance' and 'smart mobility' alone as well as combined. I did the same for the selected use cases representatives of smart mobility[10] and checked the following relevant articles using a cited reference search. Overall, I identified 310 relevant publications. The articles finally selected for our review had to meet three criteria. As in Figure 2.8, these are relevance, impact, and novelty. In addition, at least two out of three criteria had to be met. The evaluation criteria were:

[10] See Figure 1.5, Section 1.4.3 for a selection.

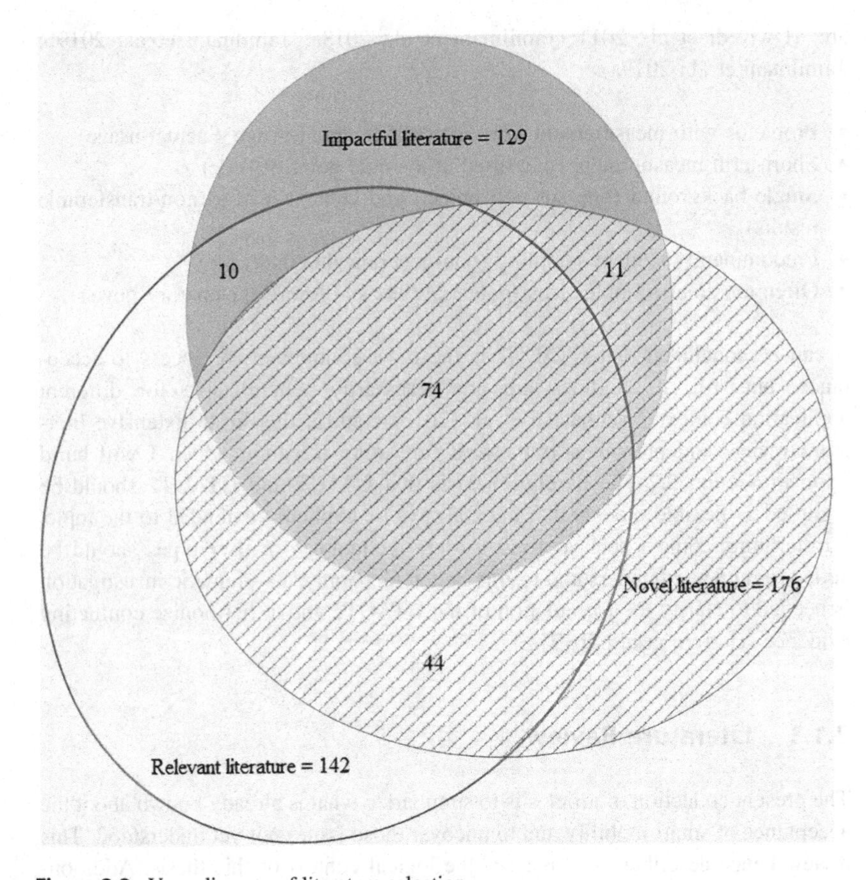

Figure 2.8 Venn-diagram of literature selection

- Impact was measured by number of citation (more than ten).[11]
- Novelty was measured by age (not older than five years).
- Relevance evaluation:
  - The article must reveal new aspects of smart mobility acceptance.
  - From a theory perspective, the scope was set on behavioral and comprehensive modeling approaches.

[11] Following van Noorden et al. (2014), more than ten citations indicate that a paper has reached the top ten percentile of papers worldwide.

- An article was more likely included if its content was unique to cover a rather seldomly investigated area.[12]

Consequently, 171 studies were no longer considered. The review, distributed as outlined in Figure 2.9, eventually included 139 articles, of which just one directly examined the acceptance of smart mobility (Ahmed et al., 2020). However, the authors consider intelligent mobility as the sum of new IoT services in the mobility environment, which does not fully correspond to the current definition in Section 1.2.

The selected acceptance research uses approaches from the social sciences, psychology, and economics, which, in the end, often correspond to each other in different fields of technology acceptance. While general questionnaires on existing attitudes are used at the beginning of the investigation in a particular field, acceptance is usually mapped using models at the end of the development process. Other socio-psychological paths, such as risk or causal analyses, are usually staggered chronologically. Economically oriented acceptance research, on the other hand, deals with forecasts, willingness to pay, or decision experiments (e.g., stated modal choice experiments). Furthermore, researchers distinguish two main types of analysis: descriptive and explanatory studies. Descriptive analyses present and summarize observations of certain aspects of a population sample without considering correlations or cause-effect relationships (Kyriakidis et al., 2015). On the other hand, explanatory studies try to model acceptance factors connectedly and find reasons for detected phenomena (Zhang et al., 2019).

In the following sections, literature is presented according to its content. Different types of studies are presented chronologically, and the most important acceptance factors are worked out against the background of holistic modeling smart mobility with UTAUT2. In addition to the selected use cases, two more categories were established. First, the electrification of the car's drive system is the most advanced and dynamic smart mobility trend at present. It has hence been considered regarding battery electric vehicles (BEVs) and fuel cell electric vehicles (FCEVs). Secondly, it can be assumed that an overarching concept such as smart mobility is similar in its system acceptance structure to related concepts such as smart city, smart grid, or smart home.

---

[12] E.g., the acceptance of eBikes is the topic of nine articles only, whereas AV acceptance is dealt with in 72 sources.

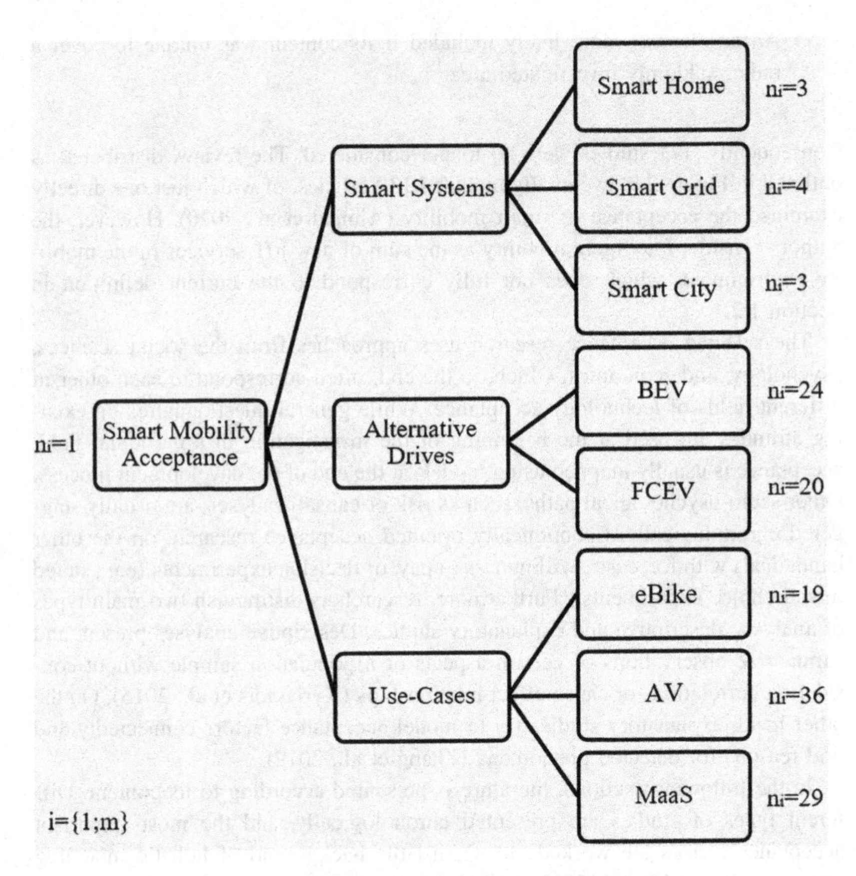

Figure 2.9 Content distribution structure of the acceptance literature review

2.1.3.1 Acceptance of Smart Systems

The concept of smart mobility refers to making the transportation sector intelligent by fundamentally connecting everything within its autopoietic systems (Luhmann, 1997) with everything else. It can be assumed that connaturally constructed concepts must be somewhat related. I discovered 11 examples of similar smart systems studies. One of these analyses looks at smart mobility acceptance itself in predicting IoT service adoption (Ahmed et al., 2020).

A well-known smart system is smart home. Its services can potentially improve daily lives while providing a modern and sustainable lifestyle. A key target group is hence 'digital natives'. In evaluating system performance, they seek convenience, health, and safety. Baudier et al. (2018) reveal that neither any UTAUT2 factor besides *habit* and *performance expectancy* (PE) nor any additional factors such as *sustainability* or *privacy* directly predict smart home acceptance for digital natives. It must thus be considered strongly utilitarian. What Baudier et al. (2018) did not factually examine was the role of *facilitating conditions* as a reflection of lifestyle compatibility, resources availability, and general feeling of control. Park et al. (2017b), on the other hand, examined exactly that with an extended TAM and found that these are most influential. They also found no hedonic motives but a larger impact of *ease of use* inconsistent with the UTAUT discoveries of Baudier et al. (2018) but in line with another TAM study (Shuhaiber and Mashal, 2019). These authors included the dimension of system *trust* and *enjoyment* and found them significant predictors of *attitude toward smart home*. Thus, theoretically speaking, the significance of *hedonic motivation* and *ease of use* may depend on whether it is intended to predict an attitude (TAM) or a BI (UTAUT). After all, acceptance of smart homes strongly depends, just like most IT, on its usefulness, its ease of use, its inception of intrinsic motivation, and the overall facilitating context. Perceived costs and risks oppose these potential benefits. Living is itself as commonplace as mobility. The acceptance factors in smart mobility could in part, therefore, be similar to smart homes.

Another related system is smart energy, i.e., smart grid technology. Smart grid describes the future energy system which integrates the behaviors of all connected users to become safely integrated prosumers. From the start, acceptance for smart grid projects has been emphasized as a socio-economic issue. In the conclusion of their vast review, Bigerna et al. (2016) find that social costs must be investigated more deeply because the perceived costs only reflect private costs, not the collective ones. This is a point Toft et al. (2014) studied in combining TAM with the norm activation model (NAM). They propose that smart grid users are only likely to accept the system if they perceive usefulness positively impacting society and the environment. These assumptions are confirmed in three countries. Hence, they conclude that the promotion of smart grid must not narrowly be based on individual but also on collective benefits. In a following study on EVs, Barth et al. (2016) confirm a significant influence of *collective efficacy* and norms on acceptance.

Acceptance of smart energy is high, yet the uptake remains slow. To investigate this problem, Gimpel et al. (2020) strived to review acceptance factors and integrate them into a comprehensive smart energy model. The model finally reflects UTAUT and extends to include *environmental concern* as a personal belief. In summary, Gimpel et al. (2020) conclude that smart energy is just like smart home, a very rational issue with strong determination through *performance expectancy* and *intrinsic motivation*. In mirroring Girod et al. (2017), *facilitating conditions, ease of use, risk, habit,* and *price value* were found to be insignificant in predicting BI of smart energy. The same applies to *environmental concern*, but in this case Gimpel et al. (2020) can astonishingly enough demonstrate that this belief influences all other factors as a second-order construct. Other studies confirm that risks for users must be reduced and benefits promoted (Park et al., 2017a). By its very nature smart energy is a concept that provides feedback that is not as directly tangible as smart home (e.g., light on/off) or smart mobility (vehicle available/not available). Hence, I assume acceptance of smart mobility might be different.

Smart city is a kind of parent concept to smart energy and smart mobility, although it is geographically limited to urban areas. Nevertheless, it describes integrating different smart systems into comprehensive intelligent surroundings. Others see smart city as the sum of e-government concepts. In this path of research, Habib et al. (2019) think that the seven significant acceptance factors of smart city technologies are: *effort expectancy, self-efficacy, perceived security, trust in technology, trust in government,* and *perceived privacy*. Their proposed model hence focused on issues related to trust and risks. The distinction between technological and administrational trust is important in understanding the cognitive processes surrounding smart services in general. Also, staying in control of what is happening (*self-efficacy*) is essential. *Price value* is also a valid predictor of acceptance. The higher people's estimates of a reasonable price, the higher the acceptance values. Survey participants of Habib et al. (2019) expect the benefits of smart city services to outweigh the costs. These results are already known from mobile services such as SMS or banking, but cannot be taken for granted in the field of smart services in general. Findings show that customer benefit perception seems to emerge uniquely in the presence of high superior functionality (*relative advantage*) in contrast to existing solutions, low *effort expectancy,* and high *perceived usefulness* (Roy et al., 2018). Sepasgozar et al. (2019) add that acceptance benefits from urban smart services being compatible with a personal lifestyle, providing high service quality, and gains in efficiency.

Finally, smart mobility as a system is the object of acceptance in the conceptual study of Ahmed et al. (2020). They build a new acceptance model by enriching the TAM with variables from a digital IoT domain (e.g., *IoT service quality*). They subsume risks associated with smart mobility under *intrusiveness concerns*, and split *social influence* into *electronic word-to-mouth* (for web interactions such as Facebook, Twitter, etc.) and *subjective norms* (e.g., overall normative pressure). The inclusion of *personal innovativeness* moreover aims at demonstrating that early adopters are especially likely to welcome smart mobility. In the end, the model achieves a decent R^2 value of 0.78 with only five significant predictors (e.g., *service quality* and *personal innovativeness*). Particularly strong influences on acceptance are exercised through *attitude, intrusiveness concerns,* and *subjective norms.* Despite the fact that, from the perspective of the present study, Ahmed's work outstandingly explains IoT services in a mobility setup, it cannot adequately explain the acceptance of smart mobility as a holistic system. Methodologically, using neural networks for best model evaluation kind of takes away the theoretical ground of previously drawn hypotheses. Thirdly, not considering factors such as experience, demographics, habits, and costs makes it impossible to recap the modeled acceptance process comprehensively. Smart mobility, by definition, means more than just IoT services and their quality. Nevertheless, Ahmed et al. (2020) provide a stimulus and positive impulse, highlighting that many traditional acceptance factors can be evaluated and positioned in relation to one another from the perspective of a digital society. One example is that systems become so complex that people hardly really understand them, and thus personal perceptions such as the *perceived intrusiveness* gain new and greater relevance. This digital society context must be considered in constructing acceptance of smart mobility. Additionally, it is interesting to find norms so highly influential for smart mobility (Barth et al., 2016), especially since, in other contexts, this influence was rather small (Baudier et al., 2018; Gimpel et al., 2020). This finding may indicate the stability of the existing socio-economic regime in transportation.

In conclusion, we know that smart systems, first of all, must be compatible and useful, i.e., in terms of efficiency gains. If they are to be accepted, they should also be fun, convenient, and reliable (Park, 2020). However, this is true for some systems (smart home vs. smart grid) and some user groups (early adopters vs. others). If tangible, *service quality* has a more significant impact as well as *attitude.* However, mixing TAM and UTAUT, as freely undertaken in some studies, is theoretically impure. Whilst it is evident that improving framework conditions can lead to a higher sense of personal control, the roles of many aspects, from environmental concerns to social impacts, remain vague. Including collective social processes could be a good starting point for further research (Bigerna et al., 2017). Finally, even if BI is positive, habitual behaviors derived from current regimes might severely hinder adoption processes. New experiences might, however, transform these habits. In general, the acceptance of smart systems varies according to the requirements placed on them by an individual. For the interactive 'smart' part of it (e.g., *ease of use, flow,* or *self-efficacy*), acceptance can be compared between smart technologies, which is not so much the case for the manifest background parts (e.g., *social influence, habit,* or *environmental concern*). Table 2.3 sums up our knowledge of smart systems acceptance.

Table 2.3 Potential strengths of smart system acceptance predictors

Predictor	Smart home ($n_1 = 3$)	Smart energy ($n_2 = 4$)	Smart city ($n_3 = 3$)	Smart mobility ($n_4 = 1$)
Performance expectancy	++	++	++[1]	++[1]
Effort expectancy	0[1]	n.s.	++	+
Social influence	n.s.	n.s.	+	++
Facilitating conditions	+	n.s.	+	–
Hedonic motivation	0[1]	+	+	–
Price value	n.s.	+ –	+	–
Habit	+	n.s.	+	–
Trust/risk	+	n.s.	++[2]	++[3]
Environmental concern	n.s.	+[1]	–	–
Sociodemographics	+	–	+[3]	+[3]
Other	++[2]	++[2]	++[4]	++[4]
	[1] significant for attitude only [2] attitude toward smart home	[1] indirect effect and social costs [2] attitude toward smart energy	[1] service quality [2] trust in technology and in authority [3] compatibility with lifestyle [4] relative advantage	[1] service quality [2] perceived intrusiveness [3] innovativeness [4] flow and attitude

++ *dominant finding*; + *weaker finding*; 0 *inconsistent finding*; n.s. *not significant*; – *not tested*

2.1.3.2 Acceptance of Alternative Drivetrains

Smart mobility covers ITS and a variety of new services for mobility that aim beyond personal ownership. Nevertheless, the driving technology of locomotion in a vehicle also remains a central challenge of the future. Electrification of the powertrain is seen as the key to sustainability in this regard, as it enables coupling between the transport and energy sector (European Commission, 2020). If there is sufficient capacity for renewable energies, transport should hence also be able to operate in a climate-neutral way. There has been a debate about onboard energy storage systems for a few years. While some people trust in the further development of battery technologies, others are counting on the advantages of using hydrogen tanks and fuel cells. Whichever side one belongs to, these two competing or, as others say, complementary technologies might face acceptance problems during their diffusional stages. This review will show which aspects can become decisive for the acceptance of battery electric vehicles (BEV) and (hydrogen) fuel cell electric vehicles (FCEV). As a recent survey suggests (Kauschke, 2020), the electrification of personal vehicles might open doors to smart mobility.

Fuel cell electric vehicles

So far, public attitudes to hydrogen mobility have been studied under very different conceptual frameworks, from simple surveys to advanced explanatory models. However, the majority of the literature initially followed a similar path; authors worked in specific hydrogen demonstrations and analyzed public reactions to vehicles in action. Reference is made here to the early lighthouse projects *AcceptH2* (O'Garra et al., 2005), *FuelCell Taxi* (Mourato et al., 2004) or *Create Acceptance* (Heiskanen et al., 2008). At this point, acceptance can be regarded as one of the most studied aspects in the research field of hydrogen mobility. Research has been conducted for a comparatively long time. Twenty relevant studies and project reports were consequently included in this review. The following section recapitulates this knowledge and carves out acceptance factors:

At the start, a larger number of studies identified a widespread lack of awareness and *knowledge*, which does not prevent a generally high level of acceptance (Altmann and Graesel, 1998). 90% of the participants in a field test were satisfied with vehicles and refueling, although 80% initially had to overcome minor technical difficulties (O'Garra et al., 2005). Ricci et al. (2008) add that although acceptance is high, people are not necessarily enthusiastic when confronted with specific vehicles, infrastructure, or large-scale roll-outs. This demonstrates a certain resistance to change, which can also be attributed to the transport sector in general due to its deep interaction with other sectors and long product life cycles of transport vehicles.

The next issue is risk perception, which has so far played a particularly important role in existing studies. Surprisingly, safety concerns appear negligible, whereas reliability concerns are serious (Mumford, 2006; Bultel et al., 2007; Zimmer et al., 2009). Hydrogen is supposed to be just as safe as diesel or petrol. The technology seems trustworthy. Ono and Tsunemi (2017) examined the acceptance of the fueling infrastructure using a risk model. Here, too, 66% of those surveyed showed no fear. Evidently, the best way to eliminate the remaining concerns in their test was to provide preventive information in the form of warning signs or notices at the H2 station.

Economic studies, in turn, suggest that the willingness to pay (as the economic counterpart to psychological acceptance) is still low, at an extra cost of around €1000 at most (Mourato et al., 2004; Chen and Chen, 2013). Hence, *price value* should be more or less level to ICEV, although factually, prices are higher. Walter et al. (2012) show at the same time that, despite the existing price structures, there are already niche markets like municipal service vehicles.

Only a few studies deal with the factors influencing hydrogen acceptance and aim to explain cause and effect. However, this type of analysis could provide a better understanding of the acceptance of hydrogen vehicles. Yetano Roche et al. (2010) present a review. With regard to two different perspectives of acceptance, namely the rational perspective (e.g., risk or benefit perception) and the behavioral perspective (e.g., intention to buy or use), the majority of studies can be assigned to the former. On the other hand, the few behavioral studies that exist concentrate on a small number of potentially influencing factors such as sociodemographics, perceived benefits, or norms. Existing knowledge was also frequently examined. The studies brought in by Yetano Roche et al. (2010) often produced inconsistent outcomes and left out some factors (e.g., habits, environmental impact, image). In addition to knowledge, experience with related technologies can influence hydrogen acceptance (Schulte, 2004). Increased familiarity with hydrogen generally leads to improved well-being and increases the acceptance of hydrogen mobility. In contrast to the NIMBY[13] phenomenon, people living near a hydrogen station show higher acceptance (Tarigan et al., 2012).

Nicole Huijts of TU Delft is the sole researcher to date to have transferred the acceptance of hydrogen vehicles into actual theory-based behavioral models. The group has thus done critical work for H2 acceptance research. Its first work was a review of existing empirical acceptance studies (Huijts et al., 2012). They demonstrated that H2 acceptance research often lacks psychological foundations and uniform terminology. Consequently, in their next research, they use the Norm Activation Model NAM (Schwartz, 1977) and the Theory of Planned Behavior TPB (Ajzen, 1985) to explain the acceptance of hydrogen filling stations more thoroughly. Subsequently, they combined different acceptance models and integrated not only cognitive determinants of acceptance (e.g., perceived benefits and risks) but also affective factors such as positive and negative feelings of fear, satisfaction, or trust (Huijts et al., 2013; Huijts et al., 2014). As a result, the moral considerations of hydrogen proved to be more important than self-interest for supporters of hydrogen mobility. These moral considerations may relate to

[13] 'Not in my backyard' is the socio-economic branding of growing hostile tendencies of residents towards projects in the vicinity (like wind power plants, highways, airports).

the expected positive environmental effects. After all, people are positive about hydrogen. This is a message worth communicating since it pioneers an even stronger *subjective norm.*

Another significant predictor was *perceived behavioral control.* To accept the technology, people want to stay in control of what is happening, where to drive, and when to refuel. Additionally, *trust in authority* and *trust in technology* are found to be significant in Huijts et al. (2014).

Research on H2 acceptance then has stopped at a time when electromobility is gaining momentum, until Schmidt and Donsbach (2016) present a media analysis showing that communication of the H2 benefits must focus on grid independence and decentralization rather than emphasize the ability of hydrogen to store surplus energy. So far, no further modeling of H2 acceptance in terms of behavioral intention has appeared. However, it is assumed that acceptance of hydrogen mobility poses less of a problem than, for example, battery electric mobility since both the refueling times and the achievable ranges are close to today's individual driving conventions.

Battery electric vehicles

In helping to regard BEVs systematically, the literature review of Geis et al. (2016) explains that research on BEV acceptance has evolved from simple usage questionnaires to grey literature behavioral approaches toward system acceptance models (Emsenhuber, 2012; Dudenhöffer, 2013; Fazel, 2013). Just like for FCEVs, the primal acceptance studies were based on empiric observations of piloted field trials (Cocron et al., 2011; Schaefer et al., 2014). Research on BEVs is generally rich. Never before have there been as many BEV publications as in the past decade (Kumar and Alok, 2020). These enclose acceptance. None of the smart mobility-related areas has been studied in such depth. One can thus expect to derive unique insights. To this end, 24 publications from the literature selection might enrich this review.

Selected key findings and known operationalizations are provided in Table 2.4. It becomes evident that limiting and promoting acceptance factors cover various disciplines such as economic, technological, and psycho-social sciences. To understand the acceptance of BEV even better, the effects of attitudes, norms, and cognitive processes were often selectively connected to discover effect sizes and reveal further interdependencies. Additionally, various studies are diving deeper into use situations, user characteristics, business models, and vehicle types.

Table 2.4 Findings on BEV acceptance

Reference	Key content	Related predictors
Sammer et al. (2008)	User requirements for an EV are equal to combustion vehicles. Range, infrastructure, and purchasing price are acceptance barriers.	Satisfaction; perceived costs
Neumann et al. (2010)	Experience with EVs increases acceptance significantly. People are optimistic about EVs when others are.	Experience; subjective norm
Cocron et al. (2011)	Safety aspects, human-machine interaction, and current personal mobility are important for acceptance.	Risk; ease of use; context
Wietschel (2011)	Too little knowledge of acceptance available. EVs are more accepted when they are environmentally friendly.	Environmental concern; perceived knowledge
Emsenhuber (2012)	An extended TAM study compares Austria and Denmark and finds hardly any cultural differences.	Price value; relative advantage; performance expectancy
Tamor et al. (2013)	Acceptance of BEV is not determined through comparing simple vehicle usage patterns but through local mobility alternatives. Plug-In EVs are often beneficial.	Facilitating conditions; economic payback
Catenacci et al. (2013)	Technical factors of battery capacities are vital issues for acceptance.	Perceived quality; perceived range fear

(continued)

Table 2.4 (continued)

Reference	Key content	Related predictors
Dudenhöffer (2013)	Firstly, tests an acceptance model with real-world data instead of survey data. Fails to predict acceptance well but finds large effects of experience on acceptance.	Experience, effort expectancy
Fazel (2013)	Combines and integrates previously discovered acceptance factors. Finds differences in acceptance for sharing or buying BEVs.	Open-mindedness; framework conditions; hedonic motivation
Paternoga et al. (2013)	Concomitant technologies like new payment systems can influence acceptance.	Facilitating conditions, context
Globisch et al. (2013)	Technology affection is good for acceptance. But change and transport constraints cause fear.	Personal innovativeness; perceived fear
Schaefer et al. (2014)	Visibility, status, and accessibility are important. Driving joy is essential for vehicle acceptance. Technology knowledge can reduce fears.	Image; perceived joy; perceived fear
Cocca et al. (2015)	New business models may increase acceptance. User requirements should thus be integrated into the development process. User knowledge can increase market share of EVs. Target groups can be clustered according to cost sensitivity and innovativeness.	User requirements; personal innovativeness; perceived knowledge

(continued)

Table 2.4 (continued)

Reference	Key content	Related predictors
Barth et al. (2016)	Different norms and collective efficacy have equal or even stronger effects on acceptance than cost-related factors.	Collective efficacy; Subjective norm; perceived costs
Nordlund et al. (2018)	Study focuses on policy acceptance. It finds that activating personal normative reasoning in individuals can have a positive influence on the acceptance of BEV policies. They must feel effective and just.	Perceived justice, personal norms, problem awareness
Mohamed et al. (2018)	Behavioral control is the most potent factor of EV adoption. Environmental concern acts as an indirect predictor. Comparative study between seven BEV body types reveals strong differences of acceptance in customer segments.	Perceived behavioral control; environmental concern, user groups.
Park et al. (2018)	An explorative study on BEV acceptance factors discovers cost as negative factor and different attitudes as positive.	Perceived costs; perceived enjoyment
Wang et al. (2018a)	Explorative study finds acceptance overall low unless technical performance improves and marketing focuses on environmental benefits.	Technical performance; environmental awareness; context;

(continued)

Table 2.4 (continued)

Reference	Key content	Related predictors
Schlüter and Weyer (2019)	Technophile, urban, and mobile users find BEV easier to use. Car owners perceive less usefulness. Carsharing experience fosters BEV acceptance.	Experience, user characteristics; ease of use; usefulness
Simsekoglu and Nayum (2019)	Risk assessment: accident risk is not a direct predictor for BEV acceptance. Best predictor is the financial risk.	Perceived risk, facilitating conditions
Wolff and Madlener (2019)	Acceptance of BEV is for the first time higher than combustion vehicles in a logistics field trial. Joy, technophilia and social norms contribute strongly to a TAM extension.	Hedonic motivation, effort expectancy, social influence, environmental concern

If one tries to set together the diversity of findings around BEV, one realizes that acceptance factors are not limited to technology, psychology, and economics alone. Contextual factors and user characteristics are particularly important. Classic socio-demography is hereby less helpful than finer measurements, such as personal mobility behavior, modernity, the wealth of experience surrounding new mobility as a whole, preferences toward certain vehicle types, or even the views on certain political interventions. Moreover, the predictors of acceptance (*actual use* or *intention to use)* are complex by nature and the nomenclature somewhat inconsistent.

Often, modifications of the TAM and TPB are applied. Overall, one significant relationship can be found for almost every factor known in acceptance research. While such individual studies provide stringent and conclusive solutions, the picture of the acceptance process for BEV becomes blurred when looking at all the studies. It may be time for a more comprehensive and statistically supported literature review at this point in time. Particularly relevant for this study is the fact that an integrative comparison of the acceptance determinants has not yet taken place. Individual attempts for holistic models (e.g. Fazel, 2013; Wang et al., 2018a) work with too many construct recreations. A comprehensive, structured investigation using UTAUT, which includes most of the constructs used thus far, has not been realized. This would be the only way to weigh up the strengths of individual effect paths. Potentially insignificant influences could be separated from the important ones and corresponding practical conclusions could be given with a clearer focus and a compelling message.

2.1.3.3 Acceptance of EBikes

Cycling is a trend that has taken off worldwide again in recent years, especially in China, but 'unexpectedly' (Reid, 2017) also in the USA and Europe. For instance, from 2002 to 2017, the share of bicycle traffic in German cities increased by 37% (Nobis, 2019). Reasons for this may, inter alia, be shifting motivations (e.g., health, nature, and youth) as well as improved framework conditions (e.g., new infrastructures, bike sharing systems, and cycling electrification). Biking provides manifest benefits to users: it is often the fastest way to travel distances between one and two kilometers, it avoids and reduces congestion and thirdly, bicycles produce zero emissions. Notwithstanding these advantages, the rise of the bicycle has lately started slowing down in some cities (McQueen et al., 2019). Electric bicycles (eBikes) may be an effective way to reboot this development. They are faster and more convenient than bicycles and can hence potentially combat the car's hegemony in urban commuting since they are comparably cheap. Finally, eBikes help users to cycle further and more often. This bears the chance to multiply the benefits already available through conventional cycling. Since the introduction of lithium batteries in 2005, the eBike market has experienced a sharp upturn. The global forecast is that the eBike market achieves a strong growth, rising from 3.3 million annual unit sales in 2016 to nearly 6.8 million units by 2025 at an 8.2% compound annual growth rate (Hung and Lim, 2020).

As in Figure 2.10, the diffusion of eBikes is going well, the market stably growing. Hence, users are of the early-majority type (Rogers, 1995). In Germany these are mainly younger commuters or older free-time users (Nobis, 2019). Not many problems with adoption and acceptance exist. According to present review, no acceptance model for eBikes is available. Nevertheless, due to its success and the high degree of market penetration compared to other new transport technologies, the eBike offers an exciting starting point to better understand smart mobility acceptance in general. Potentially, other systems can learn from the processes of successful new mobility adoption. Overall, I identified 17 acknowledgeable studies for eBike acceptance. Nine of them directly deal with eBikes, six with bike-sharing and two with innovative on-bike systems.

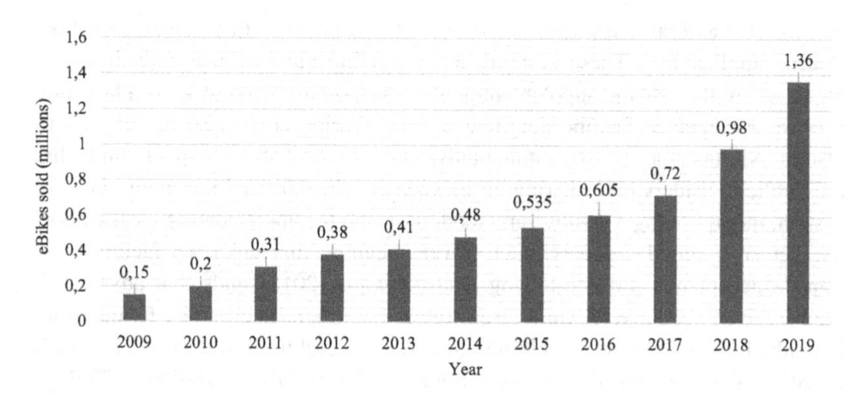

Figure 2.10 eBike sales in Germany 2009–2019 (Ahlswede, 2020)

The topics covered for eBikes range from risky on-road practices (Du et al., 2013), over market and ecologic potentials (Plazier et al., 2017; McQueen et al., 2019), to attitudes and motivations (Chaney et al., 2019; Mayer, 2019). Le Bris (2015) first describes how the adoption of eBikes is shaped by three dimensions: our social context, our relation to materiality, and our experiences and attitudes. All this leads to a habitual script which can trigger the adoption process. These thoughts are in line with existing acceptance theory. Two Dutch studies assessed experiences of the main user groups of eBikes. Plazier et al. (2017) showed that younger users value speed, fun, and ease of eBikes, while complaining about high prices. On the other side, van Cauwenberg et al. (2019) demonstrate that older users enjoy the small riding effort and the ability to bike longer distances. At the same time, eBikes' heavy weight was reported a serious disadvantage. Women show more fears toward injuries and men enjoy cycling alone for recreational reasons. Mayer (2019) interviewed 47 eBikers to understand their motivations. Improving health and savings costs were reported to be the main drivers of eBike usage. The lack of local infrastructure and social shaming from other cyclists were found to be the main barriers. The study of Chaney et al. (2019) analyses the attitudes and perceptions of electric mountain bikes. Their findings confirm certain prejudices and concerns around eBikes. Some bikers call e-biking 'cheating' and 'unnatural', while acknowledging that eBikes can be a way to mobilize unathletic user groups.

Most behavioral acceptance studies identified concern themselves with bike sharing applications. These systems are expanding quickly, especially in urban areas, where they should supplement public transportation and serve as a last-mile service. Acceptance factors identified in bike sharing could also be relevant to eBikes. Nikitas et al. (2016)'s quantitative survey with 558 participants finds that the public considers bike sharing an inexpensive, healthy and pro-environmental system, incorporating a positive and modern image for implementing local authority. Yet still, actual usage is much lower, meaning that unknown factors must impede acceptance. In a following study, Nikitas (2018) finds that advantages are perceived collectively while disadvantages are more likely to be found on an individual basis. Barriers for biking are a lack of cycling infrastructure and lack of safety. This confirms the results of Souza et al. (2014)'s application of TPB, in which they additionally found that these shortcomings are related to the perception of *facilitating conditions*. In a different stream of research, Chen (2016) used a modified TAM and TPB model to analyze the continuous loyalty to a public bike system usage. The findings reveal that social norms and perceived enjoyment have the strongest influence on continuous usage intentions. These influences are positively moderated by environmental attitude. Hence, biking might possibly have to be regarded as a hedonic system influenced by environmental values. Wang et al. (2018b) have similar findings. They state that two acceptance factors must be addressed founded on each other: (1) create a preferable biking environment for convenient, safe and quick use, (2) increase the social and personal desirability to alter social norms.

Other studies in the biking domain deal with the acceptance of innovative on-bike systems that warn of conflicts with other vehicles (Angelis et al., 2017; Prati et al., 2018). The hints these studies provide for acceptance factors are particularly interesting for intelligent eBikes. Accordingly, exemplary factors were *trust*, *attitude*, and *price value*. Generally, acceptance for these systems is high, both on-bike and in-vehicle. Trust is increased while risk perception is lowered. These two facts increase the *intention to use* a bike. With no larger models available for eBikes, the following evidence on influential factors of eBike acceptance can be captured (Table 2.5):

Table 2.5 Aspects of eBike acceptance

Positive	Negative
Compared to cycling	
Higher trust in technology	Higher weight
Less effort	Social norms (shaming)
More fun	Worse facilitating conditions (charging/weight)
Longer distance covered (performance)	Higher price
Compared to driving	
Lower price	Worse infrastructure
Healthier	Less comfort
Social norms (environmental benefit)	More risk
	Unknown negative factors

2.1.3.4 Acceptance of Automated Vehicles

The acceptance of automated vehicles has in recent years been of major interest to practitioners as well as to academics. It is still ongoing. The present review discovered more than 91 articles, of which 36 have been selected for a deeper discussion. Besides applied methodologies, studies can be subdivided according to the technology examined. The focus of present research is on the use of fully automated vehicles as alternatives to today's manual vehicles. Nevertheless, the discoveries of previous studies about related technologies such as automated road transport systems (ARTS) or other lower SAE levels can be just as interesting (ORAD, 2014). Since this discriminatory power through the object of investigation (e.g., the level of automation) is not always given in existing literature, I would rather like to present our current understanding of AV acceptance chronologically. After six years of AV acceptance research (Payre et al., 2014) it is time to see what is comprehensible so far.

Before acceptance research started focusing on AVs, some compelling findings were achieved in support systems for automated-connected driving, i.e., intelligent transport systems (ITS). Jinke van der Laan et al. (1997) first built a simple model assessing acceptance of the big variety of ITS application. This model was grounded on the constructs *usefulness* and *satisfaction* which were both proven significant predictors of acceptance in six different applications. Ten years later, Molin and Brookhuis (2007) included driver attitudes, beliefs and emotions to their model on intelligent speed adaption based on findings of Marchau (2000). In conclusion, they discovered that explaining to users clearly how ITS can contribute to attaining collective societal goals can increase acceptance. The next important work was done by Adell (2009) in first applying the UTAUT paradigm to driver support systems. Since no actual use can be observed, she highlighted the three determinants of BI. While *social influence* and *performance expectancy* were highly significant predictors of acceptance, *effort expectancy* was not. After all, the model explained as little as 20% of the variance in BI. Hence, the author concluded that more constructs such as perceived control, perceived enjoyment, or risks (technology and reliability-related) should be added. Aptly enough and more or less at the same time, Vlassenroot et al. (2010) conducted an exploratory analysis of unknown factors of ITS acceptance and arrived at 14 variables, divided into system-specific indicators (e.g., efficiency, usability, equity, affordability) and general indicators (e.g., background factors, social aims, attitude toward driving). Modeling showed that the *performance effectiveness*, the *performance equity* and the *fit with personal and social aims* had the largest effect on acceptance (Vlassenroot et al., 2011). Automated functions were proven to be welcome only if they suited our current behavior and attitudes. It is obvious and yet still important for policy development: If a driver is likely to speed frequently today, they will not voluntarily embrace an automated system that regulates their doing so as a means of achieving higher-order societal goals (like a continuous traffic flow). The model of Vlassenroot et al. (2010) revealed many new insights, but in the end, the authors themselves had to state that acceptance of automated driving assistance systems is too complex to be modeled both simply and comprehensively at the same time. One solution might hence be to regard acceptance split according to different driving-related task levels (ORAD, 2014). Burnett and Diels (2014) accordingly suggest dividing automated system into information systems (e.g., navigation), assistance systems (e.g., adaptive cruise control), and automation systems (e.g., driverless vehicle). Their overview on human factors of acceptance concludes that information systems are evaluated more by ease and usefulness, while assistance systems are perceived more acceptable if they increase the levels of comfort and safety. For the future fields of automation, reliability, reliance, and trust are considered rich areas of interest. They furthermore add that researchers need to be aware of interaction effects between different in-vehicle systems that may affect future acceptance. The findings on ITS acceptance of Larue et al. (2015) suggest that the cockpit should not, if possible, be overloaded with systems. Additionally, they found proof that on-road flashing markers for an approaching danger find higher acceptance in users than corresponding visual or audio in-vehicle systems. Hence, the AV environment, in acceptance modeling often conceptualized as *facilitating conditions*, might be of importance.

One of the first important works on actual FAV acceptance was published by Payre et al. (2014). They aimed at predicting BI via a newly developed a-priori acceptability scale and contrasted this acceptance measure with *attitudes*, *sensation-seeking* and general *interest in impaired driving*. Outcomes show that, in line with previous findings on partial automation (Nilsson, 1996; Waard et al., 1999), 68.1% of study participants would a-priori be in favor of FAVs. People are obviously positive about FAVs and highly interested. However, this *interest in impaired driving* was not proven a significant predictor of BI, while *attitudes* and a-*priori acceptability* were. The authors assume that the pleasures of driving contradict the interest in impaired driving. Hence, the role of this hedonic pleasure must be further examined. Following this stream of research, the public opinion study of Kyriakidis et al. (2015) provides hints that manual driving is the most enjoyable mode of driving. At best 33% of 5,000 international respondents would prefer fully automated driving to manual driving. The main concern about FAVs were legal issues, safety concerns, and software hacking/data misuse. Despite this, people willing to embrace the idea of FAVs value their price and so a total of 5% are willing to pay even more than $30,000 extra. These groups which are fascinated by AVs and harbor high expectations in the technology are predominantly male. In analyzing why these gender differences of AV perception exist, Hohenberger et al. (2016) add to the discussion that emotional affects (joy and anxiety) mediate the effect of biological sex on the *intention to use* AVs. Other studies could barely detect any effect of sociodemographics (Nordhoff et al., 2018a). After the extensive study of Kyriakidis et al. (2015), other survey-based studies on public acceptance of AVs, i.e. FAVs followed (Piao et al., 2016; Liljamo et al., 2018; Cunningham et al., 2019; Wu et al., 2019)). Piao et al. (2016) found that the greatest benefit of public AVs would be to lower prices thanks to not having driver costs. This seldomly mentioned argument was also confirmed for smart mobility in general in a survey done by Kauschke and Maringer (2019). This emphasizes the roles of *price value* in FAV acceptance. Most people would like to try and experience FAVs for example in carpooling or car sharing before buying one. Prior experience with autonomous transport does positively affect user acceptance (Pakusch and Bossauer, 2017). Next, Liljamo et al. (2018) presented confirmatory large-scale and representative evidence for previous results (Alessandrini et al., 2014; Kyriakidis et al., 2015; Bansal et al., 2016). Overall, they conclude in line with other authors (König and Neumayr, 2017) that FAV adoption will reflect the principles of technology diffusion (Rogers, 1995). Currently, the interest in FAV technology is highest in the segment of early adopters, namely young males who live in densely populated areas and have higher-than-average incomes. However, in another Australian study (Cunningham et al., 2019) the affiliation to the group of early adopters did not necessarily lead to a higher willingness-to-pay. Cunningham et al. (2019) furthermore found that the biggest benefit of FAVs was to mobilize people with driving impairments (e.g., fatigue) or restrictions (e.g., medical issues). People did not believe that FAVs would increase travel efficiency, but they believed they would economize on fuel. This is a contradiction that most likely shows that people are neither able to capture the full meaning of FAVs, nor smart mobility as a whole. Concerns about FAVs revolve around liability issues, ethical considerations, and—associated with both—a general lack of trust (e.g., 85% of the respondents would not let their child travel alone in an FAV). In opposition to Kyriakidis et al. (2015), data privacy was the least important concern. This finding is supported by the results on initial risk assessment of Zhang et al. (2019), who found *perceived safety risk* to strongly influence BI of FAVs via *trust* and *attitudes*, but *perceived privacy risks* as not affecting acceptance. Cunningham et al. (2019) speculate that besides cultural aspects, the ubiquity of digitalization has lowered risk perception quickly within the past few years. Acceptance is currently higher for partial automation technology than for full automation. Psychological barriers evidently exist. The mindsets of users seem to be focused habitually around contemporary modes of transportation. A stepwise introduction of AV features is hence advised (König and Neumayr, 2017). The underlying factor of user resistance toward AVs is, again, assumed to be a lack of trust toward the machine. Studies found that the keys to creating trust in AVs are to solve the existing ethical dilemmas (Adnan et al., 2018) and to increase perceived safety (Zhang et al., 2019; Zoellick et al., 2019).

Eventually reaching modeling acceptance, three approaches are identified in AV literature: stated-preference method (Haboucha et al., 2017), multiple extensions of the TAM (Panagiotopoulos and Dimitrakopoulos, 2018; Sener et al., 2019; Wu et al., 2019; Zhang et al., 2019), and applying the UTAUT model (Madigan et al., 2017; Nordhoff et al., 2018b). As in the preceding section, I want to present these findings pivoted around the question of which aspects have a positive, negative, or non-significant impact on acceptance. First of all, Haboucha et al. (2017) did an exploratory study and developed five significant variables of

AV acceptance: *enjoy driving, technology interest, environmental concern, public transit attitude*, and *pro-AV sentiments*. These attitudes were combined with current mobility habits and willingness-to-pay. Moreover, the individuals could choose whether they would prefer owning a conventional vehicle, owning an AV, or sharing an AV. In combining all these measures to a nested logit kernel model, they conclude the following for AV acceptance: (1) 44% want to remain with their regular vehicle, (2) sharing AVs is as popular as personal ownership, (3) the relative price difference between AVs and regular vehicles is important for acceptance, (4) promoting public transit incentivizes the likeliness to favor AVs, and that (5) educational campaigns on environmental impacts can foster AV acceptance. Applying these thoughts to behavioral modeling of AVs means that existing models should be extended not only to the above discussed dimensions around trust and ethics, but also to environmental, price, and contextual influences (e.g., access to AVs, multimodality) to improve our understanding. This is what several authors have subsequently tried to do by flexibly extending the TAM, i.e., the UTAUT model. Table 2.1 Table 2.6 sums up the results using the UTAUT nomenclature of variables.

We can see that, given the small number of model-based studies available, our knowledge on AV and FAV acceptance is still limited. Additionally, inconsistent findings call for further investigation. While, logically, it seems obvious that easy and convenient transport is desired (Nordhoff et al., 2018a), empirical findings do not all support this hypotheses, in addition to the fact that they do not consistently find significant sociodemographic differences.

Nevertheless, interesting new findings have been achieved in the past years. Especially the significant influence of variables such as *hedonic motivation, environmental concern* and *perceived risk* (PR) is exciting, but not sufficiently validated. On the other hand, it is certainly clear that *perceived trust* and the utilitarian *performance expectancy* impact AV acceptance strongly. Also, *social influence* is a stable variable. The roles of other potential predictors are rarely or never (e.g., *price value*) examined in the context of technology acceptance modeling. The overall explanatory power is small (Adell, 2009; Panagiotopoulos and Dimitrakopoulos, 2018), rather low (Madigan et al., 2017), or achieved by quite unspecific variables (Zhang et al., 2019). Zoellick et al. (2019) advise future researchers to use a mix of attitudes, emotions, and behavioral measures.

Table 2.6 Significance of predictors of behavioral intention to use AVs

Sources	Constructs											
	Performance expectancy	Effort expectancy	Social influence	Facilitating conditions	Hedonic motivation	Price value	Habit	Trust (+) or risk (–)	Environmental concern	Socio-demographics	Other	Explanatory power (R^2)
Adell (2009)	**	n.s.	**	–	–	–	–	–	–	n.s.	–	0.20
Madigan et al. (2017)	**	n.s.	**	**	**	–	–	–	–	n.s.	–	0.22
Panagiotopoulos and Dimitrakopoulos (2018)	**	**	**	–	–	–	–	**	–	**	–	0.44
Sener et al. (2019)	**	**	**	–	–	–	–	**	–	**	**[a]	–
Wu et al. (2019)	**	**	–	–	–	–	–	–	**	–	–	–
Zhang et al. (2019)	**	**	–	–	–	–	–	**	–	–	**[b]	0.61

** significant at $p < 0.01$; n.s. not significant; – not included; [a] perceived safety; [b] attitude toward AVs

2.1.3.5 Acceptance of MaaS

Some of the latest works relevant to smart mobility acceptance come from the area of Mobility-as-a-Service (MaaS). Acceptance research for MaaS is still in its infancy, but urgently needed, as it is a fully user-centered technology (Jittrapirom et al., 2017). In present review, only few acceptance-model-based publications were identified (Schikofsky et al., 2020; Ye et al., 2020). Some of the most impactful studies in MaaS behavioral research use stated-choice approach (Ho et al., 2018) or individual sets of exploratory methods (Polydoropoulou et al., 2018). In total, 29 publications were analyzed.

MaaS is a broad concept grounded in the connection of public transport, sharing economy, and advanced information systems to form and connect new and established mobility services (Smith et al., 2018). It is assumed that acceptance for MaaS will be formed by a mix of the same underlying mechanisms as were for instance discovered for mobile service (Escobar-Rodríguez and Carvajal-Trujillo, 2014), public transit affection (Donald et al., 2014), or open-mindedness toward sharing solutions (Fleury et al., 2017). Following up, the findings will be summarized accordingly. Their relevance for the acceptance of MaaS is discussed by contrasting them with recent literature on the uptake and potentials of MaaS technology. However, before I present the relevant studies on this, especially considering that Smart Mobility is to some extent similar to MaaS, it is important to understand that the acceptance of MaaS, as a ubiquitous and potentially disrupting technology, is multi-layered with more than one object of acceptance. People interacting with MaaS will need to accept the device, the mobile application and the associated means of transportation. Functionally, they merge. Schwanen (2015) consequently speaks of '*app-and-human entities in physical mobility*'.

Mallat et al. (2009) first studied the context in which mobile services will be accepted in the case of mobile ticketing for public transport. Their extended TAM provided two basic new insights. (1) *Use context* fully mediates *perceived usefulness*. People value mobile ticketing exclusively when no other options are available, their need for a ticket was unexpected, or they were in a hurry. For MaaS, this can mean that the best mobile application is the one not needed at all due to an intelligent system architecture incorporating people's nomadic computing. (2) *Compatibility* is one of the most important determinants for acceptance. People fundamentally reject a change in their lives and habits through the app. The app design should therefore rather be based on facilitating the mobility processes they are used to. Usefulness and experienced user-friendliness of the HMI are hereby crucial preconditions for processes like smartphone adoption (Park and Chen, 2007; Kim, 2008). Particular attention must be paid to the requirements of older people, for example by giving attention to making surfaces easy

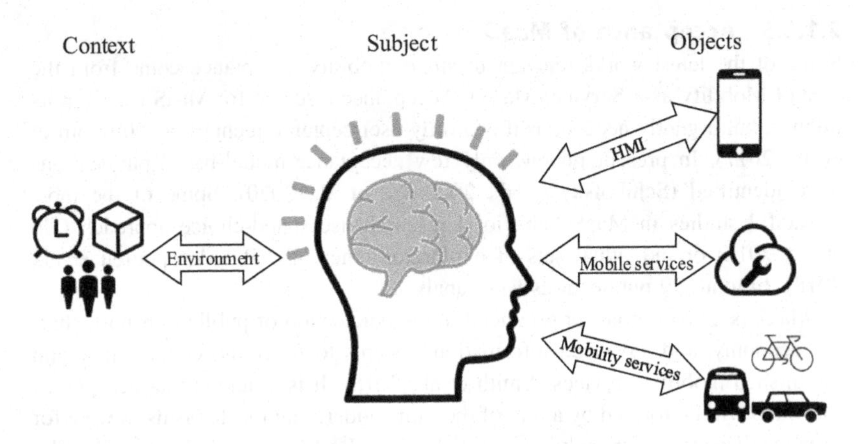

Figure 2.11 Acceptance in the MaaS environment

to interact with and enabling quick and responsive task success. This can increase *self-satisfaction* and *self-efficacy*, which according to Ma et al. (2016) are decisive for the elderly's acceptance of mobile devices.

Figure 2.11 illustrates the presented processes of MaaS acceptance. The further part of this review strives to explain our knowledge on acceptance toward mobile and mobility services since this corresponds to the level of abstraction of smart mobility.

Mobile services

Due to their constantly increasing popularity, mobile applications are of larger interest to the acceptance research community. Their key advantage for users is to provide access to information and services independent of time and place. It can hence be expected that acceptance of MaaS will to an important extent rely on the user experience with mobile applications in general.

Research on acceptance of mobile services can predominantly be found in the mobile banking and payment sector. Other research areas for mobile services are tourism, healthcare, and transportation. The most popular modeling methods are TAM and UTAUT. The review of Tamilmani et al. (2018b) analyzed 16 UTAUT studies and reveals that, of all aspects of acceptance, *performance expectancy*, *trust/risk* and *habit* were the variables with the highest predictive power for BI, i.e., *usage* (*habit*). *Effort expectancy* was least important. One can conclude that people are confident to be able to easily use mobile services, but that trustworthiness and effectiveness of the apps are paramount. Also, the insufficiently studied role of *habit* must be emphasized. Following more free modeling approaches like Yen and Wu (2016) or Saeed (2020), the focus should, in a complex mobile world, rather be on variables such as *habits, ubiquity of services*, or *perceived task control*. These aspects might reflect the reality of mobile service usage better than for instance the standard variables like the ones in UTAUT.

Ooi and Tan (2016) wanted to anticipate the complexity of mobile environments and proposed a mobile technology acceptance model. They add *risk, trust, compatibility*, and *financial resources* to the original TAM. While *financial resources* tested insignificant, *compatibility* was found to be a major predictor. It can be derived that a certain fit with the present lifestyle will foster mobile service acceptance. This coincides with Bouwman et al. (2012), who find that acceptance of a specific service depends highly on the potential user's lifestyle and the social influence of others.

Other authors focus on further details of mobile service acceptance, such as issues related to the so-called 'privacy paradox' (Barnes, 2006). This paradox describes the phenomena that people share personal information willingly even though they are normally very cautious about their privacy. Obviously, the instant benefits received from using a mobile service are often more important than potential risks associated with higher-order normative goals. Additionally, the general public's attitude toward privacy and security is changing. People are becoming accustomed to sharing their data, from names and numbers to live locations and banking accounts (Cunningham et al., 2019).

Moreover, from the outset, there have been major cultural differences in this area. While most studies show that people in Asian countries have fewer problems sharing their data, people in Germany, for example, are particularly skeptical. This might be related to a higher level of uncertainty avoidance and level of social individualism (Baptista and Oliveira, 2015). The omnipresence of data services continues to blur these boundaries. The shaping of data protection in the mobility of the future will become a significant task that requires a social debate (Kauschke, 2020). For MaaS as a new technology, these risks should be examined as an acceptance factor.

Finally, another facet of mobile services can supposedly be revealed in looking at a special kind of mobile service: games. As expected, their acceptance depends less on their usefulness than on the hedonic pleasure they provide (Hamari and Koivisto, 2015). *Ease of use* works as a predictor of the described pleasure. Ha et al. (2007) connect mobile games with Csikszentmihalyi (1988)'s flow theory. Following this theory, a psychological state of 'flow' exists, in which people are totally involved in a holistic job experience. This undertaking is so intrinsically rewarding that it needs no external stimulus. In the study of Ha et al. (2007), *flow experience* becomes a significant antecedent of BI. In terms of smart mobility, this may mean that the overall system experience should be seamless, immersive, and single-sourced.

Mobility and sharing services

The acceptance of mobile services in the transport area has its peculiarities. These depend on the technologies that can be managed by the MaaS user. New mobility services comprise for example multimodal travel information, mobile ticketing, or various kinds of sharing services. Their data- and often platform-driven interplay orchestrates an intelligent transport chain for each user. It is assumed that the same applies for acceptance. The acceptance of MaaS will be closely related to the acceptance of the integrated services. Subsequently, 14 studies are discussed.

The reduction of private car use and the simultaneous incentive of public transport have long been a major objective of public policy and thus, acceptance research. What we learn from behavior psychology is that switching intentions toward public transit are hindered by habitual behavior and the perceived enjoyment in driving as well as the experience of self-determination (Chen and Chao, 2011). This fits with the results of Donald et al. (2014). They summarize that *habit* and *intention* are the main psychological factors in mode choice. The greatest influence on these two factors in turn comes from *perceived behavioral control*. It is larger in public transport acceptance than in car-related studies. These results must be considered of major importance for mobility in general. A promising approach to overcome psychological barriers of modal change is the introduction of multimodal traveler information (MMTI). MMTI are a fundamental part of MaaS. Information presented to users of public transit might include different journey options, costs, comfort, duration, and convenience factors. This increases users' *perceived behavioral control* (Kenyon and Lyons, 2003). Yet still, subsequent research by Geis and Schulz (2017) suggests that behavioral change through MMTI is achieved solely for individuals with weaker habits.

In addition to the systemic view of public transport chains, individual service acceptance aspects will matter for MaaS. Topics covered in existing research range from flight ticketing privacy concerns in navigation apps to various new services in the area of the sharing economy. Escobar-Rodríguez and Carvajal-Trujillo (2014) analyzed the acceptance of mobile ticketing for low-cost carriers using UTAUT. The interesting part of their work is that they combine previous findings, which have been revealed by more specific studies, and comprehensively examine aspects of acceptance. They find that intention is mainly influenced by *perceived behavioral control* (i.a. *facilitating conditions*), *trust*, *price saving*, and *habit*. *Trust* is shaped by perception of safety and security. *Trust* does not matter for actual usage, whereas *habit* and *facilitating conditions* do. This provides evidence that the proposed structure of UTAUT can also be validated in mobility services applications.

Eklund et al. (2016) examined how location privacy impacts attitudes toward a transport app's usage. In pre- and post-evaluating a shuttle bus field trial they find circumstances under which users are willing to trade in their location privacy against desired features. These are: high usefulness, high ease of use, and—as a precondition—the fundamental functional need for location disclosure grounded in the app's purpose. One can assume that if MaaS provides the benefits as forecasted and expected, people will not have a problem to provide personal information.

Within the sharing economy, comparable insights exist. Users seem to be willing to share goods and services despite potential digital and, in the sharing economy, even physical privacy threats. Lutz et al. (2018) consequently call it the 'sharing paradox'. Attitudes and behavior do not seem to coincide. For most people, the monetary benefit is crucial for acceptance and hence dominates negative transaction evaluations related to privacy issues. The higher the privacy threat, however, the larger the monetary compensation necessary. It is only for few people that a *hedonic motivation* can be detected in sharing. Overall, people who are interested in sharing are influenced by *compatibility* with everyday life, *ease of use*, and *social influences* (Burghard and Dütschke, 2018). Sustainability, often operationalized as *environmental concern*, on the contrary, is only 'a nice bonus' (Hartl et al., 2018). Individuals who are interested in carsharing are often young and do not own a car themselves. Nevertheless, they are often also curious about electric or automated vehicles. These affinities seem hence to be related to personal level of innovativeness. As for AVs, this indicates that adoption patterns for MaaS might follow innovation diffusion theory.

The motivations in corporate carsharing (i.e., a company's shared vehicle fleet) are comparable. Fleury et al. (2017) use an extended UTAUT and find no *social influence*. As expected, social factors are often not relevant in mandatory settings

(Venkatesh et al., 2003). Anyway, as in normal carsharing, the *facilitating conditions* and *effort expectancy* variables were very strong determinants. It could be argued that the importance of easy access increases with the amount of estimated personal organizational effort. Making sharing simple will increase acceptance. This can be done by stimulating the *facilitating conditions* by firstly making access to the system as easy as possible and secondly implementing a high level of service for the user (Tsou et al., 2019).

Besides these well-known forms of sharing material goods, the sharing of mobility services has recently received more attention. A good example is ridesharing. This service is currently offered through platforms such as Uber. In contrast to what has just been described for carsharing, however, the TAM-based study of Wang et al. (2018c) finds no connection between BI and *ease of use*. Given the high degree of organization needed for such a service, this is surprising. Wang et al. (2018c) speculate that using intelligent mobile services is already so popular in China that the mere user-friendliness of a service is no longer a sole reason for its use. → Similarly, they detected small positive effects of *personal innovativeness*, *perceived risk*, and *environmental concern*. The other TAM constructs were significant. Finally, in a subsequent study, Wang et al. (2019) dived even deeper into the acceptability of ridesharing by examining expected risks and perceived values created through ridesharing for non-users. In conclusion, the utilitarian value of ridesharing is larger than its hedonic or its social value. Privacy, conflict, performance, and safety risks are of equal importance. Overall, *perceived risks* affect the BI less strongly than *perceived values*. Consequently, promoting ridesharing's benefits is advised as a practical implication. Be that as it may, a serious amount of variance in $R^2 = 0.44$ remains unexplained.

Findings on MaaS

Two very recent studies (Schikofsky et al., 2020; Ye et al., 2020) specifically deal with the acceptance of MaaS. Their approaches admittedly differ. While Schikofsky et al. (2020) aim at exploring motivational mechanisms from scratch, Ye et al. (2020) deductively test a UTAUT paradigm for MaaS. They first expanded the model to embody innovativeness and risk. They did not cover intrinsic motivations, habits, or pricing aspects from UTAUT2. All the hypotheses were found to be true, including *effort expectancy's* effect on BI. The strongest variable was *performance expectancy*. The strongest item within PE was convenience. MaaS can complete many tasks from simple information to booking and payment. Hence, all these processes and their acceptance are reflected in MaaS acceptance, which is then relatively comprehensive.

All hypotheses are also found to be true in Schikofsky et al. (2020). Having said that, the pathways and variables are different. After conducting expert interviews, the motivational factors included in their model were extrinsic beliefs (*perceived usefulness, perceived ease of use*), intrinsic beliefs (*hedonic motivation*), *habitual congruence*, and higher-order *compatibility*. This is eventually based on thoughts about mode choice (Donald et al., 2014; Geis and Schulz, 2017). The authors assume that, conceptualized as second-order constructs, underlying psychological wants exist that make a MaaS user want to experience *autonomy* and *control*, while at the same time feeling a certain *relatedness* to their social peer-group. This can be regarded as an interesting alternative for *perceived behavioral control* and *social influence*. Additionally, *habitual congruence* can be an interesting replacement for Venkatesh et al. (2012)'s *habit* construct, as it is better suited to measure potential future behavior. In summary, all expected relations are strong except the effect of *effort expectancy* on BI. The role of this variable remains inconsistent. In order to practically introduce MaaS, Schikofsky et al. (2020)'s advice to managers is to address autonomy and relatedness as well as lifestyle aspects, whilst simultaneously establishing a well-performing system. An advertisement could, for example, display a likeable person or a sympathetically regarded celebrity who checks their phone saying something like: 'My world. My mobility. My choice.' In that respect, Bamberg et al. (2011) remind us that it can be overstraining for people in an early phase of the introduction of MaaS to think in terms of mobility instead of normal transport. This calls for careful and stepwise marketing.

Table 2.7 illustrates support for the assumption that the acceptance of mobile and mobility services coincides with the acceptance of MaaS. On the other hand, acceptance research in general, with its variables and hypotheses, also reaches its limits in this respect, since MaaS not only has the dimension of a new technology, but also represents a new type of mobility in itself. It is an elementary part of smart mobility and must therefore always be considered fully and systemically in connection with mode choice, individual context, and overall social, psychological processes in the existing automobile regime.

Table 2.7 Potential strengths of MaaS acceptance predictors

Predictor	Mobile services ($n_1 = 13$)	Mobility services ($n_2 = 14$)	Mobility-as-a-service ($n_3 = 2$)
Performance expectancy	++[1]	++	++
Effort expectancy	+[1]	0[1]	0
Social influence	+	0[2]	+
Facilitating conditions	++	++	+[1]
Hedonic motivation	+[2]	n.s.	+
Price value	n.s.	++	–
Habit	++	++	+
Trust/risk	++[3]	+	++
Environmental concern	–	+	–
Sociodemographics	++[4]	+[3]	++[2]
Other	+[5]	++[4]	++[3]
	[1] especially for HMI [2] for games only [3] privacy threats play a minor role [4] age, modern lifestyles [5] flow experience, ubiquity	[1] depends on experience [2] depends on voluntariness [3] personal innovativeness [4] use context, compatibility	[1] indirect effect [2] innovativeness [3] psych. background processes

++ *dominant finding*; + *weaker finding*; 0 *inconsistent finding*; n.s. *not significant*; – *not tested*

2.1.4 Synthesis for Smart Mobility

After analyzing the literature, I want to logically link this study to superordinate theory. In this sense, two of the research questions require consideration:

(1) How should a model that explains the acceptance of smart mobility be designed?
(2) Which factors determine the acceptance of smart mobility, how, and to what extent?

One possibility to answer this is to conceptualize and test a new framework. Its aim must be to push forward the current boundaries of our knowledge. To do so, the things we know and the things we do not know about smart mobility acceptance must be discussed and shall be disclosed in summary. I identify theoretical and substantive research needs.

2.1.4.1 About Acceptance and its Conception

Acceptance is a complex issue. It is not surprising to see that heterogeneous approaches toward explaining acceptance of new technology exist. An enormous number of variables and models were used in the efforts of information system research to explain acceptance. This fact can, to this day, be regarded as a certain limitation that makes results appear desirable and hard to compare. Earlier, Legris et al. (2003) concluded that many of these useful models need integration into a broader approach to include variables related to human and social change processes. The key psychological factors of acceptance must hence be detected and integrated into a model which is comprehensive and yet easy to apply. To the best of the author's knowledge, no such model exists for smart mobility. It would be vital for managers and policy makers to own such a framework for all the smart mobility technologies in order to be able to make wise decisions regarding future design and system setup. Present review shows that the UTAUT model can be a good starting point for making progress on this matter as it is the most recognized comprehensive acceptance model (Sovacool, 2017; Taherdoost, 2018).

The second, more theory-driven need for research lies in the field of locating acceptance within the interplay of users with surrounding systems. While the relation of acceptance to technology is predominantly defined as an individual's use or behavioral usage intentions (Adell, 2009; Huijts et al., 2012), the implications for markets and social systems are insufficiently explored. In disassembling the smart mobility system with multi-level perspective (Geels, 2012), present work helps to understand acceptance as a catalyst and gateway for niche technology to enter and reconfigure the existing socio-economic regime. Suitable landscape pressures such as adaptive policy or social movements can moderate this process (Docherty et al., 2018). Acceptance is hence an ongoing precondition of smart mobility transition.

Thirdly, many terms and definitions are used in acceptance research, inter alia: social and public acceptance, adoption, support, or various kinds of acceptability. These inconsistent descriptions stem from different academic views and historic research focuses. A clear definition is often missing and the object, subject, and context of acceptance are somewhat vague.[14]

---

[14] See Section 2.1. for subject, object and context of acceptance.

This study proposes a new approach. Following the recommendations of Venkatesh et al. (2016) to implement further context and levels in acceptance, I develop a taxonomy of acceptance based on the thoughts of Lyons et al. (2019). They emphasize that, from a user perspective, the cognitive effort to adopt a technology shrinks with a higher level of system integration. In the field of smart mobility, one can find exactly such different levels in relation to the different sub-technologies. For example, the diffusion of the eBike is already much more advanced than that of the electric car, albeit the technology readiness is more or less the same in both cases. It can hence be hypothesized that the level of system integration influences acceptance. The proposed acceptance levels show acceptance as a dynamic process, which is merely logical, because the object, the subject, and the context change with time (Venkatesh and Davis, 2000; Wolsink, 2018). To operationalize these dynamics, our taxonomy relates to the level of user experience and is thus grounded in innovation diffusion theory (Rogers, 1995). This means that the catalyst for the aforementioned adoption must be activated by different factors level-wise, depending on the state of the regime or system. This appears highly relevant for the future, because different triggers to facilitate acceptance should be activated by decision-makers during the stages of the diffusion process. For instance, it can be argued that individuals with no exposure to the object will firstly rely on their perception of general attitudes. Later in the acceptance stages, emotional evaluations and personal norms become of larger importance, before a confrontation with practical feasibility and habitual barriers finally takes place (Farag and Lyons, 2012; Barth et al., 2016).[15] This work tests these assumptions using the example of the smart mobility use cases.

2.1.5 The Acceptance Factors of Smart Mobility

A wide range of factors is potentially able to predict acceptance. When designing adequate models, one often sees factors being used that have already proven to be significant and statistically valid in previous seminal research (Davis, 1989; Ajzen, 1991; Venkatesh et al., 2003). Such models can inherit completely novel variables and contexts that embody the current socio-political research interest. The pool of possible acceptance factors for smart mobility is thus large and hardly standardized. Moreover, even with these kinds of default variables such as *usefulness* or *ease of use*, changes in design and measurement can be observed. It requires unification and interaction detection.

[15] See Figure 2.2, Section 2.1.1.3 for the proposed taxonomy of acceptance.

The literature review collects more than 300 acceptance factors for predicting the *behavioral intention to use* smart mobility. These were derived from quite heterogeneous fields of research.[16] Hence, a method is required to distillate the variables that matter and to channel them to compare the related research with each other. From the observation that about 80% of the used acceptance constructs and moderators can be found in one variable of the model, UTAUT2 was used as a super-instance to allocate acceptance findings. UTAUT2 claims that users will ground their decision toward adoption on seven dimensions (Venkatesh et al., 2012). Variables from the review are assigned accordingly. The results illustrated in Table 2.8 are the starting point for deeper thinking on modeling smart mobility acceptance in Section 2.2. Overall, I find at least two more dimensions worth adding to the model, if, for the sake of comprehensibility, one does not want to discount new influences of social and human change. The dimensions are *perceived risk* and *environmental concern*. The choice of these variables is grounded on the observation that these two dimensions are the most frequently used extensions in literature.[17] The first is the counterpart to technology *trust* and in this fashion supports models in overcoming the criticism concerning innovation positivism (Schepers and Wetzels, 2007). The second is a personal or a collective norm and thus not part of UTAUT's *social influence*, which contains the *subjective norm*.

The acceptance factors of smart mobility have been hardly investigated so far. It is therefore worth comparing it with similar systems and sub technologies. In looking at the following paragraphs, it must be kept in mind that the deficiency of comprehensive modeling leads to results that can be difficult to reconcile due to inconsistent applied methods.[18] However, it was possible to identify the similar kind of constructs across several methods and technologies.

Present literature reveals that smart systems can offer a starting point for discussing acceptance factors although many further aspects of adoption have not yet been sufficiently discussed. All in all, both differences and similarities can be seen when comparing the acceptance of smart systems with each other. One could therefore assume that the similarities may be transferred to smart mobility and that knowledge gaps might be closed in the field of varying results. Interferences could be detected. The existing studies reveal that *performance expectancy* is

---

[16] See Figure 2.9, Section 2.1.3 for a quantitative classification.

[17] Please refer to Table 2.3, 2.7 and 2.8, Sections 2.1.3.1 ff., which have been merged by forming average values.

[18] Referred to the identified 139 most important articles, only 23 applied UTAUT and another six used an exploratory holistic approach. It is possible that factors which are actually significant lose their predictive power when the full range of predictors is considered.

Table 2.8 Potential smart mobility influences by acceptance dimensions

Predictor	Smart systems (n_i = 11)	Mobility-as-a-service (n_i = 29)	Automated vehicles (n_i = 36)	Electric vehicles (n_i = 44)	eBikes* (n_i = 19)
Performance expectancy	++	++	++	++	++
Effort expectancy	0	0	+	0	+
Social influence	0	+	+	++	++
Facilitating conditions	+	++	+	++	+
Hedonic motivation	0	0	+	++	++
Price value	0	0	–	+	+
Habit	+	+	–	++	+
Perceived risk	+	++	++[1]	++	++
Environmental concern	0	+	+	+	+
Sociodemographics	+[1]	++[1]	+	++[1]	+[1]
Other	++[2]	++[2]	+[2]	+[2]	+[2]
	[1] Innovativeness, lifestyle compatibility [2] Service quality, relative advantage, attitude	[1] Innovativeness, age, gender, lifestyles [2] Use context, compatibility, flow, psych. background	[1] Trust, ethics, various kinds of risks [2] Perceived safety, attitude, technology interest	[1] Customer segments, technophilia [2] Perceived policy justice, problem awareness	[1] Age, job [2] Health aspects, nature immersion

++ *strong finding*; + *weak finding*; 0 *inconsistent finding*; – *not tested*; **assumptions derived from non-model literature* repeatedly the key predictor of BI. Thus, comparable to most information systems from early TAM research, the systems can be regarded as utilitarian. Nevertheless, *ease of use* and the *hedonic motivation* are also important motives, at least for some user groups (e.g., early adopters) and some smart systems (e.g., smart home vs. smart grid). The same applies for *social influence* (smart home vs. smart mobility). This has not been researched to a satisfying extent, nor have the roles of *price value* and *environmental concern*. Influences such as *perceived risk*, *habit*, or *facilitating conditions* seem to be positively affecting acceptance. Other variables including *service quality* and *relative advantage* are very likely antecedents of *performance expectancy*, as they measure aspects of usefulness (Venkatesh and Davis, 2000). *Attitude toward use* is found to be a strong predictor in many studies, not uniquely in smart systems. In keeping with the predominant opinion of acceptance research theorists, this study neither discusses nor utilizes this construct as it was proven unnecessary in explaining variance of BI (Goodhue and Thompson, 1995; Teo, 2009).

The concept of MaaS can be a second whistle-blower to smart mobility acceptance. It comes closest to the idea of smart mobility as a whole and should therefore receive special attention. These encompass digital mobility services and contexts, human machine interaction technologies and sharing services. What we learn from the review is that acceptance of MaaS can relate not only to the concept, but also to its parts[19]. I hence aggregated findings from 29 studies which had been clustered to three fields.[20] Only two studies used behavioral models to comprehensively discuss MaaS acceptance (Schikofsky et al., 2020; Ye et al., 2020). They must be recognized as particularly important. Finally, it becomes visible that MaaS, just as smart systems, is evaluated strongly through its *usefulness*. Surprisingly, the *effort expectancy* is not a predictor of acceptance, whereas *facilitating conditions* have a large effect. In the MaaS environment, effects of *social influence* and *environmental concern* can be detected. It is valid to assume that these effects are also encountered in smart mobility, as they are also evident in the other mobility technologies of this study. Effects appear to become stronger, the more discernible a technology is as part of everyday life (level of system integration). Thus, risks associated with MaaS are manifold reaching from perceived privacy to conflict to performance risks. *Habit*, on the other hand, appears as a valid predictor in most studies. This could be because many people see MaaS as a counter-draft to owning a vehicle and thus the habits associated with it. Another interesting revelation is the role of psychological background processes

[19] See Figure 2.11, Section 2.1.3.5 for the MaaS acceptance environment.

[20] See Table 2.7, Section 2.1.3.5 for details.

(*autonomy, relatedness, compatibility*) as potential second-order predictors. Further investigation is recommended. Socio-demographically, the clientele of early adopters (male, young, urban, academic) tends to be the most positivistic toward MaaS.

Smart mobility contains more than MaaS. The trending areas of automation and electrification are subject to larger socio-political discussions (Parkhurst and Seedhouse, 2019). The acceptance factors of these two fields can also be sorted and evaluated in the UTAUT setup.

Electrification is the most tangible trend that is just about to break through on the market. For electric vehicles, hydrogen energy storage is preferred by customers over battery. Thus, acceptance is generally higher. Conceivably, *habit* is of exceptional importance in this regard. *Performance expectancy* (*reliability*), *facilitating conditions*, and *social influences* are of large importance to electric vehicles acceptance. The largest risk associated with electric vehicles is range anxiety due a perceived deficient infrastructure. *Price value* is found a significant predictor too. One could therefore assume that the price begins to play a role in acceptance as soon as a market appears. The role of *effort expectancy* remains conflicting. It is possible that driving is considered too ordinary for drivers to even be considered challenging. The driving joy, operationalized as *hedonic motivation* is of major relevance, typically for acceptance studies on cars (Tamilmani et al., 2019b). Further exciting evidence on electric vehicle acceptance was found looking at different customer segments and checking vehicle body types. The study of Mohamed et al. (2018) reminds us that people today articulate finely differentiated requirements toward a vehicle and its utilization profile. Thus, for being accepted, vehicles have to be designed to fit in contexts. One of them may be smart mobility. These adoption scenarios are often ignored.

The eBike displays slight differences in supposed prediction strengths in Table 2.8. Since no acceptance model exists, my assumptions will be presented. First, as is the case with all mobility technology, I believe *performance expectancy* would be a highly significant predictor. Second, it can be suggested that *effort expectancy* is of important for e-biking. Especially older users seem to value lower effort (van Cauwenberg et al., 2019). Third, the *social influence* should also be large. One must ascertain whether the effect is unexpectedly negative as investigation lets us presume that users are confronted with many emotions from social shaming to public aggression (Chaney et al., 2019). I expect the *facilitating conditions* to play a smaller role, since *self-efficacy* can be exercised directly. Although, for example, the infrastructure might be worth improving, it is already far more developed than for EVs or AVs. Moreover, eBikes are inexpensive compared to cars, but they are less safe. I thus expect a positive influence of *price value* and negative of *perceived risk*. In addition to rational arguments, many people, mostly young users, cycle for fun, for health, or simply to experience nature. *Hedonic motivation* will therefore powerfully contribute to acceptance, in my opinion. For those users willing to switch from car to bike, *environmental concern* could very well also be a valid point toward using the eBike (Nikitas et al., 2016).

Finally, the acceptance of fully automated vehicles is currently being researched at a rapid pace and discussed in important journals. Nevertheless, not all UTAUT acceptance dimensions have been satisfactorily examined in detail so far. Many authors have confirmed that *performance expectancy* is significantly associated with the *intention to use* AVs. The same accounts for *facilitating conditions* and *social influence* (Madigan et al., 2017; Buckley et al., 2018; Panagiotopoulos and Dimitrakopoulos, 2018). The construct of *effort expectancy* displays inconsistent results in AV acceptance.[21] Perhaps AV usage is considered too simple or *effort expectancy* works differently than what UTAUT assumes. Inconsistencies about *effort expectancy* are mainly found between UTAUT and non-UTAUT studies. What we also notice about AV acceptance is that driving joy seems to be declining as automation increases (Kyriakidis et al., 2015). Still people can perceive enjoyment from fascination about technology, speed, or relaxing. At the same time, a feeling of anxiety may arise, being under the control and mercy of the machine (Bansal et al., 2016). The counterpart to this risk perception would be *trust. Trust* has been subject to many investigations around AV acceptance. They concerned ethics, trust in authority, and trust in technology. Other aspects are privacy and safety issues or even social equity (Liljamo et al., 2018). Hence, since the subject of trust was already investigated in-depth, the present study rather checks for risks than for trust. In the AV environment this, inter alia, concerns machine failure, machine intrusiveness, or human-machine accidents. Two issues rather seldomly investigated are *price value* and *habit*. This seems logical since AVs are neither for sale, nor could habits develop. What's more, in the acceptance process designed here[22], their relevance grows individually at Level 4–5. Thus, these issues may remain unsolved until automated vehicles are available on a consumer market. Despite this fact, there are arguments to assume that there might be a predictive relevance. For one thing, AVs are probably expensive at the beginning, which would influence the evaluation of the price leadership. Despite this, AVs could also make mobility collectively cheaper, since fewer resources and entities of the mobility system would be in demand in total

---

[21] Compare i.a. Table 2.6, Section 2.1.3.4 for conflicting evidence on AVs.

[22] See the level of acceptance in Figure 2.2, Section 2.1.1.3.

(Piao et al., 2016).[23] In observing the influences of habit, Schikofsky et al. (2020) teach us to elaborate the construct to *habitual congruence* or beyond, when examining future technologies. Last but not least, the role of sociodemographics in AV adoption was analyzed in many studies. The characteristic as of the early adopters crystallizes here too. Berliner et al. (2019) conclude that, above all, those who perceive AVs safer than non-AVs are interested in buying an AV.

Ultimately, we can recognize many acceptance factors for different applications of smart mobility. We are also aware of numerous acceptance process variants. What we do not know is how to make use of a process that includes these different approaches and creates a sphere of comparability for the entire smart mobility system in order to understand the variety of technologies both horizontally[24] and vertically.[25] The dynamic interplay between constructs and their context must be especially recognized. Presented outcomes create a sound basis for the formulation of a new model that can help understand the shift from automobility to smart mobility.

2.2 Conceptualization

Results from the literature review and previous thoughts on embedding acceptance in smart mobility transitions led to establishing a new framework. I wanted to develop and validate an overarching and yet simple acceptance model for smart mobility. This model should offer a strong explanatory power and significant path coefficients that are transferable between use cases of smart mobility. The goal is to provide a deeper and more precise understanding of smart mobility acceptance. Three steps are necessary to achieve this. First, UTAUT2 is used as a starting point and adapted to smart mobility. Second, the model is extended on the basis of present literature analysis. Lastly, the dynamics of smart mobility context are integrated into the model as proposed by the models' theorists (Venkatesh et al., 2016).

2.2.1 Model Development

UTAUT2 is likely to be one of the most useful and robust tools for examining technology acceptance for all kind of IT-related technologies. It includes eight

[23] E.g., in AV-based transport, no driver would have to be paid.

[24] E.g., from the system to the service to the technology.

[25] E.g., between different services or technologies.

seminal theories from acceptance research and syncs them into seven salient dimensions of acceptance. In 2017, five years after its publication, it had been used for conceptualization after having already been used more than 150 times for conceptualization purposes (Tamilmani et al., 2017).

Despite this, it has not been applied to smart mobility. To date, practically no study has sought to identify the psychological determinants of the systems smart mobility use or usage intentions. The explanatory power of existing models in the smart mobility environment is often low, with as little as 20–40% explained variance in the intention of use (Adell et al., 2014; Madigan et al., 2017; Schlüter and Weyer, 2019), whereas UTAUT2 can potentially reach up to 70% explained variance in BI and 50% in use. Moreover, models are becoming more comprehensive (Huijts et al., 2012; Gimpel et al., 2020), but lack a unified structure. Furthermore, we know that for a complete picture, social standards, technology, economy, and environmental aspects must be considered in the way they interact with each other (Legris et al., 2003; Vlassenroot et al., 2010).

In filling this void, the present study will use UTAUT2 and elaborate on it with the most promising predictors from previous research. This means that foremost, and in order to retain the unified structure, I stay within the boundaries of the main hypotheses proposed in Venkatesh et al. (2003) and Venkatesh et al. (2012):

2.2.1.1 Behavioral Intention to use

Social psychology supports the notion that intentions cause actions. Some might be abandoned, some might be revised in the light of new circumstances, but they generally lead to actions (Bandura, 1977). This behavioral *intention to use* (BI) a piece of technology hence predicts its actual *use behavior* statistically significantly in a number of seminal explanatory models of human behavior (Ajzen, 1985; Davis, 1989; Venkatesh et al., 2003). BI is therefore one, if not the central measure to capture acceptance. There is no reason to doubt that this holds true for smart mobility. The construct variable is measured with four items which were adapted from original works.[26]

H1.	The behavioral intention to use smart mobility[27] positively affects its usage.

[26] Please see Table 2.11, Section 2.3.3.1 for the list of items.

[27] Respectively the sub-use cases Level 5 automated vehicles, mobility-as-a-service and e-Bikes.

2.2.1.2 UTAUT Constructs

Originally, the model contains four independent variables: *performance expectancy* (or *perceived usefulness*), *effort expectancy* (or *perceived ease of use*), *social influence* (or *subjective norm*), and *facilitating conditions* (or *perceived behavioral control*). Thus, UTAUT can explain around 40% of variance in BI and 25% of variance in usage. Definitions and origins of the variables are expounded in Table 2.2, Section 2.1.2.7. All measurement scales were taken from original sources and have eventually been extended with newly developed items to suit a smart mobility context.

Performance expectancy

Performance expectancy is the degree to which a person believes that the system will help them accomplish a task. For the mobility application, it is the assessment of the subjective benefits a smart transport technology offers to an individual. This construct is historically the strongest predictor of acceptance across various models and application areas. Different reviews, including the present one, demonstrate 90–95% significant path coefficients between with large or medium effect sizes (Schepers and Wetzels, 2007; Khechine et al., 2016). The utilitarian nature of mobility makes it self-evident to assume a strong influence of *performance expectancy* on BI.

H2.	The performance expectancy toward smart mobility positively affects the behavioral intention to use the system.

Social influence

Humans do not live in a social vacuum. Our actions are determined by the most diverse forms of norms and collective social processes, which subjectively affect the individual. Despite this, popular early theories of acceptance like TAM initially neglected these influences until the *subjective norm* construct from TRA was included in TAM2 (Venkatesh and Davis, 2000). Nevertheless, problems in psychometrics and theory remained (Taylor and Todd, 1995b). Davis observed that it is hard to distinguish if usage behavior is caused by the influence of others on one's intent or by one's own attitude. So, they also added *image* and *voluntariness* as predictors of perceived usefulness to model social influence processes comprehensively. Other theorists followed different paths. For example, Thompson et al. (1991) included the variable social factors while Malhotra and Galletta (1999) extended TAM with the construct *psychological attachment*. Finally Venkatesh et al. (2003) find that, albeit being differently labelled, all these attempts contain the implicit notion that an individual's behavior is influenced by the way in which the individual expects others to view them after using or adopting a certain technology. They hence constructed the variable *social influence* that channels these different streams of prior research. The focus of the resulting new construct is thereby compliance with others' behavior and not reaching for social status gains as a motive for adoption (as proposed by Warshaw, 1980).

Social influence (SI) has been proven a valid predictor in many studies since. Khechine et al. (2016) found significant correlations between SI and BI in 54 out of 71 cases with stable medium effect sizes. In mobility, the role of social processes seems especially strong (Barth et al., 2016). This could be since, in comparison to computer work, for example, it takes place primarily in public. Furthermore, Klöckner (2014) reminds us that the *social influence* will affect intentions particularly strongly in the pre-decisional stage with little experience available. The following hypothesis can conclusively be made for future technology smart mobility.

H3.	The social influence of others toward smart mobility positively affects the behavioral intention to use the system.

Facilitating conditions and effort expectancy

The construct *facilitating conditions* includes cognitive issues in the areas of framework conditions (e.g., access to infrastructure, regulations), compatibility (e.g., with daily routines or other technical systems), and other perceived restrictions (e.g., control beliefs, mental barriers). The construct has repeatedly been found a significant influence for usage and usage intentions in UTAUT validations (Dwivedi et al., 2011) and in smart mobility related applications (Escobar-Rodríguez and Carvajal-Trujillo, 2014; Kapser and Abdelrahman, 2020). However, it must be considered that the constructs *effort expectancy* and *facilitating conditions* intersect in the area of perceived complexity.

To understand the issue at hand, one has to look at *effort expectancy*. Schepers and Wetzels (2007) conducted a quantitative meta-analysis and found *effort expectancy* to be a significant and positive predictor of acceptance in a salient 51 of 53 TAM-based studies (96%). However, they also found that this effect was mainly mediated through *performance expectancy*. By the same token, the UTAUT review of Dwivedi et al. (2011) highlights that only a minority of 19 out of 43 studies provided evidence for a direct relation between *effort expectancy* and BI. Apart from that, Fleury et al. (2017) demonstrate that *facilitating conditions* were almost fully mediated through *effort expectancy* when it comes to corporate carsharing acceptance. Overall, the relation is unstable and partially inconsistent. Many studies found explanations for this problem. First, Davis et al. (1992) themselves find *effort expectancy* to potentially be an indirect indicator of acceptance, affecting intention via *hedonic motivation.* Subramanian (1994) adds that *effort expectancy* might lose its predictive power when systems are by their inherent nature too simple to use. Igbaria et al. (1995) twist this idea around by outlining that factors such as *performance expectancy*, which are determined by real life necessity, might have priority for acceptance. Gefen and Straub (2000) assume that *effort expectancy* directly affects adoption solely when the primary task of a system is associated with its intrinsic involvement characteristics such as providing satisfaction of interests or value-based needs (Jackson et al., 1997). In mobility, this would mean that *effort expectancy* is not directly significant for BI unless smart mobility is considered to comprise these intrinsic features. Venkatesh et al. (2003) argue differently, theorizing that *effort expectancy* can never be significant in the presence of *facilitating conditions.* This is because issues related to support infrastructure are captured in both constructs. For example, if somebody has access to knowledge (e.g., online tutorials), then this will make technology easier to use for the individual. Logically, Venkatesh (2000) found evidence for mediation effects[28] of *facilitating conditions* on intention by *effort expectancy.* Lee and Larsen (2003) generally assign *effort expectancy* a controversy role throughout literature and emphasize the essential effects of *effort expectancy* as an antecedent of variables, rather than a parallel and direct determinant of acceptance.

All these could be valid assumptions in smart mobility acceptance. Against the background of an ongoing discussion and in order to obtain a deeper understanding, this study aims to test both variables as suggested by Venkatesh et al. (2012). Here, the authors argue that, if applied outside of an organizational setting, the environmental impacts on each individual will vary strongly across consumer segments and devices used, so that *facilitating conditions* come back overtaking the role originally postulated by *perceived behavioral control* in TPB (Ajzen, 1991). For the present study, this means that the higher the level of perceived control about smart mobility, the higher their acceptance and use (adoption). The UTAUT review of Khechine et al. (2016) also supports this assessment, showing averagely large and medium effects for H4–H6.

---

[28] Mediation refers to indirect effects exerted by one variable on another through a third variable, then called a mediator.

H4.	The effort expectancy toward smart mobility positively affects the behavioral intention to use the system.
H5.	The facilitating conditions of smart mobility positively affect the behavioral intention to use the system.
H6.	The facilitating conditions of smart mobility positively affect its usage.

2.2.1.3 UTAUT2 Constructs

The second version of the model included three new predictors of BI: *hedonic motivation* (or intrinsic motivation), *price value* (or perceived value), and *habit*.[29] The inclusion of those variables helped the model to increase its explanatory power by a relatively large 0.10 for R^2 of BI and usage. Definitions and measurement scales were adapted to smart mobility from listed sources (Table 2.9).

Table 2.9 New dimensions in the UTAUT2 model

Construct	Adapted definition	Supporting literature
Hedonic motivation	The degree of fun or pleasure derived from using a technology, apart from anticipated consequences.	Davis et al. (1992), Van der Heijden (2004)
Price value	The result of a cognitive tradeoff between the perceived benefits of the applications and the monetary cost of using them.	Dodds et al. (1991), Elder-Vass (2019)
Habit (expectancy)	The extent to which an individual expects to perform behavior automatically due to learning.	Limayem et al. (2007), Schikofsky et al. (2020)

Hedonic motivation

As pointed out in the description of the hedonic approach in Section 2.1.2.4, there is a rather long discussion about the roles of extrinsic and intrinsic motivations in psychology as well as in economics (Davis et al., 1992; Brown and Venkatesh, 2005; Teo and Noyes, 2011). In technology acceptance research, Van der Heijden's (2004) view has meanwhile become the prevailing one. They distinguish between hedonic and utilitarian systems and confirm that for a hedonic system,

[29] See Figure 2.7, Section 2.1.2.7 for the UTAUT2 model.

perceived enjoyment has a larger direct effect on BI than *perceived usefulness*. Of course in reality, systems are often hybrids between hedonic and utilitarian and thus both constructs contribute to acceptance explanation at varying levels. Tamilmani et al. (2019b) call it 'the battle of brain versus the heart'. In their review, they confirmed this pattern of partially significant and non-significant (50:50) HM-BI paths throughout UTAUT2 applications. I expect *hedonic motivation* to play a major role in smart mobility acceptance together with *performance expectancy* since mobility, which by its nature is utilitarian, comprises many intrinsic characteristics, comparable to driving vehicles for fun, to relax, or to explore.

H7.	The hedonic motivation for smart mobility positively affects the behavioral intention to use the system.

Price value

Looking at things cohesively, technology is not only employed in organizational use settings, but also in consumer markets, where pricing structures matter for acceptance and the perception of value (Zeithaml, 1988). This is why Venkatesh et al. (2012) added the financial variable *price value* (PV) from marketing research (Dodds et al., 1991) to their model. They define it as 'the cognitive trade-off between perceived benefits of the applications and the monetary costs for using them'. It can be regarded as the positive antagonist of the construct *perceived costs* that is also a popular financial variable (Jarvenpaa and Todd, 1996). It has been validated in a variety of technology acceptance research (Escobar-Rodríguez and Carvajal-Trujillo, 2014; Morosan and DeFranco, 2016; Alalwan et al., 2018), but also failed to predict acceptance in some studies (Baptista and Oliveira, 2015; Aswani et al., 2018; Kapser and Abdelrahman, 2020).

Problems with PV seem to surface when customers are neither able to estimate the benefits of a service, nor to place a price tag on its utilization. Additionally, there is often no direct entry or usage cost with new IT services. Hence, no more than 32% of UTAUT2 studies in the review of Tamilmani et al. (2018a) even utilized *price value*. In the context of a future concept like smart mobility, which is cognitively not yet captured by the public (Manders and Klaassen, 2018), it is thus reasonable to measure PV as a relative-sized factor in this way. Its value is best understood as a view of an imprecise price that a good or service ought to exchange based on the currently perceived value it offers. Elder-Vass (2019) argues that views on this are shaped normatively by lay theory and related institutions advancing them. A previous study on smart mobility has shown that people expect smart mobility, above all, to reduce their mobility costs (Kauschke, 2020). Moreover, new mobility technologies are often perceived as expensive (e.g., electric or hydrogen vehicles) as long as economies of scale are not yet realized. Hence, the assessment of a better relation of price and value should have a positive impact on acceptance.

H8.	The price value of smart mobility positively affects the behavioral intention to use the system.

Habit expectancy

Habit is the most important theoretical addition to UTAUT by Venkatesh et al. (2012). It aims to challenge the role of BI and *facilitating conditions* as lone predictors of system use by demonstrating high predictive power in UTAUT2. It includes the dimension of time but may not be confused with experience or resulting longitudinal intentions themselves. So, what is the meaning and relevance of the construct?

Habit can be separated from intentions in two ways. First, it can be the cause of intentions and attitudes which are formed through repetition (Ajzen and Fishbein, 2000). This is a conscious act. If, for example, an office worker drinks coffee everyday while checking their emails in the morning, they might develop a positive attitude toward this connection. So, every time the worker sets up the office and work in the morning, they will (guided through a positive relation of intentions and attitudes) automatically perform the behavior of preparing coffee. Second, it can influence usage directly though triggering processes that are not consciously cognitively controlled. This means performing certain behavior if certain cues are encountered. Staying with our example, this means that the office worker would be finding themselves with coffee in front of their PC in the morning without even having thought about preparing the coffee. These direct influences are not captured in intention-based models such as TPB, TAM or UTAUT (Verplanken et al., 1998). Venkatesh et al. (2012) further explains experience is a necessary but not sufficient precondition for establishing a habit. If different individuals were exposed to a technology for the same amount of time, different habits would still yield due to different interactions with and use of the technology.

Research offers different operationalizations of *habit* which bear witness to different points of view. For instance, one stream of research considers *habit* as prior use behavior (Verplanken et al., 1998; Kim and Malhotra, 2005). Another stream emphasizes the automaticity of which processes are performed (Aarts and Dijksterhuis, 2000). Finally, Limayem et al. (2007) conceptualize *habit* as a survey-based perceptual construct reflecting the subjective sum of related prior experiences. Such a construct can show a stronger effect on use than BI and has worked well in many different studies. I hence adopted their definition.

Another interesting theory comes from Schikofsky et al. (2020), who try to solve a major problem related to *habit*. This issue is about the impossibility of measuring the influence of existing habits if habits have not yet consolidated themselves, simply because the object of acceptance is a fully future technology. Tamilmani et al. (2019a) found that in their review on 66 articles that 65% excluded *habit* from their UTAUT studies due to this shortcoming. What Schikofsky et al. (2020) consequently did was to describe *habit* as habitual congruence. Habitual congruence describes the fit between the perceptions of a new technology and affiliated current behavior toward related technology. This understanding suits a smart mobility context, as it will not be the habits concerning smart mobility that will eventually hinder adoption, but it will be the beliefs about the non-establishment of those habits. These beliefs are shaped by the existing habitual schemata regarding related mobility technology, such as the car. Along these lines, I conceptualize habit for future technology as *habit expectancy*.

Intentions make you start technology employment; habit lets you continue to use them. People thereby tend to forget about the initial effort in learning their behavior. At the same time, they also develop mental barriers to learning something new, if they can instead rely on their habits. Path dependencies of technology are the result. It is by these means that the automobile became part of western civilizations' DNA and a cultural habit (Sovacool, 2017). Overcoming this focus could be the biggest challenge in mobility when it comes to using smart mobility (Lyons et al., 2019). New mobility options that are congruent with currently perceived schemata about the technology could be a way out. They might be able to eliminate cognitive discomfort and reduce adoption barriers. Overall, I still follow Venkatesh et al. (2012)'s assumptions and conclude:

H9.	Habit (expectancy) toward smart mobility positively affects the behavioral intention to use the system.
H10.	Habit (expectancy) toward smart mobility positively affects its usage.

2.2.2 Model Extension

In order to obtain a deeper understanding of smart mobility acceptance, this study aims to parsimoniously extend UTAUT with dimensions that go beyond the existing psycho-economic, social, and hedonic rationales. As seen in the previous review, there are countless possibilities to enrich the model. However, I find that a large number of new constructions are pushing in the same direction. Two major streams are identified: risk associated with the use of technology, and environmental considerations surrounding new systems and services. Hence, these aspects might generally be suitable to enhance the comprehensiveness of UTAUT.

2.2.2.1 Perceived Risk

Prior research indicates that smart mobility technology may appear riskier to potential users than conventional mobility (Baptista and Oliveira, 2015; Kyriakidis et al., 2015; Mayer, 2019). This can cause anxiety and discomfort. Therefore, the supposition would be justified that this perception of risk negatively impacts system acceptance. Additionally, there are strong theoretical and practical reasons to add the well-established *perceived risk* to the model:

First, Stone and Winter (1987) view risk as loss of expectation. The more certain this anticipation of loss is perceived, the greater the value of risk for the individual. In this case, things to lose would be all the benefits incorporated in current behavior. *Perceived risk* is hence the counterpart to the positive UTAUT variables such as *hedonic motivation* or ease of use. Fear is the opposite of fun, complexity the opposite of ease. The integration of *perceived risk* can help the model to overcome criticism concerning being too optimistic about innovations and neglecting unfavorable consequences of technology (Benbasat and Barki, 2007).

Second, risk modeling is an own methodological branch of acceptance research.[30] During 50 years of research various conceptions, focuses and definitions were brought into the field, causing some confusion and inconsistencies. Discussion revolved around the subjectivity of risk, for example. Research agrees that objective risk cannot be measured and consequently, that risk is perceptual by nature. Another discussion was about the difference between risk and uncertainty (a matter of knowledge about the likelihood of a possible outcome). Mitchell (1999) further elaborates that sociological or management theory concepts such as anxiety (Venkatesh and Davis, 2000) or trust (Pavlou, 2003) also have a direct

[30] See Table 2.1, Section 2.1.1.4 for key methodologies.

relation to risk. In the course of time, various definitions have emerged. For instance, the concept of *perceived risk* often used by acceptance researchers today defines risk in terms of the consumer's perceptions of the uncertainty and adverse consequences of using a technology (Dowling and Staelin, 1994). According to Jacoby and Kaplan (1972), it can be split into five basic types of risks: financial, performance, physical, psychological, and social risk. One or more of these risks can hinder potential users to move from the desire to the action stage, i.e., adopting a technology. This all demands consideration if a model is to achieve comprehensibility.

In practice today, definitions of *perceived risk* in behavior research hence have different focuses depending on risk type and related acceptance object. Thus Susanto and Goodwin (2010) noted in their study on SMS-based eGovernment: '*Perceived risk is the degree to which a person believes that using a service may cause problems. These concerns include technology risk, risk to user privacy and security, and perceived financial risk.*' Martins et al. (2014) used the definition of Featherman and Pavlou (2003): '*Perceived risk is defined as the potential for loss in the pursuit of the desired outcome of using an e-service*'. Another way to put it is introduced by Ahmed et al. (2020), whose smart mobility study focused on aspects of personal freedom perceptions in reducing risk to intrusiveness concern, which are '*based on users' view that service provider indecently introduces in their personal life.*'

Anticipated risks in smart mobility can be found in all five categories: physical (e.g., accidents or unsuited for impaired), technological (e.g., bad performance or data safety), financial (e.g., expensive hardware or ticketing), psychological (e.g., complicated or privacy intruding), and social (e.g., lack of status symbols or social shaming). *Perceived risk* is consequently often conceptualized as a second-order multidimensional construct (Im Il et al., 2008; Martins et al., 2014). Hence, an integrated concept of *perceived risk* should be applied for the present study. I adapt the suiting original definition of Bauer (1960): *Perceived risk is the sum of negative consequences potentially arising from using smart mobility.* To construct it, I integrate items from risk, anxiety, distrust, and intrusiveness.[31]

Overall, *perceived risk* has proven to be a significant determinant of BI in many impactful studies on smart mobility technologies (Ooi and Tan, 2016; Prati et al., 2018; Sener et al., 2019; Zhang et al., 2019). It has been used either as an antecedent of *performance expectancy* or *trust* (Sener et al., 2019; Zhang et al., 2019), or as a moderator (Im Il et al., 2008; Baptista and Oliveira, 2015). In contrast, in extending UTAUT2, *perceived risk* was found to be one of the most

[31] See Section 2.3.2.1 for construct operationalizations.

promising direct predictors of BI (Tamilmani et al., 2018b). Since studies on this subject are scarce, the present study strives to explore and validate this indication.

H11.	The perceived risk of smart mobility negatively affects the behavioral intention to use the system.

2.2.2.2 Collective Environmental Efficacy

Already Yang and Yoo (2004) suggested that unknown attitude variables besides the established ones may have important effects on system use. Congruent with this, I can state from my review that such factors were repeatedly added to the TAM and the UTAUT during its development. Against the background of the necessity for a climate-friendly transformation, a large part of these attitude variables scrutinizes the environmental attitudes and views. Peculiar conceptualizations exist in smart mobility and beyond. Thus, this environmental aspect of acceptance can be understood as an individual norm or attitude, technology-specific appreciation, or perceived result of an adoption. Table 2.10 provides some insights.

In other words, what we learn is that behavior is a consequence of anthropocentric or system-specific attitudes. At the same time, we also observe a common good dilemma: the divergence between collective needs and individual action (Hardin, 1968). Where such action is missing, responsibility awareness could be one way to overcome the issue. Schade and Schlag (2003) suggest that environment-preserving behavior would be more likely if individuals recognized the damaging consequences of their own actions on the environment and others, and at the same time took responsibility for them. The question is whether to address the government or the public with the matter.

Turning these thoughts around, Barth et al. (2016) pick up another potentially major reason for missing action: no feeling of efficacy. Bandura (1977) shaped and defined the term self-efficacy as the belief in one's capabilities to organize and execute the courses of action required to manage prospective situations.[32] Since then, an overwhelming number of studies have confirmed that self-efficacy is one of the absolutely seminal variables of human decision-making. Barth et al. (2016) now realize that bringing about sociotechnical change such as establishing new forms of transportation is usually not the work of single individuals but depends on collective behavior. They hence develop *collective efficacy* as a new acceptance variable.

[32] A reminder: Following Ajzen (2002) this is the origin of strong acceptance predictors from perceived behavioral control to facilitating conditions.

Table 2.10 Selected conceptualizations of environmental aspects in technology acceptance

Reference	Name	Role	Variable type
Kinnear et al. (1974)	Ecological concern	Ecological concern is a function of attitudes and behaviors. A new index is developed.	Personality specific
Roberts (1995)	Ecologically conscious consumer behavior	Ecological conscious consumers are different for socially responsible consumers.	Personality specific
Toft et al. (2014)	Personal norm	The norm activation model is combined with TAM to include moral obligations toward sustainable technology. A medium effect is found.	Technology specific
Chen (2016)	Green usefulness	Present a modified TAM and TPB to analyze the effect of perceived green value on loyalty to a public bike system through green usefulness.	Technology specific
Møller et al. (2018)	Awareness of need/ consequences	Study detects and conceptualizes a contradiction between a general ecological awareness and personal behavior.	Personality specific
Hartl et al. (2018)	Environmental concerns	Sustainability is relevant for carsharing users, but only minor to financial considerations.	Personality specific
Ru et al. (2018)	Green travel intention	Green travel intention is a function of norms, attitudes, and self-efficacy moderated by experience.	Technology specific
Wang et al. (2018a)	Environmental awareness	The general 'greenness' of an individual helps to predict EV acceptance with a medium effect.	General belief
Wang et al. (2018b)	Green value	Perceived green value shows a small effect on behavioral intention to use bike sharing services.	Technology specific
Wu et al. (2019)	Environmental concern	General environmental concern is a major direct and indirect stimulus for AV acceptance.	Personality specific

(continued)

Table 2.10 (continued)

Reference	Name	Role	Variable type
Wolff and Madlener (2019)	Ecological attitude	Construct results from principal component analysis; works as a strong predictor of perceived usefulness of light-duty electric vehicles.	Technology specific
Gimpel et al. (2020)	Environmental concern	Influences all technology-specific beliefs in UTAUT2 for smart energy significantly.	Personality specific

It is hypothesized that if people believe their own behavior can contribute to a larger group behavior which can then enable change (such as smart mobility transition), this initial belief can influence acceptance for the better. Barth et al. (2016) define *collective efficacy* as the belief that the in-group is capable of affecting important aspects of its environment. They apply the new variable to an acceptance model for EVs and demonstrate a stronger positive influence of *collective efficacy* than all cost-related disadvantages. Following studies confirm that expected eco-societal benefits are vital factors of adoption not only for EVs but also for AVs (Manfreda et al., 2019; Zhang et al., 2019).

Jugert et al. (2016) and Fritsche et al. (2018) highlight the potential of *collective efficacy* to serve as a factor predicting sustainable behavior. I thus aim at specifying it to a personality specific belief that focuses not on a general perception of *collective efficacy*, but on the *collective efficacy* toward environmental challenges. It is quite possible that people believe that they can collectively overcome a pandemic, for example, but do not feel a common impact on the challenges of climate change. It is unlikely that every individual inherits a general belief in their *collective efficacy*. I consequently renamed it and adapted items and hypotheses originally brought in by Barth et al. (2016). In present work, collective environmental efficacy consequently takes over the role of the environmental aspect.[33]

H12.	The collective environmental efficacy of smart mobility positively affects the behavioral intention to use the system.

[33] See Table 2.10, Section 2.2.2.2 for alternative conceptualizations.

2.2.3 Model Context

This work aims to apply UTAUT for two reasons: to deepen our knowledge on smart mobility transitions and to refine associated theory. In this endeavor, adding further context proves helpful. Venkatesh et al. (2016) did a review on this case and identified six contextual dimensions in UTAUT research: (1) user groups[34], (2) organizational and societal settings, (3) location and cultural settings, (4) type of technology[35], (5) type of task[36], and (6) time of technology use[37]. These dimensions can be specified by choosing the respective sample group, the survey locations, and periods, or by describing model interactions such as moderation. The effect of context was in this regard mainly operationalized as moderations between BI and the external variables, but also between BI and use behavior (Tamilmani et al., 2020).

In the original UTAUT model, Venkatesh et al. (2003) included gender, age, experience and voluntariness of use. In UTAUT2 they dropped the latter to open the model from organizational to consumer behavior research. In consumer markets the use of technology is normally completely voluntary. Consequently, there would not be any variance in the construct, which would make it redundant. Overall, these context variables helped to predict acceptance significantly better. By modeling multiple moderation (Braumoeller, 2004), the R^2 increases from 35 to 56% in UTAUT and from 44 to 74% in UTAUT2. What is unfortunate and 'disappointing' (Venkatesh et al., 2016) for the authors is the fact that these context findings could initially hardly be replicated in subsequent studies and were eventually no longer used in research practice (Tamilmani et al., 2018b). The reason for this could be rooted in an erroneous study design (not longitudinal), statistic methods (higher-order interactions) that were too complex, or an insufficient theoretical foundation (cause-effect relation). So, while UTAUT is a robust tool, there is not enough evidence to draw conclusion for the generalizability of the model's marginal conditions. Venkatesh et al. (2016) finally offer fundamental advice for further investigation. In consequence, one needs to distinguish between individual-level (i.e., user personality, technology types, time, and task) and higher-level contextual factors (i.e., the techno-economic environment of adoption). Second, one must conceptualize the acceptance of a technology at

[34] These are often based on sociodemographic characteristics.

[35] These reach from general examination of 'the Internet' Gupta et al. (2008) to specific applications such as in 'SMS- based eGovernment' in Susanto and Goodwin (2010).

[36] Things that are enabled by the technology like communication, banking or here: mobility.

[37] E.g., pre- or post-adoption.

a feature level and retest the current context factors (age, gender, experience). Third, higher-order context effects could be discovered. All this is to make substantial contributions to UTAUT in particular, and to transport acceptance research in general. Accordingly, I will adhere to these pieces of advice as follows:

(1) The new variable collective environmental efficacy is a personality trait and not a technology-specific belief. This is in line with the requirement to consider the user personality.
(2) The study's design, in the way that both smart mobility as an overarching concept and selected sub-use cases are considered allows for exploration of both the boundary and core conditions of the model in desired depth. Therefore this work conceptualizes *levels of system integration* (Lyons et al., 2019) as a higher-order context.
(3) Present study retests age, gender, and experience to shed more light on the issue of possible moderation in UTAUT. This is based on understanding acceptance as a process. In the upcoming paragraphs, potential effects will be reviewed. I conclude that no trustworthy effect pattern exists and thus instead of hypotheses, an exploratory test is indicated.
(4) This study explores the role of level of system integration, i.e., technology type to discover dynamic changes in the acceptance setup.

2.2.3.1 Age and Gender

The role of sociodemographic variables is a central area of investigation in acceptance research. Essential representatives of this genre are age and gender, which are usually measured as categorial variables.[38] Gender schema theory (Bem, 1981) suggests that cognitive differences in processing information exist. These variations probably originate from trained socialization processes, rather than having a biological root, as in genetics, for example. Empirical studies on acceptance have consequently tested and confirmed strong and enduring effects of gender in the adoption process (Gefen and Straub, 1997; Venkatesh and Morris, 2000; Wang et al., 2009). Bluntly summarized, recent literature assumes that men are more usefulness-, fun-, and status-oriented than women. These, in turn, can be better convinced of a new technology through social influences, perception of behavioral control, and simplicity of use. Since usage of IT is often rather based on

[38] In keeping the research efficient, other conceptualizations of sociodemographic values like lifestyles (Schlüter and Weyer, 2019) or customer groups (Mohamed et al., 2018) will not be considered in present work.

the former factors, men are often found to be the early adopters. However, for some IT like social networks, women display higher acceptance (Hwang et al., 2019). In other studies, no difference exists at all (e.g., in online shopping (Lian and Yen, 2014)).

If one pursues this argumentation, one can expect to see a higher smart mobility acceptance in men due to its utilitarian and technology-related character. Some researchers also put forward the hypothesis that it is primarily men who shape mobility and thus tailor it to their own needs.[39] All in all, hypothesizing about the interaction effects of gender on the elements of the acceptance process is complex. Initial insights are provided in our review. In e-biking, for example, women show a higher risk assessment concerning injuries and men seem to be more influenced by intrinsic motivations (van Cauwenberg et al., 2019). MaaS, Ye et al. (2020) tested UTAUT with multigroup analysis and found a single gender effect for the relation *effort expectancy* on BI. The relation was only significant for females. Furthermore, gender differences also exist in the adoption patterns of electric vehicles. Fazel (2013) elicits that an electric vehicle can through its invasive shape or pushy design noticeably lose its appeal to women. O'Garra et al. (2005) show that willingness to pay for FCEVs is higher among males, which makes it reasonable to believe *price value* might have a larger influence on them. Gender might also influence the way *performance* and *effort expectancy* cause change in BI, such as proposed by Gefen and Straub (1997). They find men's acceptance to rely more on performance and women's to rely more on ease of use. The clean vehicle incentive study of Potoglou and Kanaroglou (2007) indicated this for smart mobility too. They demonstrate that high speed performance is of greater importance to male respondents, while females value the simplicity (ease) of vehicles. Teo et al. (2015) speculate that women might find technology more challenging. With automated vehicles, too, it is (younger) men who are particularly attracted to the technology (Sener et al., 2019). However, other surveys (Morris et al., 2005; Panagiotopoulos and Dimitrakopoulos, 2018) discuss that this gender gap is progressively closing in younger samples. Until then, I expect to see some gender particularities and interactions in acceptance schemata. After all, not so much knowledge is yet available for new mobility systems.

Similar to gender, human cognitive functioning is affected by age. Attention, memory, and perception are permanently subject to biological and psychological change (Verhaeghen and Salthouse, 1997; Murman, 2015). In acceptance research, several assumptions were made about the influence of age. These were initially based on the empiric observation that older people are usually rather

[39] Please see Section 2.4.3.3 '*The Thomas Principle*'.

reluctant to adopt new technologies. Potentially, younger individuals are snoopier and thus more excited about using new technology (Lian and Yen, 2014; Zhou et al., 2014). Also, habits may not yet be as ritualistically practiced. On the other hand, applying IT means elderly people could often actually reap more benefits than the young in terms of increasing quality of life (Niehaves and Plattfaut, 2014). For some technologies, the effect of age on acceptance might hence be reversed. For instance, Mitzner et al. (2010) find that older adult's positive assessments of healthcare, home and, work IT systems largely outweighs the negative attitudes toward the system. This contradicts the stereotype of the old, afraid, unwilling user. In this manner, Burnett et al. (2011) for example propose generally empowering older adults with digital technology. Seeing these diverging results, it was sometimes also not possible to prove the influence of age on acceptance in prior studies (e.g. in Chung et al., 2010). Chen and Chan (2011)'s review provides guidance, suggesting that exploring the technology-specific impacts of age on acceptance in whatever way must be considered valuable for our understanding. In smart mobility-related acceptance, results are as fragmented as in general acceptance research. The universal effects of age, as suggested by UTAUT, could not be replicated (Gupta et al., 2008; Khechine et al., 2014). Nonetheless, in selected studies that take age into account, there are some findings to date.

Generally, acceptance in the field of smart mobility seems to be higher among younger people (Sener et al., 2019; Schikofsky et al., 2020). Ye et al. (2020) find that this effect may be based on higher *performance expectancy* impact on acceptance for the young. Kauschke and Schulz (2017) on the hand demonstrate how age influences the perception of *ease of use*. However, Rödel et al. (2014) show that age effects differ between various technologies and ergo need further investigation. In most studies in present review (e.g. Kyriakidis et al., 2015; Madigan et al., 2017; Panagiotopoulos and Dimitrakopoulos, 2018), age had no influence at all. In the case of e-bikes, on the other hand, even higher acceptance was found among older people, who might appreciate performance and ease of use (van Cauwenberg et al., 2019).

In conclusion, the significance of demographic variables in smart mobility acceptance appears uncertain and requires clarification. To put it simply, it is assumed that younger men overall have a higher affinity for adopting IT systems. However, due to social change, especially within the western industrial nations, these boundaries appear to be increasingly blurring and it is possible that some technologies work differently (Morris et al., 2005).

H13.	The influence of age and gender on smart mobility acceptance will primarily <u>not</u> follow the patterns proposed in UTAUT.

2.2.3.2 Experience

Salient studies confirm the large influence of experience on acceptance (Ajzen and Fishbein, 1980; Bagozzi, 1981; Venkatesh et al., 2003). However, one can see different views of what experience entails, as well as varying proxies of experience on acceptance.

The Cambridge dictionary defines experience either as *'getting knowledge or skill from doing, seeing or feeling things'*, or as *'something that happens to you that affects how you feel'* (Heacock, 2009). Thus, experience is the idea that, starting from one or more prior event(s), a change in the cognitional and emotional reactions toward and beyond a matter is caused. This was acknowledged early on in acceptance research. The first psychological studies dealing with experience (e.g. Regan and Fazio, 1977; Triandis, 1977) suggested that knowledge gained from past experiences will contribute to shaping intentions. This is supposed to happen because experience makes knowledge more accessible in memory and because prior experience makes low-probability events more conspicuous, serving to ensure that they are taken into account in the formation of new intentions. In a similar vein, Bandura (1986a) confirmed that individuals who have had the opportunity to test a particular technology tend to exhibit lower levels of internal resistance. In this regard, established theory (Karahanna et al., 1999) suggests that additional to knowledge, the skills acquired and the associated usage confidence obtained through experience also have an important role to play; for instance through reducing the influence of *subjective norm* on acceptance. Gefen et al. (2003) call it a barrier based on experience that can be torn down by building *familiarity*. Using this theory, one could say that at a certain level of experience, internal resistance gives way to acceptance. Consequently, there would be a breaking point to identify. So, different approaches to the functioning principles of experience coexist.[40] What unites all of them is the conjecture that there may be differences between experienced and inexperienced users. Such disparity may suggest alternative ways to effectively manage the implementation of new systems.

Also in mobility, experience seems to be positively targeting the cost-benefit of a technology and the relationship between trust and risk (Rousseau et al.,

[40] Please see Section 2.3.2.1 for the different measurement models.

1998; Schuitema et al., 2011). This, inter alia, translates into possible effects on *perceived risk*, *price value*, and *performance expectancy* for present UTAUT extension. In the original model, Venkatesh et al. (2012) moreover found effects on FC, SI, EE, and habit. With growing experience, *habit* and *performance expectancy* became more important in the formation of intention and use behavior, while the influence of the others shrank or disappeared completely. This is potentially because the efficacy and capabilities of new technology can be assessed more precisely by experienced users. Though these moderations in the UTAUT were not sufficiently validated in follow-up studies, this view nevertheless tends to reflect today's common understanding (Venkatesh et al., 2016). Further mobility studies found links between growing EV experience and a reduction of the influence of *effort expectancy* (Fazel, 2013), trialability (Nordhoff et al., 2021), car sharing experience, and *EV acceptance* (Schlüter and Weyer, 2019). Overall, the influence of experience is a much-discussed acceptance factor in the environment of smart mobility, but an integrated study is not available as of today's knowledge.

In line with other authors (Taylor and Todd, 1995a; Blanca Hernández et al., 2009; Nordhoff et al., 2019), present work elaborates an understanding of smart mobility acceptance as an individual, experience-based process, with the factors at play altering between the beginning and end. That said, there are unfortunately some white spots about how experience might manipulate the patterns of acceptance (Venkatesh et al., 2016). Consequently, UTAUT assumptions on experience effects have not proven robust enough to build hypotheses from them. Hence, in order to provide a feasible alternative for practice and theory, this paper conducts an exploratory investigation into the issue of experience. Figure 2.12 therefore depicts experience, as is the case with age and gender, as a construct outside the research model, with arrows pointing at all constructs in the research model.

H14.	The influence of experience on smart mobility acceptance will primarily not follow the patterns proposed in UTAUT.

2.2.3.3 Levels of System Integration

Prior studies have highlighted the importance of dynamizing the classic conceptual understanding of acceptance. This, for instance, implies multiple measurement time points. Kollmann (2013) in this regards exemplifies that acceptance, before, while and after purchase work differently. Similarly, the psychological model of Bamberg (2013) conceptualizes mobility-related decision making as a transition through the time-ordered sequence of four stages: pre-decision, pre-action, action, and post-action. Comparable approaches exist in mobility acceptance research (e.g. Im Il et al., 2008; Nordhoff et al., 2019). Not least of all because this study will also examine a dynamization of the acceptance process via six acceptance levels developed by splitting *acceptance* and *acceptability* apart.[41] It is assumed that the various smart mobility technologies are integrated to varying degrees into today's mobility system. This makes it difficult for an individual to gain cognitive access to the topic (e.g., through practical experience). Furthermore, it is expected that individual acceptance characteristics will emerge depending on the stage. For example, the evaluation of less integrated technologies such as MaaS could take place more on an emotional level than on the basis of a performance evaluation. Alternatively, a very tangible technology like the eBike will already rely strongly on fun and habits, while AVs, being at an earlier stage, do not.

These considerations are currently still relatively experimental, but what will be investigated in any case with these hypotheses is whether the relative importance in the prediction of intention varies across use cases. The conceptualization of this model context is thus also close to Im Il et al. (2008)'s *technology type* (e.g., hedonic, utilitarian, job related).

H15.	a. The higher the level of system integration, the larger a technology's experience and acceptance. b. The influence of system integration level on smart mobility acceptance will be based upon the schemata drawn in Figure 2.2.

2.2.4 Research Model

The following structural model was conceptualized as a starting point for analyzing the acceptance of smart mobility. It contains nine exogenous variables whose effects on *use behavior* are all mediated by the endogenous BI. Three dimensions of model context are to be explored. *Age* and *gender*, and *experience* originate from UTAUT, while *system integration* was newly included to analyze related technology types in different innovation phases.

[41] See Figure 2.2, Section 2.1.1.3 for the proposed levels of acceptance.

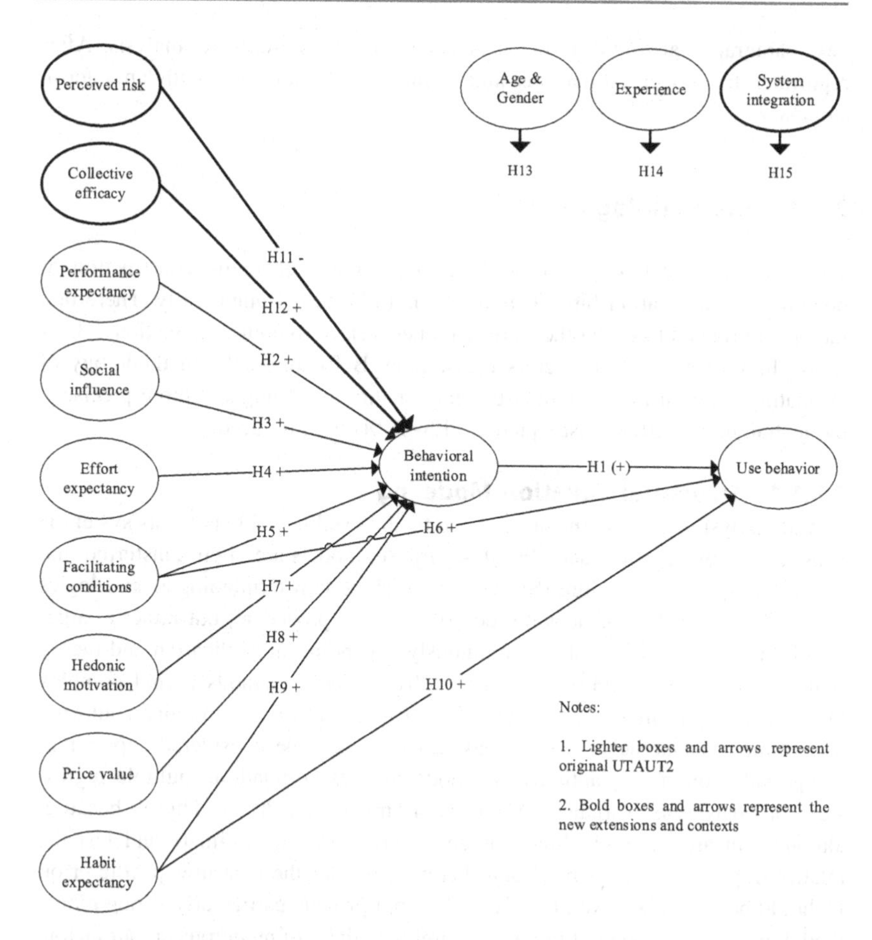

Figure 2.12 Research model with hypotheses

2.3 Results

Model assessment is carried out in accordance with the guidelines of Hair et al. (2017) and selected advanced methods from Hair et al. (2018). SmartPLS3 proved to be an ideal software to perform these calculations. This section presents the considerations of selecting variance-based SEM for the analysis of the model as well as the methodological background of its application. Moreover, these next paragraphs also outline the data collection and descriptive analysis. After reporting the model estimations and evaluations, I conclude with an interim summary.

2.3.1 Methodology

The research model in preceding Section 2.2.4 is a theoretic construction of potential causal relationships. This model is to be tested empirically. Therefore, factors are related to each other through a network of hypotheses. Section 2.3.2.1 shows how to make these factors measurable. Beforehand, the methodology of evaluating a causal model with structural equation modeling and more precisely, the technique of partial least squares (PLS-SEM) is introduced.

2.3.1.1 Structural Equation Modeling

Causal analysis controls whether a theoretically established hypothesis system is consistent with empirical data. For this purpose, models are often transferred into structural equation modeling (SEM), for which certain terminologies and logics apply. Today, SEM itself does not describe a fixed procedure, but rather a larger set of analysis methods that is continuously expanding in application and theory (Ullman and Bentler, 2006; Hair et al., 2017). These methods permit complex phenomena to be modeled and tested in a quantitative and initially confirmative fashion (Schumacker and Lomax, 2010). It can be considered superior to comparable common quantitative methods such as correlation, multiple regression, or analysis of variance (ANOVA) (Ramli et al., 2018). This is because, although all are generally linear models, only SEM can estimate and test the relationship between constructs and thereby consider the measure-specific error. It should be noted, however, that SEM does not provide particularly strong global model fit indices. Instead, it requires a careful analysis of numerous output factors depending on the goal of the investigation (see Henseler, 2018).

As in Figure 2.13, SEMs are routinely visualized with path diagrams. Ovals (Y_1 to Y_4) represent constructs, also known as latent variables.[42] These are not directly observed but inferred from direct measures through varying mathematical equations. Those manifest observations are called indicators or items (Kmenta, 1990). The path diagram graphs them as rectangles (X_1 to X_{10}). Single-headed arrows, which generally represent a directional relationship, connect indicators and latent variables. In SEM, no such thing as a bidirectional or non-directional

[42] Derived from Latin: *lateo* ('lie hidden').

relationship exists[43], although it can be assumed that hypotheses often have an effect in both directions. It is essential here to find the properly directed hypothesis through a strong and sound theoretical underpinning. Then, and only then, can such a predictive relationship also be interpreted as a causal relationship. And that is what is at the core of SEM: providing empirical support for theoretic cause-effect assumptions. To find this, SEM's key characteristic in terms of conceptual logic is the distinction between a structural and a measurement model.

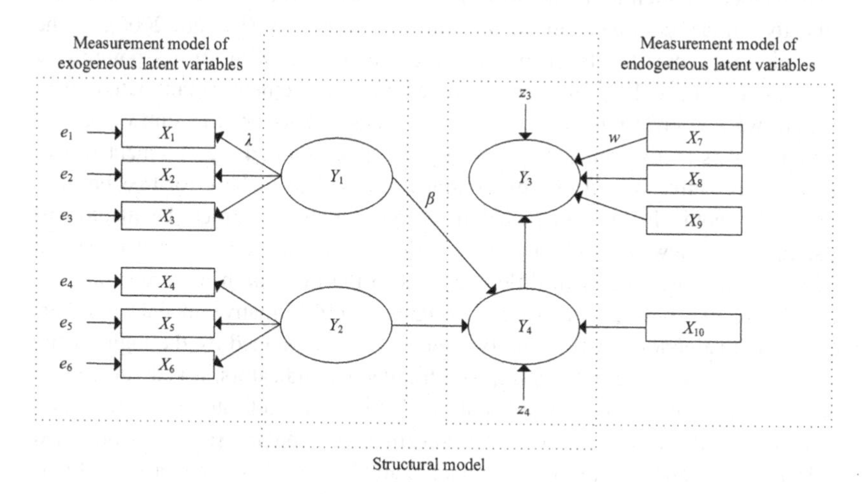

Figure 2.13 An exemplary path diagram

The structural model is the inner part of the model that contains the relationships between the latent variables. We can describe direct effects or indirect effects through interactions, equal to moderation and mediation. The path coefficient β of every directed relationship can be interpreted just as a regression weight and describes the strength of the relationship. The constructs without predecessors (e.g., Y_1 or Y_2) are called independent, or exogeneous latent variables. The constructs with arrows being pointed at are consequently dependent, or endogenous variables (e.g., Y_3 or Y_4). They are predicted by the model. It is of course also possible that a construct acts in both ways (e.g., Y_4). In this case, they are still endogenous. It is worth mentioning that the error z is included in SEM. From a

[43] Figure 2.13, Section 2.3.1.1 displays a single-item measurement for the latent variable Y_4. This must not be confused with a non-directional relationship but a special case, in which the construct and the item are simply the same. No measurement model is necessary.

classic test-theory view, variance of any observed measure consists of true scores and errors (Weston and Gore, 2006). The true score is the variance explained through the predicting variable; the error term z accounts for all the unexplained variance due to unmodeled relationships. The coherence of true score(s) and z indicates the explanatory quality of the model for a certain construct.

The measurement model represents the 'outer' part of a SEM. It specifies how latent variables are measured. Two measurement modes exist. The first, the formative mode, is often found in business studies, whereas the second, the so-called reflective mode, dominates in the domains of psychology (Fassott, 2006). In the example, Y_3 shows the formative measurement variant with arrows pointing at the construct. Y_1 and Y_2 represent the reflective measurement variant with opposing arrows and error-term e attached to it. These errors are the sum of natural random errors and systematic errors due to a miscalibrated measurement instrument.[44] Formative measuring, by contrast, is assumed to be error-free because indicators should include all potential reasons for the construct (Diamantopoulos, 2011). In view of this assumption, different methods apply to reflective and formative models. One main difference lies in the evaluation of factor loadings λ for reflective measures and outer weights w for formative models. Furthermore, the distinction between the two modeling is explained by the logic of the underlying cause-effect relationship. With reflective indicators, it is assumed that the construct causes the measurement. With formative indicators, it is the other way around: the measures cause the construct. A popular example is provided in Hair (2014) and visualized next on an example originally brought in by Ebert and Raithel (2011). Accordingly, drunkenness is to be measured or understood either as a consequence of past consumption or as a derivative of actual behavior (Figure 2.14).

[44] See assessment of the measurement models in Section 2.3.4.2.

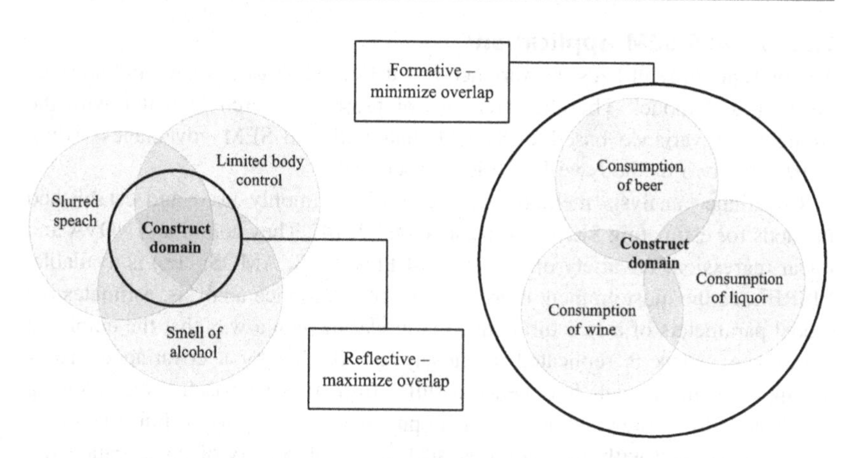

Figure 2.14 Reflective and formative measurement on the example of drunkenness

The formative approach minimizes the overlap between complementary indicators. For this, no item could be removed without the danger of changing the nature of the construct. The reflective measurement approach on the other hand maximizes the overlap between indicators. Thus here, items can be omitted and are more or less interchangeable, as long as content validity is ensured. If it is not clear which measurement model is active, confirmatory tetrad analysis (Bollen and Ting, 2000) can be used to determine the measurement type.

The estimation of hypothesized effect relationships of latent variables in SEM is commonly carried out using second-generation multivariate analysis procedures. These combine regression and factor analysis methods. Given their ability to handle numerous variables parallelly, they are superior to univariate methods (e.g., description of data). They hence allow for estimation and testing of correlations between dependent variables (DV) and independent variables (IV). In the course of the development of SEM, more and more precise methods have been developed to make the insights more stable and to even detect hidden structures in the data (Backhaus et al., 2018). After all, two highly recognized procedures for solving SEM exist: covariance-based (commonly referred to as CB-SEM) and variance-based (commonly referred to as partial least squares PLS-SEM) methods. Which one to choose boils down to the goal of the research.

2.3.1.2 PLS-SEM Application

The present study utilizes the variance-based PLS-SEM to estimate and advance the proposed model. The characteristics of present research fit better with the strengths of variance based PLS-SEM than with CB-SEM advantages (Nitzl, 2010). The two methods will be briefly discussed.

Covariance analysis methods are the most commonly used and established methods for estimating SEM (Astrachan et al., 2014). They combine ANOVA and linear regression. A variety of software (M-Plus, EQS, AMOS, etc.) is available. LISREL, as the most eminent representative of covariance analysis, estimates the model parameters of a structural equation model in such a way that the empirical covariance matrix is replicated as closely as possible by a covariance matrix arising from the model. It is hence a full information approach, even allowing conclusions to be drawn about a basic population. This quality of information is, in turn, associated with preconditions, such as a high quality of linear data and a large sample size. Thus, CB-SEM is primarily used to confirm or reject theories (Hair et al., 2017).

Variance-based SEM, contrastingly, focuses on expanding and exploring theory. It is specialized in explaining the variances of the endogenous variables. An iterative sequence of ordinary least square (OLS) regressions is employed to maximize this value. It is accordingly rather predictive than confirmatory. Hair et al. (2011), in line with PLS theorist Wold (1982) and the review on non-arguments of Rigdon (2016), therefore suggest preferring PLS-SEM to the popular CB-SEM, if the goal of the research lies in exploring new relationships or extending current theory. It is also recommended for small samples sizes, non-normally distributed data, and generally for complex models with many variables. CB-SEM should be used if the research is confirmatory, if the model includes circular relationships, or if the goal of the research implies the need for a global-fit indicator. This may, for instance, be the case if one assesses different theoretical approaches for the same data. In the discussion about choosing the right method, Henseler (2018)'s notion makes a point that is often neglected today: PLS, like other estimations, is in essence just a tool for solving statistical tasks. The evaluation of the results might differ depending on which output factors are emphasized. Four examples can demonstrate how to interpret results depending on the research scope:

- If the goal is explanatory, such as explaining a unique phenomenon as well as possible, analysts should predominantly examine the coefficients of determination and effect sizes.

- If the analysis is exploratory, quick graphical modeling and a profound inspection of path coefficient and proxy constructs is what researchers should concentrate on.
- If the goal is to predict behavior, the predictive relevance and the individual prediction errors are the relevant output sizes of SEM techniques.
- If the research is conducted for confirmatory purposes, the focus must lie on model fit indices.

In practice, many application situations are hybrid by nature. Often, the research goal requires a combination of theory-testing (confirmatory and explanatory) and theory-building (exploratory) approaches. Hence the boundary between confirmatory and exploratory is not always clear-cut. This is also the case in the present modeling and in a growing number of related research (e.g. Fazel, 2013; Ooi and Tan, 2016; Schikofsky et al., 2020). I therefore choose to apply PLS-SEM. In summary, the following points are decisive for this choice:

- As a second-generation technique, it is superior to first-generation techniques such as ANOVA or regression in terms of parallel processing (Chin, 1998). In contrast to CB-SEM, PLS-SEM is highly recommended by experts (Hair et al., 2017) if the goal of the research is primarily not to confirm established theories. Typically, few rigid rules complicate its application.
- PLS-SEM algorithms optimize toward prediction and the overall explanatory value. At the same time modeling remains flexible. Hence, alternative specifications may easily be tested. The focus in the present study, which is not laid upon specific relations, but on discovering patterns of acceptance, is thus reflected nicely. Or as PLS inventor Wold (1980) puts it, models should be built and evaluated by a researcher '*on the joint basis of his rudimentary theoretical knowledge, his experience, and intuition about the problems explored, and the data that are at his disposal*'. Exploratory research must thereby be followed by confirmatory research, as a principle.
- After all, PLS corresponds to the quintessence of applied sciences, which is to provide concrete recommendations and offer practical value (Nitzl, 2010).

SEM experts share consensus on the six steps of model testing. Besides data collection, these are the steps of model specification, identification of data, estimation, evaluation, and modification of the model and its parameters (Hoyle, 1995; Weston and Gore, 2006; Kline, 2015). In present examination, these steps are to a large extent performed based on software. Thanks to its user-friendliness, the richness of implemented methods, and the ongoing development in a large scientific community, SmartPLS3 was selected from other software available (LVPLS, PLS-Graph, WarpPLS, etc.) as the computing solution for present examination.[45]

2.3.1.3 Model Specification

The process of setting up a structural and a measurement model is the so-called model specification. As the term already suggests, it is about defining the relationships between all latent variables, error terms, and measurement indicators. Equipped with this information about positive, negative, or zero relationships, the PLS algorithm can estimate the model parameters from the raw data (Backhaus et al., 2018). The basis for present testing is the model conceptualized and specified in Figure 2.12, Section 2.2.4, which includes 108 parameters (including error terms). The key concept behind the model is an extended UTAUT2 theory (Venkatesh et al., 2012). I initially stick to the majority of these common specifications and measurements but test the model under different circumstances and with multiple data sets. The models are implemented in SmartPLS.

2.3.2 Data

This section describes how to proceed from theory to application of PLS-SEM. This includes a discussion of model identification and the necessary sample sizes. Additionally, the operationalization of the measurement constructs is described. Next, the design and implementation of the survey sets out how the data for the acceptance model was collected. Finally, a quality assessment of the data is provided. It can be summarized that the collected data is of high quality and indisputably well-suited for PLS analysis.

The first and second steps of SEM, model specification and model identification respectively, should ideally be scheduled prior to data collection (Weston and Gore, 2006). To achieve meaningful results from an SEM analysis, a model must be overidentified. This means that fewer variable relationships are hypothesized than there are elements included in the correlation matrix. To test this, the number of *degrees of freedom* dF can be calculated using the following formula:

$$dF = \frac{no.variables(no.variables + 1)}{2} - no.parameters$$

[45] For details on software settings please refer to Section 2.3.4.1.

Degrees of freedom refer to the maximum number of logically independent values, i.e., scores that have the freedom to fluctuate in the data sample. If $dF > 0$ the model is overidentified. In addition, the higher the value of dF, the more parsimonious the model. For the present model, the calculation results in $dF = 728$. Hence, the model is well identified.

Unlike other statistical analyses, the degrees of freedom have little influence on sample sizes requirements in PLS-SEM, because the algorithm does not compute all relationships at the same time, but applies sequences of OLS-regressions (Hair et al., 2017). Concerning the appropriate sample sizes, a history of controversy can be seen in literature (Hair et al., 2011; Tarka, 2018). Today, the old ten-times rule is regarded outdated and would hardly serve as a guideline. It states that the minimum sample size should be ten times the largest number of arrows pointing to a single construct in the model. Meanwhile, alternative and potentially more precise methods have been developed to obtain the sample size threshold. Examples include the inverse square root and gamma-exponential methods (Kock and Hadaya, 2018). The overall goal of all approaches is achieving a satisfactory degree of statistical power. Simply spoken, statistical power is the probability that a test of significance will pick up on an effect that is actually in place. The theorists of SmartPLS hence suggest relying on rules of thumb originally developed in Cohen (1992)'s seminal work on power analysis. It considers model complexity and significance levels. For a common statistical power of 80% at a significance level of 1%, present model would consequently need at least 204 observations to discover even weak effects with a minimum R^2 of 0.10.

2.3.2.1 Operationalization of Constructs

In behavioral economics, operationalization is the systematic process of defining a latent construct through several indicators (Hair, 2014). Ergo, operationalization is the link between reality and the model. Commonly, statements are made about specific topics and the degree of individual agreement of the study participants is queried. In principle, this yields in quantitatively observable measures, the so-called items.

Researchers in the social sciences often find it difficult to develop corresponding statements, because usually a great deal of effort (e.g., surveys, pretests, focus groups) is required. In this context, the quality criteria of validity and reliability are of particular importance. Only with valid and reliable indicators can scales and indices be meaningfully formed, meaning more complex theoretical concepts can be empirically captured. Some authors (e.g. Markus and Borsboom, 2013) critically note in this regard that there are no generally valid rules for the discovery, development, and formulation of items for a latent variable scale. Therefore, a lot of the developer's subjective perception may be embedded in a construct. Due to all these challenges, it is common practice to utilize existing and already well-tested scales. This is also the case in the present study. Nevertheless, this study claims to be capable of selectively adapting and expanding the constructs for the purpose of mapping the smart mobility reality better than the original items form information system research were able to do. The measurement quality was hedged by an adequate pre-testing procedure with 30 participants (Perneger et al., 2015). Pretesting means applying the survey in a small pilot study to see how well it performs outside the academic environment. Backstrom and Hursh-César (1981) jovially write: 'No amount of intellectual exercise can substitute for testing an instrument designed to communicate with ordinary people.' Since most of the scales have already been validated extensively during the last decades, only minor changes had to be undertaken after the pre-test.

All UTAUT related item adaptions are listed in Table 2.11. As with the original UTAUT, the present measurement model is mainly reflective. The one exception is *actual usage*, which is operationalized as a single-item variable. Although generally a multi-item design of constructs with three to five items is advised for reasons of validity (Ringle et al., 2012), single-item conceptions are not without their advantages (Loo, 2002). Their application is easy and cost-effective. Additionally, they reduce mental fatigue of respondents and thereby help to increase the response rates. In some cases, a single question might even be better suited to sense a certain phenomenon. Conversely, they do not offer 'more for less' (Hair et al., 2017). Single-item designs are sometimes not suitable because reality might be too complex to be understood in one term. In addition, problems of a statistical nature can arise: single items reduce a model's degrees of freedom and inherit lower reliability because measurement errors cannot be removed. Thirdly, choosing one item in isolation increases the risk of bias due to the author's limited perception. This does not imply, however, that single items should not be used at all. Research shows that they can also be valid (Diamantopoulos, 2011). It is just a matter of considering exactly when and for what you use them. Following Fassott (2006), such is the case, for example, when there is a common understanding of a subject that is on the one hand simple and on the other hand hard to disaggregate (e.g., customer satisfaction).

Table 2.11 Operationalization of UTAUT2 constructs for smart mobility

Construct	Code	Item	Supporting literature
Behavioral intention to use	BI1	I intend to use smart mobility.	Warshaw (1980), Davis (1989)
	BI2	I plan to integrate smart mobility into my daily life.	
	BI3	I expect to use smart mobility in the future.	
	BI4	I can well imagine using smart mobility.	
Performance expectancy	PE1	I find smart mobility useful.	Davis (1989), Compeau et al. (1999), Escobar-Rodríguez and Carvajal-Trujillo (2014), Kauschke and Schulz (2017)
	PE2	Smart mobility helps me to travel more efficiently.	
	PE3	Smart mobility increases my safety.	
	PE4	Smart mobility improves my comfort.	
	PE5	Smart mobility enhances flexibility in my everyday life.	
Effort expectancy	EE1	Learning how to use smart mobility is easy for me.	Davis (1989), Thompson et al. (1991), Herrero et al. (2017), Kapser and Abdelrahman (2020)
	EE2	I find smart mobility easy to use and organize.	
	EE3	It is easy for me to become a smart mobility user.	
	EE4	My interactions with it would be clear and understandable.	
Social influence	SI1	People who influence my behavior think that I should use smart mobility.	Ajzen (1985), Venkatesh and Davis (2000), Barth et al. (2016), Madigan et al. (2017)

(continued)

Table 2.11 (continued)

Construct	Code	Item	Supporting literature
	SI2	People whose opinions I value prefer it for themselves.	
	SI3	Experts and media agree on a positive assessment of smart mobility.	
	SI4	Most people who mean something to me would approve of me using smart mobility.	
	SI5	I feel social pressure to use smart mobility.	
Facilitating conditions	FC1	I have the necessary resources to use smart mobility.	Bandura (1977), Ajzen (1991), Venkatesh et al. (2008), Fleury et al. (2017)
	FC2	I have the necessary knowledge to use smart mobility.	
	FC3	Smart mobility is compatible with systems that I use.	
	FC4	In case of difficulties with smart mobility, I can get help.	
Hedonic motivation	HM1	Smart mobility is fun.	Davis et al. (1992), Van der Heijden (2004), Teo and Noyes (2011), Pfeiffer et al. (2016)
	HM2	I enjoy using smart mobility.	
	HM3	Smart mobility is exciting.	
	HM4	Smart mobility reflects my inner values.	
Price value	PV1	I think smart mobility is reasonably priced.	Dodds et al. (1991), Wu and Wang (2005), Ramírez-Correa et al. (2019)
	PV2	I think smart mobility has a good cost-benefit ratio.	

(continued)

Table 2.11 (continued)

Construct	Code	Item	Supporting literature
	PV3	At the current price, smart mobility offers great value.	
Habit (expectancy)	HE1	Smart mobility has become a habit for me.	Limayem et al. (2007), Baptista and Oliveira (2015), Schikofsky et al. (2020)
	HE2	Using smart mobility is normal for me.	
	HE3	I do not think twice about using smart mobility.	
	HE4	I cannot do without smart mobility anymore.	

scale: linear, ordinal Likert scale from *I do not agree* (1) to *I agree* (7)

The adapted and extended operationalization of the constructs of the smart mobility acceptance model are discussed below. In the process, different characteristics within the use cases to be investigated are also dealt with. In addition, the pooling (negative or positive expression of items) and exact scaling (phrasing of Likert scales) of the indicators are addressed. It emerges that the original sources could be adapted fairly consistently. All constructs that are employed in this study are theoretically adapted from Venkatesh et al. (2003) and Venkatesh et al. (2012). Within the context of the four use cases, selectively further literature was consulted. Scaling finally relies on seven-point Likert scales, which are the state-of-the-art measurement tool (Likert, 1932; Kulas and Stachowski, 2009; Willits et al., 2016). From a theoretical perspective, these represent multi-level interval scales. Odd numbering is advised if the research object in question is uncertain future technologies so that respondents have the possibility to retain a neutral position (Fazel, 2013). This is important for the present study because it is assumed that a part of the population has not yet formed meaningful opinions. Table 2.11 ff. list the construct ©tems and related references for the use case of smart mobility. Other use cases were operationalized as similarly as possible. Nevertheless, small variations emerged. For instance, for the eBike use case, a health-related item was added to PE ('An eBike improves my health.') and for the more visionary use cases of MaaS and fully automated vehicles, the conditional form was applied (e.g., for *habit expectancy* 'Using a fully automated vehicle would be normal for me').

Additionally, to the independent variables of UTAUT and UTAUT2, the dependent target construct *use behavior* must be measured. Acceptance literature therefore provides diverse and arbitrarily complex concepts of operationalization. In this study, I use a simple single-item assessment. The rationale will be given in the following.

The first question to clarify is whether to collect real usage data or survey information. In studies on the utilization of computer software, model development frequently involves focus groups whose usage behavior can be precisely recorded with system logs (Davis et al., 1992; Venkatesh et al., 2003). However, this measurement of exact behavior is costly and hardly feasible in mobility practice for larger samples. In transport acceptance studies for example, Dudenhöffer (2013) tried to measure electric vehicle usage with an experimental research design with 232 participants over the course of one year. Although the approach was very innovative and advanced our understanding of electric vehicle adoption in many dimensions, it failed to apply TAM because of data inconsistencies. Other model-based mobility acceptance studies (Madigan et al., 2016; Madigan et al., 2017) were eventually able to achieve significant results with 'real' usage data from large-scale field studies. However, this is not possible in present examination because, on the one hand, it would exceed the scope of this work and, on the other hand, smart mobility is simply not yet available in its currently conceived form. Due to these issues, some authors simply omitted the variable *use behavior* from their models (Xu et al., 2018; Shaw and Sergueeva, 2019). Present study chooses a different path. Instead of neglecting *use behavior*, self-reported measurements for each of the use cases 1–4 are applied.[46] Even if this form of operationalization is sometimes criticized because of potential bias (e.g. in Donaldson and Grant-Vallone, 2002), it offers the advantages of being cheap and simple. Moreover, newer evidence suggests that self-reported data are not necessarily bad data if they are accurately implemented (Chan, 2009). This involves, for example, paying attention to the wording, the position in the survey and circumventing the dangers of social desirability (Miller, 2011). Indeed, self-reported data have become common practice today. For gauging usage, the instruments of measurement are the subject of ongoing enhancement. Venkatesh et al. (2012) explain that authors are struggling to integrate the extent of use (e.g., the no. of uses), the breadth of use (e.g., the no. of different applications/features used)

[46] Since Mobility-as-a-Service in general, and fully automated vehicles in particular, are expected to be not accessible to the vast majority of survey participants, the use of digitally supported mobility services (as a substitute for MaaS) and the use of automated driving functions (as a substitute for Level 5 vehicles) are deployed as fallback operationalizations.

and the depth of use (e.g., the duration per use). Thus, one can see that an initially simple phenomenon such as the use of a technology becomes seriously complex due to the nature of the technology and the anticipated implications of the responding individual in an effort to eliminate methodological biases. As a result, *use behavior* scales expand more and more in size, again reducing ease of measurement. To counteract this trend, some authors in mobile technology acceptance studies started falling back to single-item scales (Chen et al., 2002; Wu and Wang, 2005), originally brought in by Ajzen and Fishbein (1980). These researchers advocate the fact that measures comprising one item can be roughly as effective as multi-dimensional items as long as the problem of common method bias can be ruled out (Sharma et al., 2009). Since the results achieved in the supporting literature were very good, I follow their approach in scaling for reasons of cost-effectiveness (Table 2.12).

Table 2.12 Operationalization of the construct use behavior

Construct	Code	Item	Supporting literature
Use behavior	1_USE	How often do you use smart mobility on average?	Chen et al. (2002), Wu and Wang (2005), Escobar-Rodríguez and Carvajal-Trujillo (2014), Baptista and Oliveira (2015)
	2_USE	How often do you use automated driving functions on average?	
	3_USE	How often do you use digitally supported mobility services on average?	
	4_USE	How often do you averagely use an eBike?	

scale: never (1); less than once a year, but I have tested the technology (2); more than once a year (3); more than once every six months (4); more than once a month (5); more than once a week (6); more than once a day (7)

Next, measurements for the model extensions *perceived risk* and *collective efficacy* were developed. While the scales of *collective efficacy* were adapted in dialogue with the original authors, the operationalization of *perceived risk* has not been as straightforward.[47] This arises from the fact that *perceived risk* is

[47] See Section 2.2.2.1, p. 64 for risk operationalization.

conceptualized in many different forms throughout literature (e.g., as institutional distrust, privacy risk, safety risk, technological risk). Occasionally, it is also designed as a so-called second-order construct to integrate different reflective risk measurements (e.g. in Wang et al., 2019). Overall, risk thus displays its influence on acceptance inconsistently in literature depending on conceptualization and time of measurement (e.g., before and after market diffusion of a new service). Faced with this problem and against the backdrop of the explorative character of the present work, the solution was to create and test a single, but holistic construct. Its aim is to reflect the manifold facets of risk in one reflective form. Table 2.13 presents corresponding items and sources.

Table 2.13 Operationalization of model extensions

Construct	Code	Item	Supporting literature
Perceived risk	PR1	Using smart mobility is risky.	Bauer (1960), Susanto and Goodwin (2010), Wang et al. (2019), Zhang et al. (2019)
	PR2	I do not trust smart mobility technologies.	
	PR3	Smart mobility may not work as well as conventional mobility and cause problems.	
	PR4	There are too many open questions about smart mobility.	
	PR5	Smart mobility is beyond my control.	
	PR6	I have a certain fear of smart mobility.	
Collective efficacy	CE1	The people in our region can pull the strings to significantly reduce CO_2 emissions.	Barth et al. (2016), Jugert et al. (2016)
	CE2	If all people in our region participate, we can help solve environmental problems through smart mobility.	
	CE3	I believe that, together, we can create a sustainable future.	
	CE4	If everyone participates, smart mobility will help to significantly improve air quality.	
	CE5	Together, we can achieve that Germany will remain the location for technical innovations in Europe.	
	CE6	Together, we people in our region can achieve the change in mobility.	

scale: linear, ordinal Likert scale from *I do not agree* (1) to *I agree* (7)

Finally, control and moderation variables for model context had to be captured. These comprise sociodemographics, experience, and the level of system integration. Since the level of system integration[48] is already immanent in the survey design, only the first two had to be operationalized. The sociodemographic variables serve two purposes in the present study: to enable group specific acceptance analysis and to perform quota sampling. To gauge them, a standard set of scaled questions on age, gender, education, job, and living place with optimized wording was applied (Hoffmeyer-Zlotnik and Warner, 2014). Additionally, I asked about ownership of an eBike using a nominal yes and no scale.

The operationalization of the construct experience can be compared to the one of *use behavior*. Early research like Taylor and Todd (1995a) even equaled *use behavior* and experience. Overall, the discussion about the right conceptualization can be conducted in great depth. For example, the question arises as to what people understand by experience; theoretical knowledge, practical experience, or mental confrontation with both? Is experience a state of an individual's present understanding or a retrospective on past behavior?

Table 2.14 Operationalization of the construct experience

Construct	Code	Item	Supporting literature
Experience	1_EXP	How much experience do you have with smart mobility?	Taylor and Todd (1995a), Castañeda et al. (2007) Suki and Suki (2017)
	2_EXP	How much experience do you have with automated driving functions?	
	2_EXP_a	How much experience do you have with fully autonomous driving?	
	3_EXP	How much experience do you have with digitally supported mobility services?	
	3_EXP_a	How much experience do you have with Mobility-as-a-service?	
	4_EXP	How much experience do you have with eBikes?	

scale: none (1); little (2); some (3); a lot of (4); very much (5)

Typically, experience is operationalized as a passage of time from initial use, as a counting of experiences or a summary of experience intensities (Kim

[48] See Section 2.2.3.3, p. 71 for further information.

et al., 2005). Mixed methods exist. Irani (2000) for example employs a formative four-item scale that includes time as well as sum of expositions, whereas Kim (2008) applies a two-item reflective approach. Originally, in developing UTAUT, Venkatesh et al. (2003) utilized a three-stage research scheme that intelligently derives experience from the point of time at which recurring individuals were asked to fill out the same survey. Such a longitudinal research design usually delivers excellent data yet drastically raises survey costs, due to repeated inquiries and the rigid experimental framing. For UTAUT2, Venkatesh et al. (2012) thus kept in line with Limayem et al. (2007)'s more efficient work on habit and experience. Experience was therein measured as a single item asking for the number of months of technology exposition. This worked well for mobile internet experience due to the subscription involved that commits customers to the technology. Potentially, in smart mobility, voluntariness of use must be considered, and such a scale cannot be transferred. In another example of single-item experience measurement, Chen et al. (2011) simply asked whether online games have been played or not and formed two groups accordingly. Fazel (2013) similarly operationalized experience for electric cars. However, for a new concept that is not yet quite tangible, such as smart mobility, a yes/no distinction appears to be just partially effective. Cases in which users have, for example, already gained experience with a particular mobility app but have not yet done so in the context of an overarching understanding of smart mobility, would fly under the radar. Consistent with the *use behavior* construct, the present study hence applies a single-item measurement with a five-point scale reflecting intuitive levels of experience (Liébana-Cabanillas et al., 2014; Suki and Suki, 2017). The evaluation of what experience denotes, like theoretical or physical interaction, is thus left with the individual. For instance, I assume that participants will consider themselves to some extent experienced when just working in a specific area (e.g., automotive or IT), whereas others might regard themselves experienced at best when they have physically been in touch with smart mobility technology. Any type of experience might matter. Deviations in the data are anticipated to even out across a larger sample (Hair, 2014). To receive more reliable data, experience with MaaS and automated vehicles are backed up with exposure to alternative technologies (EXP and EXP_a in Table 2.14). Liébana-Cabanillas et al. (2014) demonstrated that experience with related technologies can help to predict intention and usage.

2.3.2.2 Data Collection

A self-administered survey was developed to collect the empirical data for this study. This type of paneling entails the benefit of eliminating bias from professional responders. It was implemented and published in *soscisurvey*.[49] The open-source platform places particularly high demands on data protection, allows for mobile optimization, and offers extensive customization possibilities. The questionnaire was published bilingually in German and English. Having first been developed in English, the items were then translated into German. I initially relied on the artificial intelligence of DeepL for this task. Next, a group of university staff checked the translations. Last but not least, a blind reverse translation into English using the same procedure with fellow researchers was applied. Across all use cases and acceptance factors, almost no variance ($s^2 < 0.006$) between the two language groups[50] was found, confirming the quality of the measurement approach.

Substantially, the survey consisted of two main parts: a technology-acceptance part A with specific preliminary information and a general part B with questions about sociodemographics, the latter including captured age, gender, or job. Moreover, Part B offered the possibility to leave qualitative feedback and participate in a lottery. Ten 50€ shopping vouchers served as an attractive incentive to participate since comparable surveys on relevant panels[51] paid less on average. In part A, each participant was randomly assigned one to four of the smart mobility use cases. The number of draws, and thus the number of item batches to be answered, was based on the participants' individual preferences. Prior to Part A, participants had therefore been able to select how much time they wanted to invest (eight minutes per use case) in the survey and correspondingly received one to four lottery tickets for the shopping vouchers. This was integrated in order to obtain as high a level of participation as possible. As a result, 25% of the participants chose to complete only one questionnaire. 10% selected two and 5% three. A majority of 60% answered the surveys for all four use cases. In terms of the pure number of completed questionnaires per technology, this procedure consequently generated around 400 additional responses[52] that might otherwise not have been received.

---

[49] https://www.soscisurvey.de/

[50] See Table 2.16, Section 2.3.2.4 for sample characteristics.

[51] E.g., SoSci Panel (https://www.soscipanel.de/), SurveyCircle (https://www.surveycircle.com), SurveyMonkey Audience (https://www.surveymonkey.com/surveymonkey/audience).

[52] From a total of 2143 filled surveys valid after data cleaning.

Reaching for objectivity

The central question of objectivity is whether identical results can be generated when using different people for the same research task (Bühner, 2011). While reliability and validity can be statistically verified and structural changes are still possible, objectivity must be inherent in the survey design. In addition to naturally occurring measurement errors, the main goal in this context must be to minimize systemic error. This type of bias is within the responsibility of the researcher not only because systemic error includes a researcher's own faults (e.g., misleading wording or scaling), but also because it is the researcher who selects who to query. Finally, after data collection and cleansing, systemic bias can be partially assessed with common method variance (CMV) tests. However, it is no longer possible to draw any causal conclusions about a lack of objectivity from CMV, since other processes (e.g., technical errors) can contribute to systemic error occurrences. Various countermeasures have been implemented in the current study to minimize systemic errors, of which the most important will be presented:

- *Multiple eyes principle:* The subjective influence of the researcher can be very effectively objectified by involving a number of third parties in the research process. Collectively, these are more likely to detect errors or other sources of biases such as the threat of non-response bias (Kalton and Schuman, 1982). For this research, external experts were coherently consulted for cross-checking the translation of the items, piloting the survey logic or ensuring legally compliant anonymity, which can, inter alia, reduce response bias due to social desirability.
- *Pooling:* Constructs should be thoughtfully arranged within the survey. For example, it is sensible not to position items with overlapping interpretive spaces such as *effort expectancy* and *facilitating conditions* directly next to each other. Participants should after all not be guided through a survey with a cognitive map, second guessing what they are giving their opinion on (Chang et al., 2010). Additionally, the inclusion of both positively and negatively worded items for the same scale can help to prevent extreme responses and acquiescence response-style biases. This allows, for example, to test the attention of the subjects and can be extremely useful in data aggregation. Negatively pooled items were therefore randomly interspersed in the questionnaires.
- *Multi source data:* Gathering data from multiple sources helps to even out individual biases and gain a richer picture of the situation as a whole (Rahm

and Do, 2000). This was accounted for by addressing different social groups via different media at different times.[53]

- *Minimum of a-priori knowledge:* Overall, smart mobility still seems to be a largely unknown and unfamiliar topic that many people cannot relate to their daily lives and experiences (Mukhtar-Landgren and Paulsson, 2020). So, given this widespread lack of consciousness and familiarity with smart mobility, collecting meaningful opinions on a-priori acceptability is challenging. A common solution practice is to provide short descriptions beforehand along with pictures and examples (e.g. in Payre et al., 2014; Schikofsky et al., 2020). This information was expected to level citizens' knowledge and encourage engagement on the topic, which might then lead to more stable opinions and fewer non-responses.

Implementation and sampling

The data collection took place from October 2018 to January 2019. I decided to survey exclusively online because it is efficient and far-reaching. Since the total population size of potential smart mobility users is unknown, several non-random convenience samplings were made successively to fulfill a predefined quota (San Martín and Herrero, 2012). This quota intends to represent the population strata in Western industrialized nations that is of relevance for smart mobility. Given the small amount of knowledge about specific target groups for smart mobility (Flügge, 2016), the mean characteristics of users from comparable studies* were utilized to define the quota. These characteristics mainly stem from innovation diffusion theory (Rogers, 1995) and commonly relate to a particular distribution of age and gender. In the present case, this yields a mix of the actual demographics of Western nations with a slight skew toward *early adopters* (younger, higher education, and more males). Table 2.15 confronts the compiled quota with the results at a raw data level.

In total, 1148 interviewees evenly participated in 2.99 questionnaires on smart mobility use cases. The predefined minimum sample requirement of 204 (Cohen, 1992) hence safely holds true for each use case. Kaiser-Meyer-Olkins above 0.95, moreover, confirm the adequacy of the sampling for the four scenarios (Kaiser, 1974). To reach the quota requirements, students were surveyed first. These usually provide high-quality data (King and He, 2006) and are considered early adopters (Lee, 2014). Following this, convenience sampling was carried out because, at this point, the sample was too young and too well educated

[53] See upcoming paragraph for the applied sampling procedure.

Table 2.15 Quota results

Criterion	Specification	Sampling (n = 1148)	
		Quota population* (%)	Raw data (%)
Gender	Male	55	58.8
	Female	45	39.2
Age	0–30	50	49.3
	30–60	40	41.1
	60–90	10	7.9
Education	Non-academics	40	38.0
	Academics	60	61.0

* age and gender derived from Fazel (2013), Barth et al. (2016), Xu et al. (2018); education derived from Soong (2000)

for the quota. Therefore, selected samples were added stepwise, as displayed in Figure 2.15. Overall, representativeness in terms of the quota was achieved quite well through this process. Solitarily gender ratio fell short of the targeted values. Hence, a weighting vector was included in PLS to compensate for the surplus of men (Becker and Ismail, 2016). Hence, supplementary data collection was not necessary.

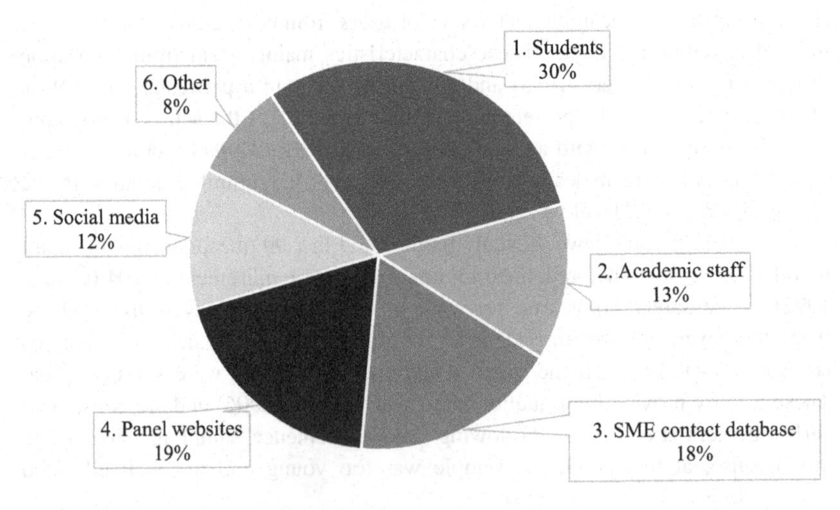

Figure 2.15 Data sources in the adjusted sample (N = 713)

Further sources included a small and medium enterprise (SME) contact database, professional social media, and open self-panel websites. The origin of the respondents was tracked by means of reference IDs, so that an evaluation of the individual data quality per source was enabled. The analysis revealed that an average of just over 35% of student data were insufficient in terms of the data-cleaning criteria. In comparison, other data performed worse. For instance, 42% of academic employee data and as much as 50% of social media data had to be removed. The best data were the 'other' data, which mainly consisted of individual targets. Potentially addressing respondents personally meant that only 10% of the data had to be removed in 'other' data.

2.3.2.3 Data Cleaning

In order to improve the quality of the information, rigor data cleaning was performed on the raw data. At the end of the process, 713 of the original 1148 cases (62%) remained for further analysis.

In essence, I followed the recommendations of Field (2011). Thus, first, the average number of missing responses was evaluated. As Hair et al. (2017) suggest, once more than 15% of the responses were missing in a single survey, the case was omitted (n = 233). For the remaining missing values, –9 was imputed as a dummy and other 'dirty data errors' (Rahm and Do, 2000) such as misspellings, slipped lines, etc. were corrected. Next, the mean response times were evaluated (~8 min/questionnaire). Data from 113 participants who finished unreasonably fast (2 SD below mean) were removed. 17 participants averagely needed more than 2 SD above mean. Nevertheless, these data were kept, since a maximum response time of 18 minutes still seemed quite realistic. In any case, such and other anomalies were always carefully inspected and investigated for suspicious response patterns (e.g., people who said yes and no, diagonal lining). Inverse pooling and forced data entry also helped in this regard to check for potential fatigue (Sue and Ritter, 2012). This case-by-case review led to the exclusion of another 89 cases. Outliers and so-called influential cases (Cook and Weisberg, 1980) were tested for in SPSS and, as expected, were not found, due to Likert scaling. Finally, 713 total cases remained. These include 2143 responses, among them 517 for smart mobility, 558 for fully automated vehicles, 537 for eBikes along with 531 responses for MaaS.

2.3.2.4 Sample

Table 2.16 breaks down the demographic composition of the adjusted sample. Due to a large overlap, the 'education' grouping is omitted from further analysis in favor of the 'occupation' grouping, as this target group is considered to have a higher practical value. Present work is consequently a mixed consumer and student study with a focus on Germany. It can be observed that data cleaning affected cases evenly across all groups. Use case-specific characteristics are not presented here because there was nothing conspicuous in terms of age or gender detected in chi-square tests ($p < 0.01$), potentially thanks to the larger sample sizes (Greenwood and Nikulin, 1996). The subsamples are thus similarly distributed.

Table 2.16 Adjusted sample characteristics (N = 713)

	No. of surveys	Sample (%)
Gender (missing = 27)		
Men	409	57.4
Women	277	38.8
Age (missing = 25)		
0–20	44	6.2
20–30	302	42.4
30–40	150	21
40–50	63	8.8
50–60	80	11.2
60+	49	6.9
Occupation (missing = 25)		
Student	294	41.2
Professional	348	48.8
Other	46	6.5
Education (missing = 25)		
Non-academic	262	36.8
Academic	426	59.7
Language (missing = 0)		
German	604	84.7
English	109	15.3

2.3.2.5 Goodness of Data

PLS-SEM is a robust tool that does not require data to be normally distributed. Nonetheless, they should not be too far from normal. Extreme distributions increase the probability of non-significant paths because they inflate standard errors in bootstrapping (Henseler et al., 2009). This effectively prohibits researchers from evaluating data precisely. Standard procedures to check for normality of distributions are the Kolmogorow-Smirnow and Shapiro-Wilk tests. Both were applied to all indicators surveyed and have unanimously held true. However, Field (2011) notes that these tests become unreliable in larger samples like the present one. To investigate the normal distribution in more depth, it is therefore recommended to examine the skewness and kurtosis of the distribution curves. Since an ideal normal distribution is uncommon in practice, data are usually considered good in empirical research if the skewness is within the range of –3–+3 and the kurtosis is in the range of –8–+8 (Kline, 2015). Present data are approximately normally distributed. Only one item[54] exceeded the kurtosis threshold and will consequently be the subject of special attention. Due to the fact that PLS-SEM can theoretically handle such data, there is no initial need for removal. The distribution across all 7-point items is plotted in Figure 2.16. One can well recognize the approximated Gaussian normal distribution with a moderate tilt toward positive indicator ratings.

Besides normality of distributions, the judgment of data quality also depends on how independent the data are from the collection method. In this context, CMV refers to variance that is attributable to the method of measurement rather than to the constructs the measures are supposed to represent (Ylitalo, 2009). This, in turn, is a severe source of methodological bias, which is a major issue in all behavioral research. Especially survey-based TAM studies have in the past been subject to controversial discussions in this regard (Podsakoff, 2003). While some authors believe that CMV represents a completely exaggerated phenomenon, others believe that studies involving CMV are entirely non-informative (Richardson et al., 2009). For TAM and UTAUT research, especially the relation between the predictor variables and acceptance (intention and use) appears to be at risk for CMV (Sharma et al., 2009). In any case, data including a serious amount of CMV are not desirable, especially when the data from the dependent and independent variables have been obtained with the same tool.

Thus, to double check, I employed two approaches to assess CMV. First, as a standard method, Harman's single factor test was performed in SPSS with exploratory factor analysis (EFA). This method involves looking at the unrotated factor solution to determine the number of factors needed to explain the variance in the variables. If a single factor emerges or a common factor explains most (more than 50%) of the covariance between the measured variables, it is deduced

[54] Too many people agreed on 3_EE1: 'Learning how to use an eBike is easy for me.'

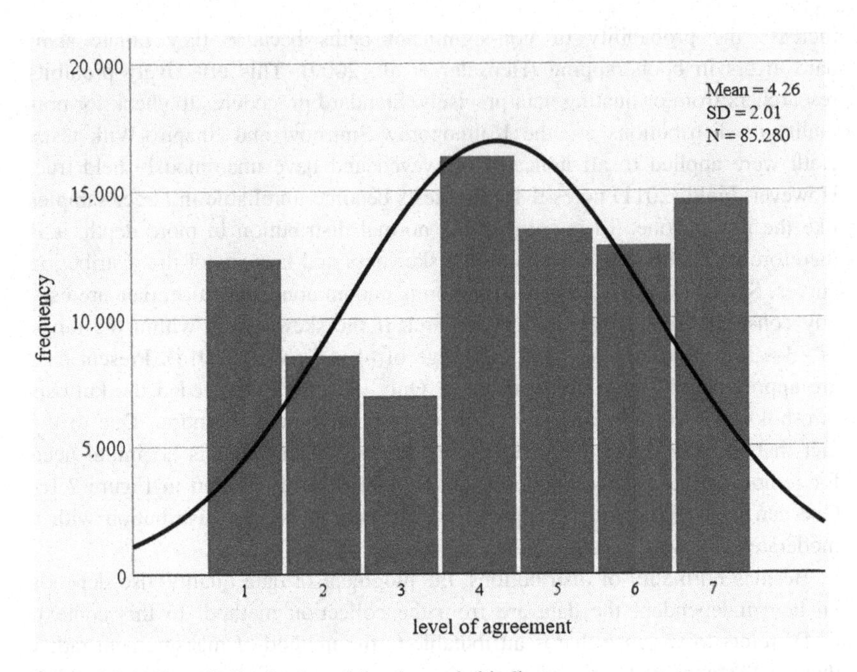

Figure 2.16 Overall distribution of 7-point-scaled indicators

that there is a significant amount of common method variance (Podsakoff, 2003). Since the extracted principal component for present data explains at best 31.04% of the variance, however, I cannot detect a problem of CMV.

Second, an advanced and PLS-SEM-specific method developed by Kock (2015) was applied. Here, the so-called variance inflation factors (VIF) are utilized to do a full collinearity assessment. These are provided by SmartPLS. The results of testing each measurement construct against each other are illustrated in Table 2.17.[55] VIF values enable the detection of multicollinearity in regression analyses. According to Kock (2015) VIF values should not surpass 3.3. Hair et al. (2017) are more liberal, on the other hand, stating that values below 5 do not indicate any substantial collinearity. Anyhow, I did not find CMV to be a problem in the present study. Despite being able to see the tendency of TAM method bias for BI, to which Sharma et al. (2009) drew attention, all values consequently fell

[55] For reasons of clarity, the illustration is limited to the evaluation of the data of the main use case smart mobility.

Table 2.17 VIF matrix of the structural model for smart mobility (N = 517)

	CollEff	Effort	FaCon	Habit	Hedo	Intent	Perform	Price	Risk	Social	Usage
CollEff		1.16	1.17	1.18	1.15	1.18	1.18	1.17	1.17	1.18	1.18
Effort	1.76		1.44	1.87	1.88	1.86	1.84	1.86	1.90	1.87	1.88
FaCon	2.21	1.83		2.35	2.35	2.22	2.33	2.30	2.39	2.30	2.42
Habit	2.02	2.09	2.07		1.99	2.05	1.98	2.02	2.05	2.04	1.83
Hedo	2.29	2.42	2.37	2.32		2.19	2.19	2.40	2.36	2.36	2.42
Intention	3.23	3.27	3.06	3.28	2.99		3.12	3.25	3.24	3.05	3.13
Perform	2.21	2.23	2.26	2.10	2.07	2.14		2.24	2.26	2.18	2.23
Price	1.22	1.22	1.22	1.21	1.23	1.22	1.22		1.19	1.22	1.23
Risk	1.40	1.40	1.42	1.42	1.39	1.41	1.43	1.38		1.44	1.42
Social	1.89	1.88	1.83	1.88	1.87	1.77	1.87	1.87	1.91		1.89
Usage	1.79	1.82	1.83	1.62	1.82	1.72	1.82	1.83	1.81	1.83	

below the threshold, after item 4_BI2 was removed for multicollinearity. Survey measurement of *use behavior* also affirms its legitimacy.

2.3.3 Descriptive Results

To begin with statistical analysis, it is useful to describe the data using univariate and bivariate analysis methods (Hair, 2014). Therefore, this section presents descriptive statistics on acceptance and usage behavior. In addition, group- and context-specific influences are assessed with exploratory data analysis. Apparently, acceptance of smart mobility is a linear or logarithmic function of the underlying experience values. In contrast to this, no distinct patterns can be discerned in the inclusion of the level of system integration.

2.3.3.1 Acceptance

As Figure 2.17 illustrates, BI of automated vehicles ($M = 3.92$, $SD = 2.09$) and eBikes ($M = 3.92$, $SD = 1.89$) is rather mediocre on the 7-point scale. Levene's test also shows that the ratings of FAVs and eBikes are significantly more heterogeneous than the other smart mobility systems. In contrast, MaaS ($M = 5.06$, $SD = 1.67$) and smart mobility ($M = 5.04$, $SD = 1.59$) find higher levels of approval and perform strikingly similarly overall, $t(1046.7) = 0.29$, $p = 0.77$.

The second key parameter of acceptance is actual usage behavior. Here, a one-way analysis of variance (ANOVA) reveals clear differences between all measured technologies and substitutes, $F(3, 2144) = 126.7$, $p < 0.001$. Figure 2.18 visualizes frequencies of responses and normal distribution curves. The mean value across all cases is 3.06, so that it can be stated that smart mobility is still rarely used, even despite the more tangible *use behavior* substitutes for MaaS and Level 5 automated driving. Concerning the different use cases, one can observe that the normal distribution builds up in waves from a I to d I. Following this scheme, the mean and median values increase as indicated below. The eBike is not yet used by many people on a regular basis. Only 15% utilize it more than once a month. After all, 8% of those surveyed own one. Standard deviations (SDs) for *use behavior* are highest for automated driving functions (e.g., automatic parking, distance control system, lane assistant). Thus, the responses are more homogeneously distributed. While 31.5% have never been in touch with any automated

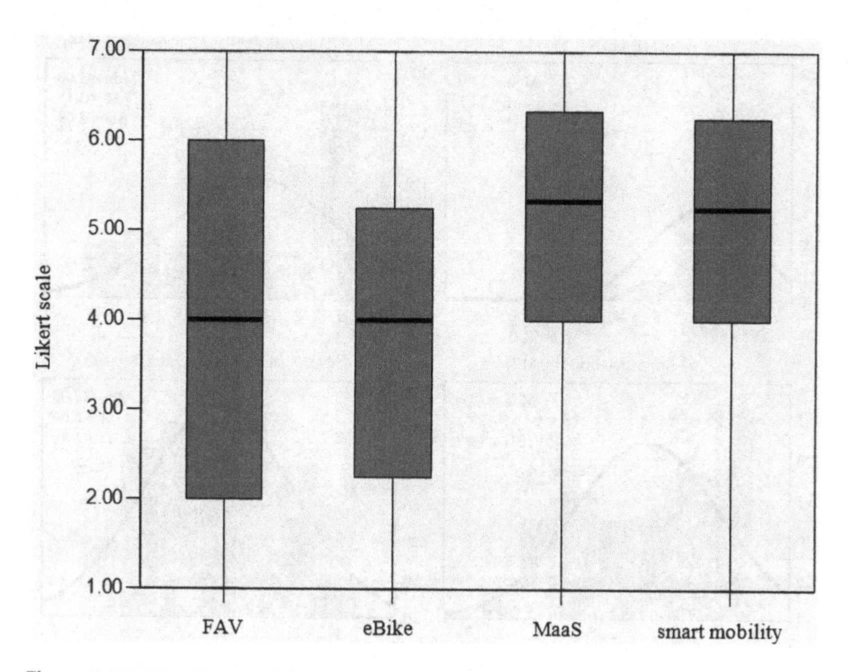

Figure 2.17 Distribution of the behavioral intentions to use

driving, 28.3% use the systems more than once a month. Smart mobility, introduced as an overarching concept of electrification, automation, and digitization, displays a slightly higher intensity of use in almost all levels. However, it is noteworthy that as little as 3.5% state that they use smart mobility on a daily basis. The most frequent answer remains having never used smart mobility. Unlike BI, for *use behavior*, smart mobility and today's digital mobility services (e.g., travel and sharing apps) are not the same, $t(1048.1) = -6.5$, $p < 0.001$.The majority already utilize these services more than once a week.

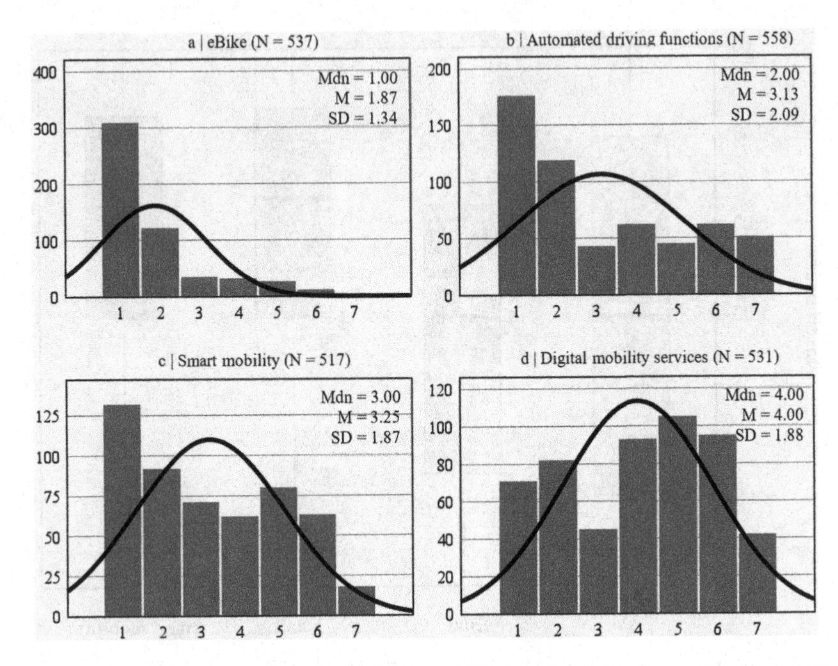

Figure 2.18 Histograms of use intensity (never (1); less than once a year (2); more than once a year (3); more than once every six months (4); more than once a month (5); more than once a week (6); more than once a day (7))

2.3.3.2 Acceptance Factors (Variables)

Figure 2.19 illustrates mean average scores of acceptance factors across use cases. On this basis, users seem to predominantly assume that new mobility technologies are or will be easy to use and that *facilitating conditions* exist that may encourage their adoption. Respondents state believing that the collective can bring in positive change for the environment through smart mobility, while on the other hand these beliefs are not reflected in social norms at the same strength. Furthermore, people expect good performance as well as enjoyment from using smart mobility.

In addition to this majority of positively rated acceptance factors, however, there are also values that contain a rather bad or moderate rating. First, these are *habit* and *price value.* Presumably, people do not perceive smart mobility a common affair and attribute a poor cost-benefit ratio to it. Second, *social influence* is medium-strong on the scale or rather low compared to the average evaluation level. Whether smart mobility is socially desirable could be interpreted as an unanswered question. Finally, risk is rated lowest of all factors, indicating that risk could be less of a concern overall.

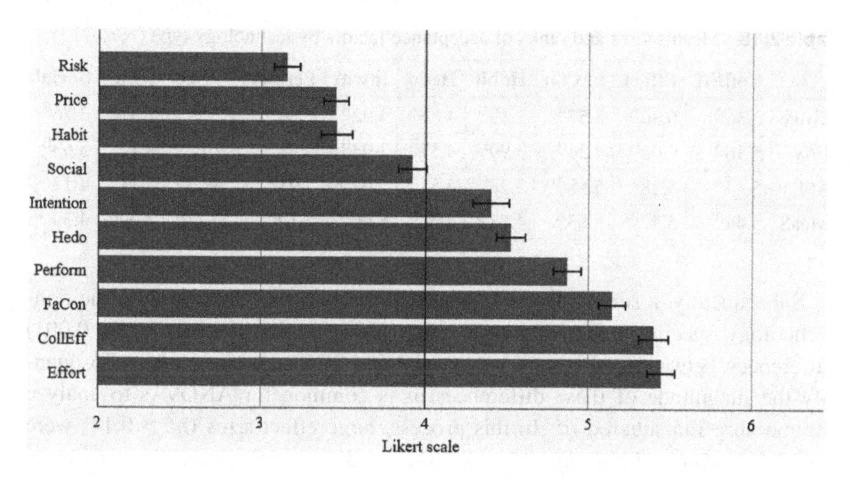

Figure 2.19 Mean average scores and 95% confidence intervals of acceptance factors

The scores in Table 2.18 were derived within performing a one-way multivariate analysis of variance (MANOVA) in SPSS. Mean value scores can be interpreted as a status quo perception of the different facets of smart mobility. MANOVA can, unlike univariate methods, include two or more dependent variables in the model. Additionally, it helps to protect following ANOVAs and post-hoc comparisons against inflation caused by Type 1 errors.[56] Prior to their application, a series of Pearson correlations were performed between all dependent variables to verify that they were meaningfully and moderately correlated with each other. The majority of correlations were observed to lie between 0.2 and 0.6, thus confirming the appropriateness of the approach. Since present group sizes were largely above 30, MANOVA can also be considered robust against violations of homogeneity of variance-covariance matrices assumption (Allen and Bennett, 2008). In applying MANOVA to test the hypothesis that there are differences in the mean acceptance factors values between technology types, a significant multivariate effect was obtained, Wilk's Lambda = 0.22, $F(33)$ =

[56] In statistics, a Type I error signifies that the null hypothesis is rejected when it is actually true, while a Type II error signifies that the null hypothesis is not rejected when it is actually false Howell (2010).

115, $p < 0.001$. The effect size was 0.4, which is regarded large (Cohen, 1988). Hence, technology type strongly affects the observed factors of acceptance.

Table 2.18 Mean scores and ranks of acceptance factors by technology type (N = 713)

	CollEff	Effort	FaCon	Habit	Hedo	Intent	Perform	Price	Risk	Social
eBike	5.36^A	6.36^A	5.57^A	2.19^D	4.75^A	3.92^B	4.30^C	2.79^C	1.91^A	3.76^B
FAV	5.36^A	5.01^D	4.24^B	3.99^B	4.57^B	3.92^B	5.32^A	2.92^C	4.16^C	3.63^B
SM	5.44^A	5.28^C	5.58^A	3.28^C	4.51^B	5.04^A	4.91^B	4.23^A	3.11^B	4.12^A
MaaS	5.44^A	5.47^B	5.63^A	4.61^A	4.38^B	5.06^A	5.19^A	3.77^B	3.09^B	4.34^A

Subsequently, a series of one-way ANOVAs confirmed that, besides the non-technology specific personality trait *collective efficacy*, significant ($p < 0.001$) differences between technology types exist for all acceptance factors. To quantify the magnitude of these differences, it is common for ANOVAs to analyze the measure Eta-squared η^2. In this process, large effect sizes ($\eta^2 > 0.14$) were found for *perceived risk*, *habit*, *price value*, and *facilitating conditions*, whereas medium-strong effect sizes ($\eta^2 > 0.08$) were observed for BI, as well as for *performance* and *effort expectancy*. Next, in order to better understand how crucial each of the individual factors are for the differences in the technology-specific acceptance processes, the canonically derived discriminating function with the highest Eigenvalue, $E = 1.37$, $\eta^2 = 0.58$, was examined (Enders, 2003). The values of the discriminant function coefficients matched those of the effect sizes, confirming risk and habit as being by far the most discriminative factors between smart mobility use cases.

Finally, a series of post-hoc analyses ($\alpha = 5\%$) based on Fisher's LSD (least square difference) were run in SPSS to examine disparities between all technologies and acceptance factors. This revealed an overall heterogeneous picture of significant and non-significant differences between the mean values. I added a superscript rank A-D to Table 2.18 to see which of the means differ significantly from each other. This must be understood such that groups with the same rank within one column do not differ. Although the mean average values of the acceptance factors evolve linearly to BI, no meaningful pattern was discovered in mapping the means and their ranks. Each acceptance process thus keeps its own peculiarities, which are stylized below:

- The *eBike* is viewed by individuals as particularly easy and low risk (A). *Facilitating conditions* are perceived good (A) and eBikes are considered being more fun than all other smart mobility (A). On the negative side, performance and prices are rated comparatively poor (C). By far the least habituated behavior is associated with eBikes (D). The large discrepancy between average BI and strikingly low *use behavior* scores, leads one to hypothesize that *use behavior* might be downshifted by variables other than BI, such as habit or prices. In practice, this results in a high potential to attract new users if only specific acceptance factors improve.

- The acceptance factors of *fully automated vehicles* exhibit a different pattern, although they bear some similarities to eBikes. For example, both technologies are significantly less socially desirable than MaaS or SM (B) and both are expected not to offer good *price value* (C). FAVs alone are considered the least easy to use with the worst *facilitating conditions* (B). Consequently, respondents indicate the highest level of risk in smart mobility being associated with FAVs (A). However, on the plus side, people in the present study value FAVs in terms of performance (A) and expect that they will become used quite routinely in the future (B).
- *Mobility-as-a-Service* achieves consistently good ratings spanning acceptance factors (e.g., in performance (A), *facilitating conditions* (A), or effort (B)). Nevertheless, some conspicuous points are found in Table 2.18. Respondents can, for instance, best imagine utilizing MaaS to become a habit in their daily lives (A) while they do not believe MaaS is much fun (B). When comparing the acceptance structure of MaaS and FAVs, it is noticeable that, although neither technology is available yet, acceptance of MaaS is consistently higher. It is the most popular technology in this study.
- *Smart mobility* can be simplified to representing ubiquitous MaaS with the accompanying introduction of electrified and automated systems. This is reflected in the acceptance structure with mean scores frequently falling in between values of Maas, eBike, and automated vehicles (e.g., for effort I, risk (B) or habit (C)). There is an exception to every rule; despite these partly reduced acceptance factors, the price-performance ratio for smart mobility is rated best (A) along with the largest BI (A).

2.3.3.3 Acceptance Factors (Items)

To now dive deeper into acceptance, an item analysis was conducted on the smart mobility use case. A group of researchers from ITS research that was unaware of the study's context, was asked to select the 15 most interesting items. Finally, these were complemented with those indicators that displayed an importance above 0.10 according to importance-performance map analysis IPMA[57] (Ringle and Sarstedt, 2016).

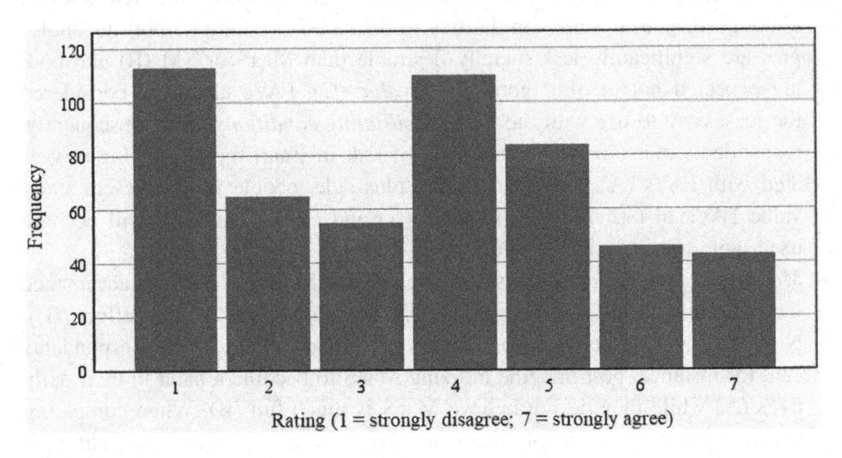

Figure 2.20 Distribution of responses for the item 'Using smart mobility is normal for me'

Figure 2.20 reviews the extent to which smart mobility is already experienced as something ordinary today. A comparatively large number of respondents evaded the question by giving moderate answers. Albeit the ratings are well distributed across the scale, the highest bar is found at value 1. Thus, there is also a large number of people for whom smart mobility is not normal.

As visualized in Figure 2.21, respondents indicated that smart mobility is perceived as something easy (Q18) that they feel comfortable with in terms of compatibility (Q20), knowledge (Q19), and availability of personal resources (Q22). They strongly believe that smart mobility can have a positive collective effect on the environment (Q16, Q17, Q21). Moreover, people appreciate the idea that smart mobility will offer them personal benefits. The analysis of the confidence intervals reveals an order: the most important benefit was comfort (Q15), the second flexibility (Q13), and the third was safety (Q10). In connection with

[57] Please see IPMA procedure in Section 2.3.4.4.

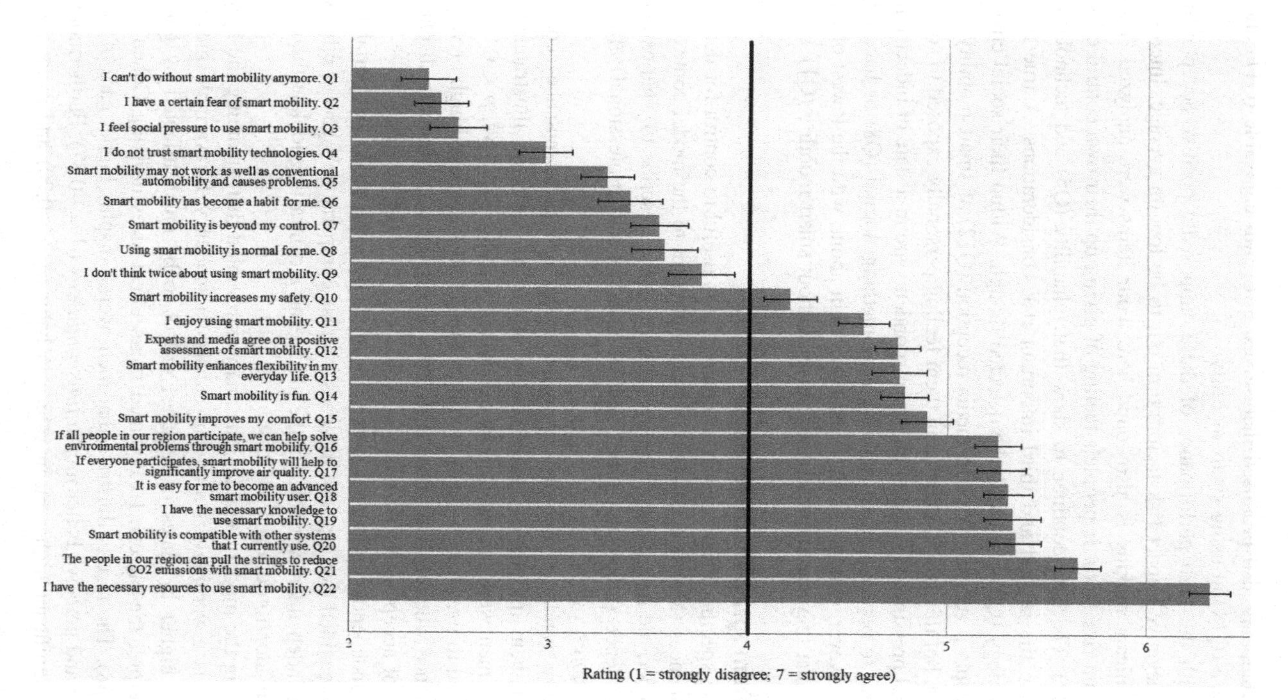

Figure 2.21 Item performances for smart mobility, means and 95% confidence intervals

this, the participants also promised themselves an overall enjoyable (Q11) and fun experience (Q14) in using smart mobility.

In line with the high performance of these supposedly positive acceptance factors, the negative-impact risk factors turn out to be low on average. Interestingly, a significant ranking is also found here. Four items were analyzed. The most important risk factor is people's feeling of giving up their own control over mobility (Q7). The data continue to show that reliability (Q5) and technology trust (Q4) are the second and third important risk considerations. A true fear of the technology (Q2) can hardly be detected though. Within their social environment, people perceive a positive media reception (Q12) of smart mobility on the one hand, but this is not reflected in them feeling personally exposed to some kind of social pressure (Q3). After all, the respondents are in a state of indecision as to whether to perceive smart mobility as something normal (Q8) or habitual (Q6, Q9). However, what those surveyed are certain about, with the lowest of all rankings, is that they can definitely still travel without smart mobility (Q1).

2.3.3.4 Demographic Differences

For detecting specific differences in acceptance, it is useful to control for demographic influences (Morris et al., 2005). In complement to the model context of age and gender, the results presented in this section also describe to what extent place of residence (urban and rural) and job (student and professional) affect acceptance scores.

To analyze *gender*, I initially calculated mean average acceptance scores and divided them into the six value ranges of the BI scale. As illustrated in Figure 2.22, men were observed to be significantly $t(684) = -3.89$, $p < 0.001$, more attracted to the offer of smart mobility. Almost 50% displayed high or very high acceptance whereas little more than 25% of women were found to display these levels of acceptance.

Second, independent sample *t*-tests were applied to the individual variables scores. This enabled two types of analyses: a cross-comparison of gender effects for the nine independent variables in the model and a contrasting look at the role of gender in adoption of each use case.

Concerning the model variables, the analysis revealed that, spanning the four cases, there is a varyingly strong but consistent tendency for variable means to be either higher in males or females, or not to be much affected by gender. To interpret the effects, I calculated mean effect sizes utilizing Cohen's d (Cohen, 1988). The results illustrate that men overall display higher ease of use ($d = -0.43$) and perceive better *facilitating conditions* ($d = -0.37$). Furthermore, they show significant higher mean values in habit ($d = -0.20$) and experience

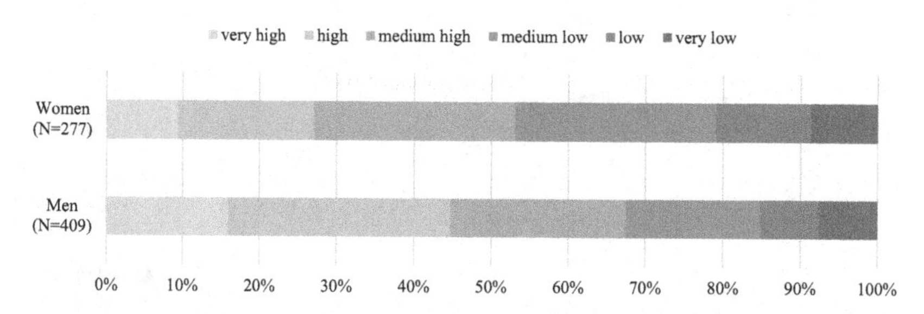

Figure 2.22 Acceptance of smart mobility technologies by gender

(d = 0.35), which may be due to causal interference of the two. For women, it can be observed that while they perceive a higher risk (d = 0.29), they also have stronger faith in the collective environmental effectiveness associated with smart mobility and especially with MaaS (d = 0.21). *Price value*, *performance expectancy*, *hedonic motivation*, and *social influence* were not meaningfully influenced by gender across cases.

Regarding the role of gender in influencing the perception of the use cases, rather specific differences were identified. As plotted in Figure 2.23, men and women rated the technologies eBike and MaaS largely equal, with solely one acceptance factor each (Effort and CollEff) standing out. Exactly these two are also found in the smart mobility use case, which is why an additional presentation was omitted. In contrast, a striking perceptual difference was found in automated vehicles, for which men evaluate almost all acceptance factors more highly (N = 558). Particularly striking is the medium-strong effect (Cohen's $d > 0.5$) of gender on FaCon and Effort of automated vehicles.

Next, a series of ANOVAs and post-hoc tests revealed no significant differences in acceptance between *age* groups. Yet the group comparison shows that younger people currently use smart mobility more often, $F(2059) = 2.90$, $p = 0.03$. This fact led to a deeper investigation about the role of age in smart mobility adoption. Eventually, some trends could be derived. As depicted in Figure 2.24, acceptance does arguably change with age if dividing the respondents into clearly above and clearly below scale-mean groups (likely and not-likely), although its mean may not vary significantly. For instance, approximately the same number of younger respondents (Age<30) indicate liking and disliking the eBike, whilst in older participants (Age>45), notably more like than dislike them. This suggests that older people might favor the technology to some extent. For FAVs, a different pattern can be observed. Throughout the age categories, the approval rate for

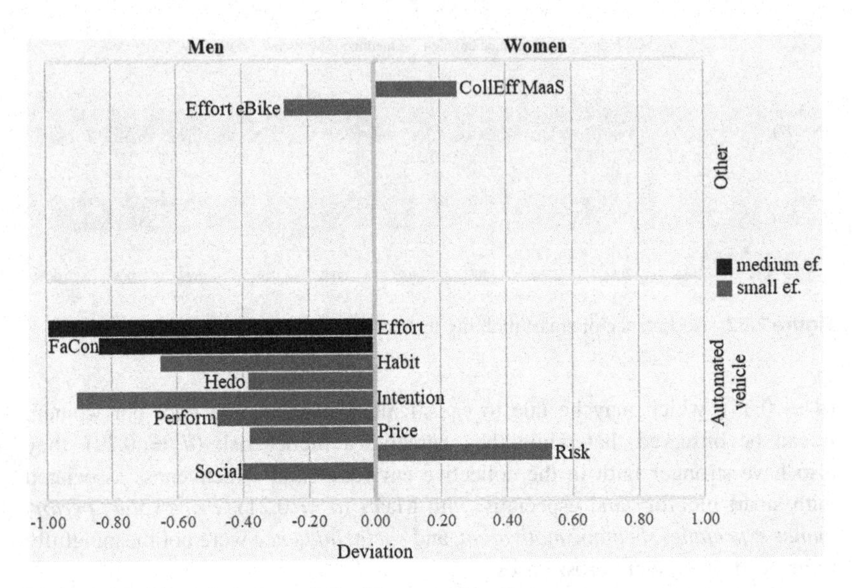

Figure 2.23 Mean deviations of significant gender effects across use cases

FAVs remains about the same, whereas disapproval grows with age. This can be seen as a sign that older people may possess greater resistance to FAVs than the young. In general, a smaller share of neutral views (not clearly above or below scale-mean) exists in intensions to use SM and MaaS compared to eBikes and FAVs. Respondents seem to have formed a more meaningful opinion for those concepts.

As shown in Figure 2.24, smart mobility (SM) and MaaS acceptance emerges as roughly the same throughout groups. Among the youngest respondents, there is a high rate of approval, which even grows in 30–45-year-olds. Then, with increasing age, likeability shrinks whilst the rejection rate grows. Results also suggest this trend might be stronger in MaaS than in smart mobility. Older people may thus reject aspects of smartphone-based MaaS more than smart mobility as a whole. In conclusion, young people tend to welcome smart mobility, with eBikes being an exception that attracts older generations more strongly. In FAVs, neither the young nor the older have formed stable opinions yet, but within the older age groups, more people dislike than like the technology.

In comparing *students and professionals* in the present sample, little overall difference in mean variance ($s^2 = 0.01$) came to light. A significant effect at

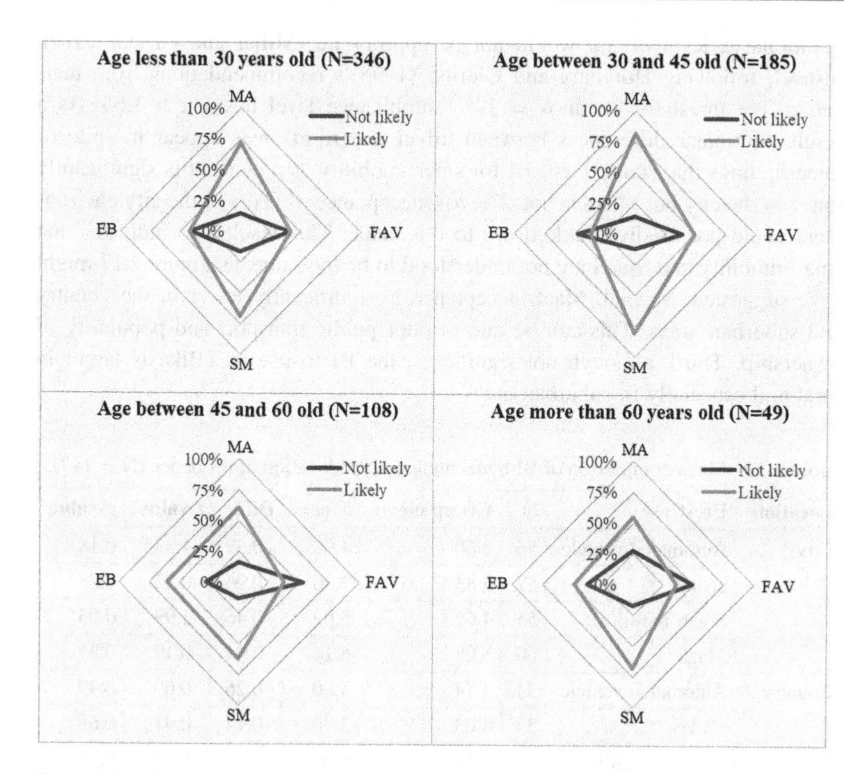

Figure 2.24 Intention to use smart mobility technologies by age

$\alpha = 5\%$ between the two groups was found in price perception of automated vehicles (d = –0.20) and eBikes (d = –0.20), indicating that students perceive lower willingness to pay, since other value indicators such as performance are not statistically different for the two groups. Additionally, students displayed significant lower *social influence* toward utilizing eBikes (d = –0.22).

The final demographic influence to consider is *location of residence*. This is carried out in a rather exploratory manner, as only 147 valid cases were available. Hence, this factor was not used to describe the sample but can be employed to better define target groups for smart mobility and improve marketing. Table 2.19 provides a comparison of the group mean values with the mean values of the sample. The table also illustrates the results of a series of t-tests that have been performed to determine the significance of mean differences. Due to the comparatively weak sample and a research area fraught with uncertainties, using a traditional α level of 5% would not be appropriate (Miller and Ulrich, 2019). Instead, following Homburg and Giering (1998)'s recommendations from marketing, the threshold for an $\alpha = 10\%$ significance level lies at $t > 1.65$. As a result, few clear differences between urban and rural areas appear in spite of three findings that stand. First, BI for smart mobility as a concept is significantly lower in the city but MaaS is not. The low acceptance of FAVs in the city environment could potentially trickle down to this value. This result also indicates that smart mobility and MaaS are not understood to be the same, as Figure 2.17 might have suggested. Second, MaaS acceptance is significantly lower in the country and suburban areas. This can be due to poor public transport and popularity of ownership. Third, although not significant, the BI to use an eBike is larger in rural and especially in suburban areas.

Table 2.19 Mean comparison of BI by technology type location of residence (N = 147)

Location	Use-Case	N	Group mean	Mean	Diff.	*t*-value	*p*-value
City	Automated vehicle	56	3.60	4.00	–0.39	–1.33	0.18
	eBike	57	3.83	3.90	–0.06	–0.22	0.83
	Smart mobility	55	4.62	5.09	–0.46	–1.99	0.05*
	MaaS	60	5.08	5.14	–0.04	–0.19	0.85
Country	Automated vehicle	34	3.74	4.00	–0.26	–0.69	0.49
	eBike	33	4.03	3.90	0.14	–0.41	0.68
	Smart mobility	30	5.19	5.09	0.11	–0.39	0.70
	MaaS	34	4.62	5.14	–0.51	–1.73	0.08*
Suburban	Automated vehicle	16	3.23	4.00	–0.76	–1.43	0.15
	eBike	15	4.65	3.90	0.76	1.59	0.12
	Smart mobility	12	5.04	5.09	–0.04	–0.08	0.93
	MaaS	17	4.38	5.14	–0.74	–1.82	0.07*

2.3.3.5 Contextual Influence

This section explores the roles of individual experience and societal system integration for smart mobility acceptance with univariate and descriptive methods. Therefore, Figure 2.25 plots BI as a function of experience. We can see an almost linear relationship between the two measures, which exponentially decreases toward reaching the scale maximum. The individual measurement points of acceptance form about the same sequence and keep a certain distance. This holds true for all use cases except automated vehicles, whose starting acceptance level is still above that of eBikes, but whose course then changes and thus encounters a small bump at Level 3 ('some experience'). One can speculate about the reasons for this. In controlling the change in means for the specific acceptance factors, SPSS finds consistently large effects ($\eta^2 > 0.14$). Sole *price value* shows a statistically minor change ($\eta^2 < 0.06$) as experience increases in present data.

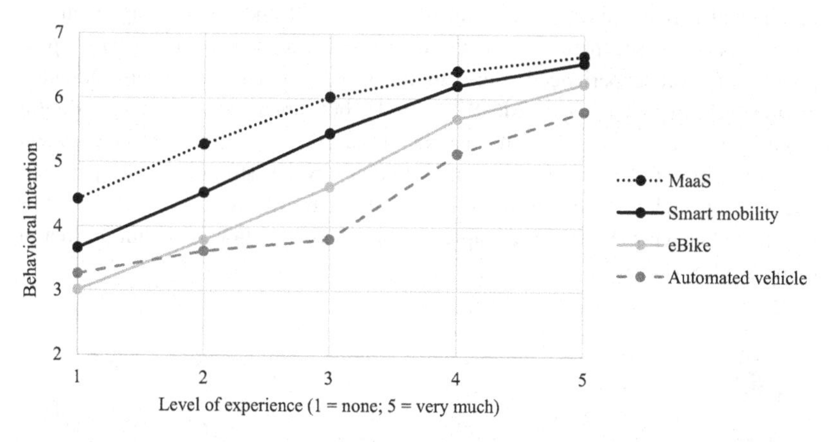

Figure 2.25 Acceptance of smart mobility by technology type and level of experience

In Section 2.1.1, I assumed that the level of system integration of a mobility technology influences the distribution of experience because it contributes to determining the cognitive effort to realize this experience (Stradling, 2006). As a result, and since experience and acceptance seem to be linearly connected, levels of acceptance may be assigned according to experience. Figure 2.26 hence combines acceptance and distributions of experience with the hierarchy of cognitive efforts as an emulation of SAE-level taxonomy from Lyons et al. (2019). In addition to the experience values of the use case technologies, empirical data for automated driving functions and digital mobility services were also available for this purpose. The eBike is in some ways a special case because, unlike the other technologies that generally address a broader target group, it is not limited to (future) users but includes current owners (N = 41). This encourages an alteration in acceptance behavior (Chen, 2016), which is why a separate analysis is useful in order to better understand smart mobility as a combined service and product concept (Manders and Klaassen, 2018).

As mapped in Figure 2.26, one can find representatives for most established acceptance levels. Prior to drawing the distributions with the specified means as fading bars, a one-way ANOVA on experience[58] confirmed ranking as meaningful, $F(2146) = 60.4$, $p < 0.001$. Furthermore, a Levene test approved that no significant differences in the associated variances exist, $p = 0.32$, which allows for a comparison. By the logic of this approach, fully automated vehicles are thus found at Level 0 of acceptance with an overlap to a-posteriori acceptance. Automated driving functions are already at Level 2 and are thereupon on the cusp of personal acceptance, even reaching social acceptance for a smaller proportion of more experienced users. In the same manner, respondents indicate having little experience with MaaS (Level 1), but larger experience with digital mobility services (Level 3). An unavoidable larger gap is found between users and non-users of eBikes. The latter are found at Level 1, struggling with personal acceptability, while users, on the other hand, have already incorporated eBikes into their behavior (Level 4). Complete habitual assimilation is within the realm of possibility (Level 5).

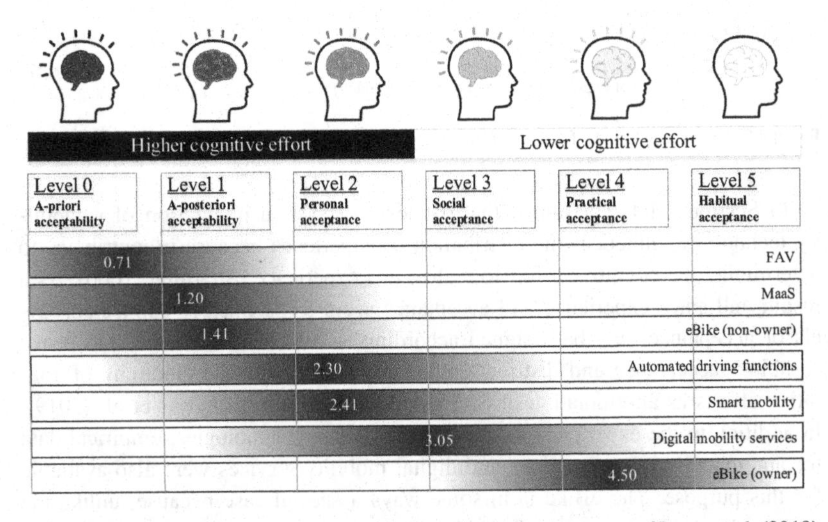

Figure 2.26 Mean average experience and allocation in the taxonomy of Lyons et al. (2019)

58 Adjusted to a six-point scale.

2.3.4 PLS-SEM Results

Model estimation and evaluation were performed using SmartPLS 3 (Ringle et al., 2015). I first assessed reliability and validity with confirmatory composite analysis (CCA) before testing the models.

Quantitative results are presented below. To judge their quality in present research, a series of local statistical criteria can be interpreted as visualized in Figure 2.27 (Nitzl, 2010; Kline, 2015; Hair et al., 2017; Hair et al., 2018). It is pivotal to understand that such an assessment is carried out individually for the measurement, structural model, and overall model. However, the interpretation of outcomes must be realized in an integrated way (Henseler, 2018). The specific procedures and criteria will be elucidated in the upcoming section.

2.3.4.1 Software Settings for Model Estimation

The parameterizations to be selected for setting up SmartPLS software correctly were adopted from the guidelines in Hair et al. (2017). For present scope, four main types of calculations are performed. Besides the PLS algorithm, these foremost include the so-called bootstrapping method, the blindfolding procedure, and the permutation algorithm. In the pursuit of result comparability and reproducibility, the next paragraphs explain the settings of the software that are used in all of the following calculations. Advanced method settings are discussed where applicable.

The PLS algorithm represents an integrated procedure within the framework of which the model estimation is carried out (loadings, weights, and path coefficients). In addition, the procedure provides information about the quality criteria of the measurement and structural model. It is thus the key method of this work. A variant of the PLS algorithm, the so-called consistent PLS (PLSc) algorithm, performs a correction of the correlations of the reflected constructs to make the results consistent with a factor model. Although PLSc promises more robust results, usage is only advised if the models are fully reflective, which is not the case for present models due to the single-item measures they contain (Dijkstra and Henseler, 2015b). I hence keep in with normal PLS and employ the settings listed in Table 2.20. The 'path' weighting scheme was chosen because, unlike 'centroid' and 'factor' weighting, it takes into account the direction of the hypotheses in the model, leading to somewhat more precise outcomes (Henseler et al., 2009). The maximum number of OLS-regressions was set to 500 and the stop criterion was set to 10^{-7}. This means that the iterative process of the PLS algorithm will either stop when it reaches 500 runs or when the change in outer

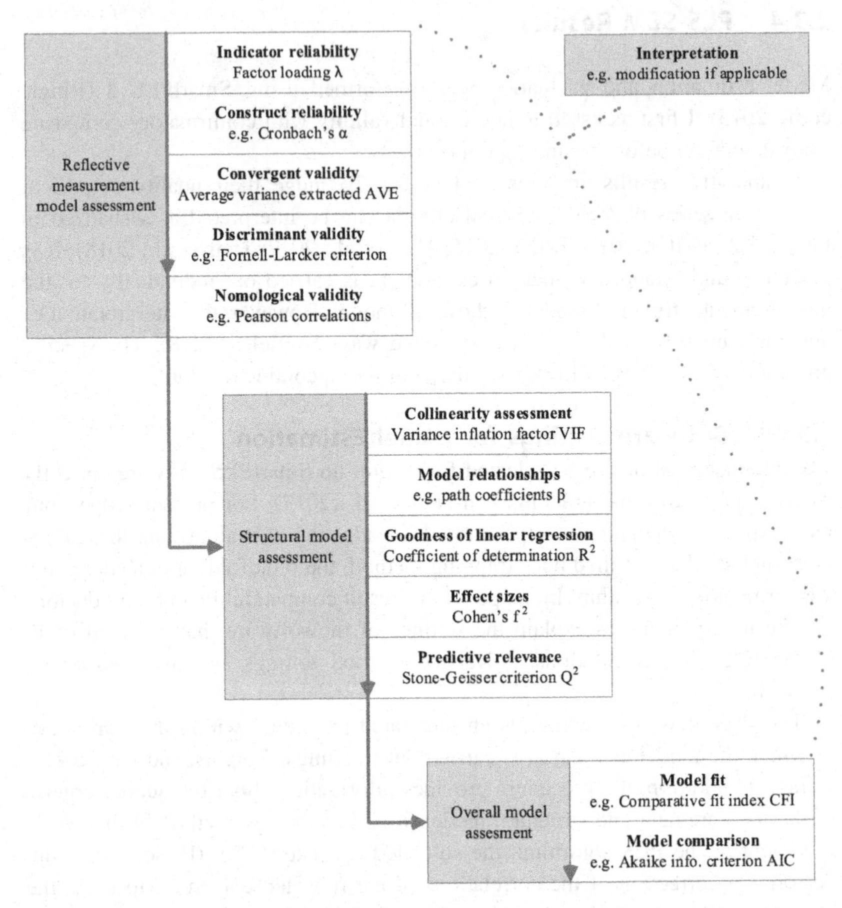

Figure 2.27 The PLS-SEM process

weights for consecutive iterations is smaller than 10^{-7}. Both values represent conservative research standards. In line with this, the initial outer weights were also left at the default value of +1. Next, the treatment of missing values must be specified. This is of particular importance because the selection has a relatively large influence on the results. From the three methods available in SmartPLS,, because it is Mean value replacement method, recommended for data that exhibit

Table 2.20 Parameter settings in SmartPLS3

Procedure	Parameter	Setting
PLS algorithm	Weighting scheme	Path
	Maximum iterations	500
	Stop criterion	10^{-7}
	Initial weights	+1
	Missing values	Mean replacement
	Weighting vector	0.8 (male), 1.2 (female)
Bootstrapping	Subsamples	5000
	Parallel processing	Yes
	Amount of results	Complete bootstrapping
	Confidence interval method	Bias-corrected and accelerated
	Test significance	Two-tailed, $\alpha = 5\%$
Blindfolding	Omission distance	7
Permutation algorithm	No. of permutations	5000

less than 5% missing per indicator[59] (Sarstedt and Mooi, 2014), was chosen from the three methods available in SmartPLS. Other methods, such as casewise deletion, may promise more precise results, but involves some pitfalls. If missing values were primarily found within a particular group, for instance, casewise deletion would systematically exclude this group and hence bias calculations. So as to reduce such bias related to the target population outside of the missing value issue, all calculations were performed using an a-priori established vector that corrects imbalances in gender ratio.[60]

By means of the so-called bootstrapping procedure, the significance of the path connections of the structural model is tested, whereby iterative subsamples of a specified size are drawn to calculate corresponding test statistics. The number of subsamples needed depends on the construction of bootstrap, its confidence interval method, and the range of null hypothesis rejection (Streukens and Leroi-Werelds, 2016). A common and efficient recommendation in terms of computer processing time, is to use 5000 subsamples for complete bootstrapping with the

[59] Within the cleaned data, we find a maximum of 3.5% missing data per indicator (for collective efficacy, which was located at the end of the survey and might thus be a victim of respondents' fatigue).

[60] See Section 2.3.2.2 for details on data collection.

default bias-corrected and accelerated bootstrapping confidence interval method at a 5% Type–1—error probably level.

Blindfolding denotes the third main procedure to be utilized, which allows the predictive validity of the model to be investigated. To this end, the procedure performs a stepwise suppression of a part of the data matrix and simultaneously reconstructs the suppressed data using the estimated parameters. In comparing the estimated data with the actual data, predictive validity can be established. Typically, an omission distance of seven has proven to deliver consistent results. This value implies that every seventh data point of a construct's indicators are eliminated in a single blindfolding round. Since every point must be predicted, seven blindfolding rounds are necessary before all data are replaced by predicted values. This equals the omission distance (Hair et al., 2017).

Eventually, the permutation algorithm is used to investigate group differences and allows for application of multigroup analyses. It can test if pre-defined data groups significantly differ in parameter estimates (e.g., loadings or path coefficients). For this purpose, the test randomly permutes (i.e., rearranges) observations between groups and reestimates the model accordingly to receive a test statistic. Following Edgington and Onghena's suggestion, setting SmartPLS to 5000 permutations ensures stability of results (Edgington and Onghena, 2007).

2.3.4.2 Measurement Model

In PLS-SEM, naturally occurring measurement errors must be proven to be within an acceptable range. To assess the quality of the measurement model, reliability and validity of the empirical measurements are examined. Reliability of the measurement describes how well certain indicators represent the same thing. Reliability is a prerequisite for validity. Validity refers to how well the measurements constitute the subject they are supposed to represent (Sarstedt and Mooi, 2014). Thus, a prerequisite for achieving reliability and validity is a well thought out research design. Following literature, the validation of the reflective measurement model is carried out in five evaluative steps: (1) indicator reliability, (2) construct reliability, (3) convergent validity, (4) discriminant validity, and (5) nomological validity. For this purpose, various test methods and associated cut-off values are available as provided in Table 2.21. This integrated procedure has recently been termed confirmatory composite analysis (CCA) and is fully conductible in SmartPLS (Hair et al., 2020). It represents an PLS-alternative to confirmative factor analysis (CFA), which originates from CB-SEM.

The goal of CCA is to achieve low error measurement in and between composites. While the research design, data collection, and cleaning can serve to reduce systemic errors, the aim of measurement model validation is to exclude significant measurement errors and biases at item and composite level. Hence, problematic items can be detected and removed. However, pure data-driven elimination of indicators is not permissible, since it would sacrifice content validity for a homogeneous measurement instrument (Hildebrandt and Temme, 2006). Content validity is crucial, describing whether the construct logically targets the actual subject in reality. Content validity was ensured in the present work through utilizing established scales and expert testing in advance of the survey.

Table 2.21 Goodness criteria for validating reflective measurement models in PLS-SEM

Goodness measure	Local criterion	Symbol	Target value	Reference
Indicator reliability	Factor loading	λ	≥ 0.7 / ≥ 0.4	Bagozzi et al. (1991), Hair et al. (2011)
Construct reliability	Cronbach's alpha	α	≥ 0.7 / ≤ 0.95	Cronbach (1951), Nunnally (1978)
	Rho alpha	$\rho\alpha$	≥ 0.7 / ≤ 0.95	Dijkstra and Henseler (2015b)
	Spearman-Brown	P	≥ 0.7	Eisinga et al. (2013)
Convergent validity	Average variance extracted	AVE	≥ 0.5	Fornell and Larcker (1981)
Discriminant validity	Fornell-Larcker criterion	$\sqrt{AVE}$	≥ 0.7	Fornell and Larcker (1981)
	Heterotrait-monotrait ratio	$HTMT$	≤ 0.90 / ≤ 0.95	Henseler et al. (2015)
Nomological validity	Pearson correlation	(r)	< 0.01	Cohen (1992)

Reliability of measurement

Indicator reliability tells us how well an indicator (e.g., an item) explains the variance in a factor (e.g., a latent variable). The factor loading λ is employed as a measure of indicator reliability. In the best-case scenario, it should be so large that more than 50% of the variance in the factor can be described by the indicator. This yields a threshold value of $\lambda > \sqrt{0.5} \sim 0.7$ for the single path connection (correlation), which will also be found frequently in following contexts. Commonly, values that do not exceed this threshold would have to be successively removed from the model. However, this threshold and its liberal or conservative interpretation are subject to ongoing debate. For example, Fazel (2013) uses a threshold of $\lambda > 0.6$, arguing that low values are also permissible when developing new constructs in the social sciences, as long as they are only sufficiently significant (5% level). Other authors propose even lower limits for latent psychological variables (e.g., 0.4 in Bagozzi and Baumgartner (1994)). Finally, Hair et al. (2017) suggest keeping all indicators with factor loadings above 0.7 and deleting all with loadings below 0.4. For all values in between, the recommendation is to remove the indicator particularly when a substantial increase in construct reliability can be achieved. Indicator and construct reliability must hence be tested in tandem. In the case of psychological latent variables that are measured with the multi-item approach, it must be questioned whether the deletion of an item does not in fact compromises the quality of the construct's overall logic or whether it no longer measures exactly what should actually be observed.

Construct reliability is the more important reliability criterion (Nitzl, 2010). In contrast to indicator reliability, it does not refer to the individual correlations of the indicators with the factor but to the correlation of the indicators with each other. These should be particularly high, as too much total variance would otherwise be generated in the associated factor, which in turn—and this is what this test is about—the associated factor would not be able to cover (Götz et al., 2010). Again, thresholds of 50% explained total variance apply, which differ depending on the applied quality criterion and the algorithm behind it. For the popular quality criterion Cronbach's alpha, the minimum value is $\alpha > 0.7$ again (Nunnally, 1978). If this value is not reached, indicators with a comparatively low correlation can be removed iteratively. Values above 0.95 are also not desirable, since this would indicate that all items measure the same thing. This often occurs if items are semantically redundant. The use of Cronbach's alpha is becoming increasingly restricted to confirmatory studies because the value is relatively sensitive to a large number of items and, from a technical point of view, tends to systematically underestimate reliability when PLS-SEM is used (Chin, 1998). Hair et al. (2017) therefore recommend considering an alternative, such as the recently developed Rho alpha (Dijkstra and Henseler, 2015b), which tends to slightly overestimate reliability. Rho alpha is a quality criterion of confirmatory factor analysis and indicates the shared variance among the observed variables used as indicators for a latent construct. Again, a conservative threshold is 0.7 and a minimum practical claim in exploratory research is 0.6 (Bagozzi and Yi, 1988). According to the authors, the true value of construct reliability lies somewhere between the two criteria. Furthermore, as outlined in Section 2.3.2.1, scales representing a construct should normally consist of more than two items. However, scientific reality means that some items have to be omitted sometimes, leaving just two items remaining in the construct domain. Due to computational biases, Cronbach's alpha is unsuitable in this case (Eisinga et al., 2013). This is where the Spearman-Brown coefficient comes into play. The coefficient tests if scales inherit internal consistency by controlling the change in reliability after manipulating the test length, i.e., the number of items. As with the other indicators, Spearman-Brown should exceed 0.7. There is often no need to test single items, since reliability is 1.

Validity of measurement

A common measure involved that tests for *convergent validity* is average variance extracted (AVE). The AVE is the sum of squared loadings divided by the number of indicators. If this value is greater than 0.5, it indicates that the construct explains more variance than remains in the error terms of the indicators. Thus, while indicator reliability indicates how much the items have in common and construct reliability observes how well those items describe a latent variable, convergent validity assesses to what extent the above correlations did not happen by chance when considering the interaction of all model indicators. Table 2.22 present results of the first part of CCA for smart mobility.

Overall, the performance of the measurement model is good. The vast majority of local criteria lie within the proposed quality ranges. Solely items FC1 and FC4 displayed a somewhat reduced factor loadings across use cases. Although omitting them could have led to slight construct reliability gains, I decided to include both for content validity. To clarify: *facilitating conditions* textually describe perceived cognitive barriers to adopting technology. This is operationalized as the reflective overlap of the four dimensions of availability of resources (FC1), knowledge (FC2), compatibility (FC3), and perception of support (FC4) in Venkatesh et al. (2003). Unlike other constructs, as for example *hedonic motivation*, which is operationalized through fun (HM1), joy (HM2), or excitement (HM3), the FaCon items appear quite distinct. Thus, removing FC1 and FC4 would manipulate the construct domain[61], in such a way that the construct would no longer have the ability to measure what it is supposed to measure. This was not the case for other indicators, meaning that a total of five items were omitted in the evaluation process. This was mostly because the requirement for a factor loading > 0.7 had not been met and leaving the corresponding indicator out

[61] See Figure 2.14, Section 2.3.1.1 for construct domains in reflective and formative measurements.

Table 2.22 Goodness of reflective measurement model for smart mobility

Variable	Item	Indicator reliability	Construct reliability		Convergent validity
Name	Code	λ > 0.7	Cronbach's α > 0.7	Rho α > 0.7	AVE > 0.5
Collective efficacy	CE1	.74	.88	.92	.66
	CE2	.88			
	CE3	.78			
	CE4	.88			
	CE5	*			
	CE6	.79			
Effort expectancy	1_EE1	.73	.84	.95	.75
	1_EE2	.92			
	1_EE3	.93			
Facilitating conditions	1_FC1	.56**	.73	.88	.56
	1_FC2	.83			
	1_FC3	.88			
	1_FC4	.69**			
Habit expectancy	1_HE1	.94	.95	.93	.78
	1_HE2	.94			
	1_HE3	.87			
	1_HE4	.77			
Hedonic motivation	1_HM1	.89	.87	.88	.72
	1_HM2	.90			
	1_HM3	.82			
	1_HM4	.77			
Behavioral intention	1_BI1	***	.94	.94	.88
	1_BI2	.93			
	1_BI3	.96			
	1_BI4	.94			
Performance expectancy	1_PE1	.77	.87	.88	.66
	1_PE2	.89			
	1_PE3	.78			

(continued)

Table 2.22 (continued)

Variable	Item	Indicator reliability	Construct reliability		Convergent validity
Name	Code	λ > 0.7	Cronbach's α > 0.7	Rho α > 0.7	AVE > 0.5
	1_PE4	.86			
	1_PE5	.82			
Price value	1_PV1	.84	.79	.72	.63
	1_PV2	.80			
	1_PV3	.75			
Perceived risk	1_PR1	.77	.85	.86	.63
	1_PR2	.87			
	1_PR3	.78			
	1_PR4	*			
	1_PR5	.79			
	1_PR6	.76			
Social influence	1_SI1	.90	.87	.87	.79
	1_SI2	.90			
	1_SI3	*			
	1_SI4	.88			
	1_SI5	*			

* Items omitted because λ < 0.7 and removal improved reliability ** retained for content validity *** redundant

improved model performance without compromising content validity to a serious extent. One item was removed to lower reliability below 0.95 (1_BI1).

A cross-comparison with the remaining scenarios demonstrates that the measurement model performs similarly well in all cases. Invariably, five to six items had to be removed to ensure full validity. Across samples, problematic items were SI3, SI5, and PV2. The latter was negatively pooled, which might have induced some bias. SI3 and SI5 might relate to descriptive instead of the subjective norms, intended by the construct's TAM roots (Barth et al., 2016). Sporadically, some insufficient loadings were found within the personality trait *collective efficacy* and technology-specific *perceived risk* constructs. Since these were kind of new scales, this is not a surprise and also not a problem with regard to reliability, since the reflective scales contained a sufficient number of interchangeable items.

One item (4_BI2) was eliminated to rule out common method bias in the MaaS scenario.

Supplementary to convergent validity, *discriminant validity* is another important criterion of overall construct validity. This type of validity aims to verify that two constructs are actually measuring something different (Diamantopoulos, 2005). For variance-based structural equation modeling, the dominant approach is to examine cross-loadings and the Fornell-Larcker criterion. This assumes that, on average, a latent construct being estimated shares a higher proportion of the variance with the respective indicators than with any other latent construct within the model. The criterion can be described as $\sqrt{AVE} > 0.7$. At the same time, no other construct may have a higher correlation to this one construct. However, Henseler et al. (2015) suggest that the above methodology cannot detect a lack of discriminant validity fail-safe. They thereupon propose utilizing the heterotrait-monotrait ratio of correlations (HTMT) that SmartPLS provides. Technically, HTMT is an estimate of what the true correlation between two constructs would be, if they were perfectly reliable. The closer to 1 HTMT is, the worse the discriminant validity between two constructs. As long as the value is below 0.9, though, discriminant validity is considered established. For values ranging from 0.90 to 0.95, a full bootstrap becomes necessary to control that HTMT confidence intervals do not include 0. Values above 0.95 would indicate a lack of discriminant validity.

As reported in Table 2.23, discriminant validity in terms of a Fornell-Larcker > 0.7 (bold) is confirmed for all use cases. Moreover, none of the inter-variable correlations surpasses 0.7. Table 2.24 presents supplementary results on HTMTs. These are consistently in the ideal range < 0.9, with the largest HTMT between FaCon and Effort (0.81). This ensures that each construct has the strongest relationships with its own indicators and not with others (Hair et al., 2017). Thus, discriminate validity is fully established.

Nomological validity is an additional method to establish construct validity (Hair et al., 2020). It is the third aspect of validity in CCA. It refers to a measure of the degree of theoretical correspondence between the scale and other constructs. Nomological validity must be demonstrated above all for new scales (Mentzer and Flint, 1997). This is achieved by comparing the construct scores with those of existing scales. The selection of existing scales must be based on previously confirmed effect correlations in the environments of the construct domain. In the present case, the scales of *collective efficacy* and *perceived risk* can be described as new or highly modified scales. In prior research, *collective efficacy* has shown to significantly correlate with social norms, *price value*, *performance expectancy*, and BI (Barth et al., 2016; Wolff and Madlener, 2019),

Table 2.23 Inter-variable correlations and Fornell-Larcker criterion

	CollEff	Effort	FaCon	Habit	Hedo	Intent	Perform	Price	Risk	Social	Usage
CollEff	**.81**										
Effort	.21	**.87**									
FaCon	.23	.66	**.75**								
Habit	.21	.39	.47	**.88**							
Hedo	.33	.38	.43	.57	**.85**						
Intent	.28	.5	.63	.57	.67	**.94**					
Perform	.28	.41	.47	.57	.65	.66	**.81**				
Price	.22	.23	.23	.3	.28	.25	.33	**.80**			
Risk	–.25	–.34	–.37	–.33	–.44	–.45	–.41	–.32	**.79**		
Social	.25	.42	.53	.47	.53	.63	.53	.29	–.33	**.89**	
Usage	.12	.42	.47	.59	.45	.57	.43	.19	–.35	.40	**1.00**

Notes: Square root of the average variance is along the diagonal. Below: the estimated correlation between the factors

Table 2.24 Heterotrait-monotrait ratio for smart mobility

	CollEff	Effort	FaCon	Habit	Hedo	Intent	Perform	Price	Risk	Social	Usage
CollEff											
Effort	.25										
FaCon	.29	.81									
Habit	.22	.42	.54								
Hedo	.36	.41	.50	.62							
Intent	.29	.54	.73	.61	.74						
Perform	.30	.45	.55	.64	.74	.72					
Price	.27	.28	.32	.37	.35	.30	.41				
Risk	.27	.39	.44	.36	.50	.49	.45	.40			
Social	.27	.47	.65	.53	.61	.70	.61	.37	.37		
Usage	.12	.44	.52	.60	.47	.59	.45	.23	.37	.42	

while *perceived risk* has frequently negatively affected *perceived usefulness, ease of use*, or BI (Featherman and Pavlou, 2003; Martins et al., 2014; Ooi and Tan, 2016). Additionally, gender differences in risk perception as well as in environmental behavior are known to exist. Women seemingly experience higher levels of risk and share more environmentally friendly beliefs (Gustafsod, 1998; Lee et al., 2013). Therefore, this work investigated nomological validity by examining the corresponding correlation levels. The numbers in Table 2.25 indicate that all variables correlate significantly in the presumed way, confirming nomological validity. In summary, the measurement model thus holds satisfactory reliability and validity and is suitable for the analysis of the structural model.

Table 2.25 Correlation matrix for nomological validity of new variables

Variable	Perceived risk	Collective efficacy
Gender	$-.11^{**}$	$-.08^{*}$
Behavioral intention	$-.38^{**}$	$.29^{**}$
Performance expectancy	$-.25^{**}$	$.33^{**}$
Effort expectancy	$-.47^{**}$	–
Social influence	–	$.32^{**}$
Price value	–	$.23^{**}$

** Correlation is significant at $p < 0.001$ * at $p < 0.005$

2.3.4.3 Structural Model

The main goal in assessing the goodness of the structural model is to test the hypotheses presented in Figure 2.12. For this task, path coefficients need to be computed and tested for significance with bootstrapping. Furthermore, the endogenous variables' R^2 and effect sizes are estimated to understand the model's performance in terms of explanatory power. Finally, to control if the model is not only explaining but predicting acceptance, the Stone-Geisser criterion Q^2 can be calculated through the blindfolding procedure.

Prior to interpreting the PLS-algorithm's results, Hair et al. (2017) request a screening for multicollinearity. For this, the relationships in the inner model should best be below 3.3. As I had already calculated the VIFs during investigation of CMV, these can be derived from Table 2.17, Section 2.3.2.5. A maximum VIF of 2.19 confirms that multicollinearity is less of a concern in the present study. All goodness measures elaborated, are presented in Table 2.26 together with their relevant thresholds.

Table 2.26 Goodness criteria for validating structural models in PLS-SEM

Goodness measure	Local criterion	Symbol	Target value	Reference
Multicollinearity	Variance inflation factor	*VIF*	≤ 3.3	Kock (2015)
Path relationships	Path coefficient	β	[–1,1]	Backhaus et al. (2018)
	Significance level	$p(\beta)$	< 0.01	Homburg and Giering (1998)
Fit of regression	Coefficients of determination	$R^2(BI)$	> 0.44	Venkatesh et al. (2012)
		$R^2(UB)$	> 0.35	Venkatesh et al. (2012)
Contribution	Effect size	f^2	> 0.02	Khalilzadeh and Tasci (2017)
Predictive relevance	Stone-Geisser criterion	Q^2	< 0	Stone (1974), Geisser (1974)

Hypothesis testing

Path coefficients are standardized values for describing the hypotheses set up in the structural model. Their values are usually between –1 and +1. The closer to zero they are, the weaker the relationship. A positive value indicates that a change in the independent variable results in a positive change in the dependent variable. For negative signs, vice versa applies. The size of the path coefficients, however, does not directly tell us anything about the predictive power because a path coefficient is a relative measure related to the SD of the underlying distributions. Thus, if the indicator changes by one SD, the predicted construct changes by β SDs (where β is the path coefficient). Thus, a high path coefficient is a good indication of the presence of a true relationship, but can initially be purely interpreted as a relational size (Hair, 2014). For a correct assessment of the quality of the structural model, it is hence also necessary to consider the *significance level* of the relationships. The number of observations and the data structure matter in this. Significance refers to the actual meaningfulness of the statistical relationship. It denotes the fact that a relationship between two variables with predefined probabilities (α levels) does not result from a random distribution of the data, but from an actual difference in the database (Backhaus et al., 2018). Whether a coefficient is significant ultimately depends on its standard error which is obtained in bootstrapping.[62] The use of three critical α levels for a two-tailed *t*-test in path coefficient bootstrapping is common: 5% level (significant), 1% level (very significant), and 0.1% level (highly significant). The corresponding *p*-value (e.g., $p < 0.01$ for 1% level) describes the probability that the null hypothesis[63] cannot be rejected.

From this knowledge about path coefficients and associated significance levels, an assessment of the structural model and the relevance of the individual relationships can be carried out as displayed in Table 2.27. As soon as a path is significant, a hypothesis can be assumed valid. However, a relative interpretation of β values is also required since, as described above, not every significant relationship has the same importance. The results show that except for relations of *effort expectancy, price value*, and *collective efficacy* on intention, all hypotheses can be accepted at $\alpha \leq 5\%$. Moreover, β values indicate that *performance expectancy, facilitating conditions, hedonic motivation*, and *social influence* are the most important predictors for BI. On the other hand, *habit* and BI seem more vital than perception of *facilitating conditions* for actual usage of smart mobility.

Table 2.27 Results of hypothesis testing and effect sizes for smart mobility

Path	Proposed effect	β	Hypothesis	f^2	Meaningful effect ($f^2 > .02$)
Perform → Intent	+	+.200***	Accepted	.057	Yes
Effort → Intent	+	+.030	Rejected	–	–
Social → Intent	+	+.197***	Accepted	.067	Yes
FaCon → Intent	+	+.247***	Accepted	.130	Yes
Hedo → Intent	+	+.258***	Accepted	.092	Yes
Price → Intent	+	–.063*	Rejected	–	–
Habit → Intent	+	+.080*	Accepted	.011	No
Risk → Intent	–	–.083*	Accepted	.016	No
CollEff → Intent	+	–.001	Rejected	–	–
Intent → Use	+	+.289***	Accepted	.075	Yes
FaCon → Use	+	+.116**	Accepted	.014	No
Habit → Use	+	+.367***	Accepted	.156	Yes (strong)

*** $p < 0.001$; ** $p < 0.01$; * $p < 0.05$

[62] See Section 2.3.4.1 for further information.

[63] No difference between actual distribution and random distribution is observable.

After examining significance, it is crucial to consider the relevance of significant relationships by examining their *effect sizes* f^2. These describe the change of the goodness-of-fit R^2, if a certain exogeneous construct were omitted from the model it was originally included. Effect size can be calculated as follows:

$$f^2 = \frac{R^2_{included} - R^2_{excluded}}{R^2_{included}}$$

Effect sizes reveal the actual practical and managerial relevance of relationships. Sullivan and Feinn (2012) thus state they are 'the main finding of a quantitative study'. While the *p*-value informs whether an effect exists, f^2 illustrates the extent to which one independent construct causes changes in the dependent variable. Following the guidelines of Cohen (1988), an effect can be regarded meaningful once $f^2 > 0.02$. From $f^2 > 0.15$ onwards, one can even refer to a medium to strong effect. As outlined in Table 2.27, meaningful effects can be found for six of the nine significant paths. The effects of risk and habit on intention, as well as *facilitating conditions* on use are negligible compared to the others. One effect (*habit* → *use behavior*) is significantly stronger than the others.

Explanatory power

The *coefficient of determination* R^2 indicates the proportion of the explained variance of a dependent variable in relation to the total variance. It is the most commonly used measure to evaluate a structural model's quality. Graphically speaking, R^2 describes how well the data points of the independent variables predict the data points, i.e., the curve of the dependent variable (Nitzl, 2010). The dependent variable in UTAUT-based research is usually BI and sometimes also *use behavior* (Tamilmani et al., 2020). In the area of future technology, measuring actual usage behavior does not come without its drawbacks.[64] This leads to a focus on intention in this work as well, whilst the inclusion of the more experimental measurement of use is somewhat exploratory in nature. Figure 2.28 plots the actual and the predicted data of BI. We can see a cloud of points that forms quite nicely around the linear regression curve.[65] R^2 in the present model is 0.66, suggesting that our model can explain 66% of variance in BI.

For interpreting R^2, no generally applicable thresholds are in place. Similarly, rules of thumb exist at best within specific disciplines. For example, Hair et al.

[64] See Section 2.3.2.1 for a discussion on measurement issues.

[65] Due to the linear appearance of the data, one can refrain from investigating non-linear effects (Hair et al., 2018).

(2011) propose that in marketing, R^2 values of 0.75, 0.50, and 0.25 should be viewed as substantial, moderate, and weak, respectively. In contrast, Ferguson (2016) suggest that in medicine 0.64, 0.25 and 0.04 are equivalents. In a more general, PLS-specific approach, Chin (1998) claims that a value $R^2 > 0.67$ represents a sound and substantial explanation for models with multiple influencing factors. Depending on the discipline and method, values below this limit can also be evaluated positively. Anyhow, it is useful to compare R^2 only for models specified in a similar manner. As shown in Table 2.28, benchmarks from both UTAUT's theory development and contextually comparable acceptance studies are available. The original UTAUT2 accounted for 44% of variance in BI and is therefore considered a baseline in present examination.[66] Within my review[67] the average R^2 for UTAUT2 application lies at 0.60. The R^2 values for BI in our model lies at 0.66 and is thus well in line with other research.

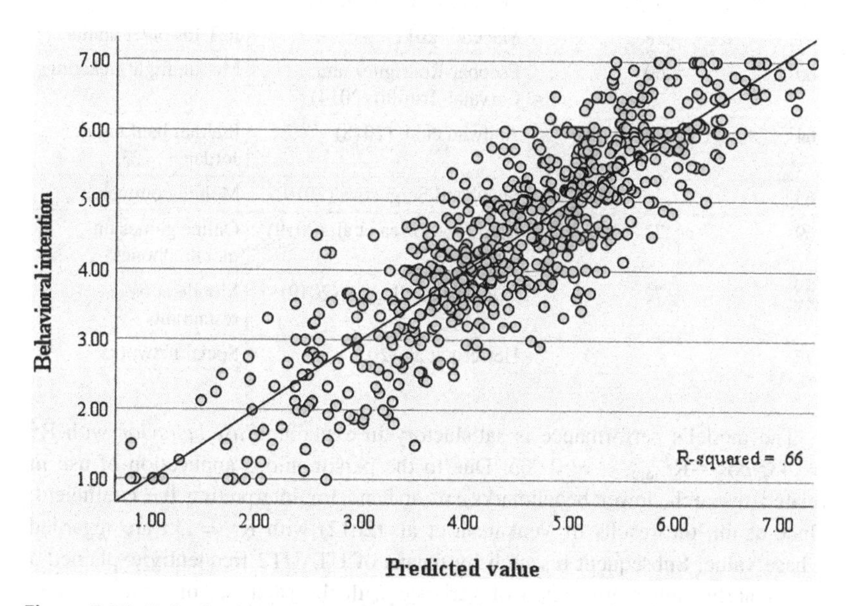

Figure 2.28 Behavioral intention prediction

[66] See Table 2.26, Section 2.3.4.3 for a goodness criteria overview.

[67] See Section 2.1.3 for literature review.

To manifest results, $R^2_{adjusted}$ can be obtained. This extended coefficient of determination considers the number of variables and data points in its calculation. It is always smaller than or equal to R^2. If it is distinctly smaller than an associated R^2, one can assume that the model contains useless data for the prediction (Everitt and Skrondal, 2010). However, with a minor Δ value of R^2-$R^2_{adjusted}$ = 0.004, this threat is averted. Thus, I conclude that the predicative accuracy of the model toward BI is quite substantial.

Table 2.28 Evaluation of the coefficients of determination R^2

R^2 benchmarks		**Reference**	**Subject**
Intention	Use		
.37	.47	Aswani et al. (2018)	Public Wi-Fi in India
.44	.35	Venkatesh et al. (2012)	Mobile internet
.57	.38	Macedo (2017)	ICT for older adults
.60	.60	Escobar-Rodríguez and Carvajal-Trujillo (2014)	Mobile flight ticketing
.64	.32	Alalwan et al. (2018)	Internet banking in Jordan
.65	–	Shaw and Sergueeva (2019)	Mobile commerce
.69	.72	Ramírez-Correa et al. (2019)	Online games on mobile phones
.72	.72	Palau-Saumell et al. (2019)	Mobile apps in restaurants
.77	–	Herrero et al. (2017)	Social networks

The model's performance is satisfactory in explaining *use behavior* with R^2 = .43, ΔR^{22}-$R^2_{adjusted}$ = 0.003. Due to the parsimonious application of use in related research, fewer benchmarks are at hand for interpreting the coefficient. Once again, the results of Venkatesh et al. (2012) with R^2 = .35 are regarded a base value. Subsequent research by means of UTAUT2 frequently explained a significantly higher proportion of variance with the same set of variables (e.g. Palau-Saumell et al., 2019), while others did not (e.g. Alalwan et al., 2018). Since *use behavior* has been measured with survey-based self-reports in all studies in Table 2.28, one can speculate that this is linked to the character of the technologies investigated. One out of many possible explanations for the intermediate explanatory power of our model would therefore be to assume that for technologies which are not yet widespread in the area where they are studied, the

UTAUT's constructs (e.g., habit) do not (yet) perform as strongly as they would in situations where technologies have already become more tangible to and able to be experienced by the public.

Model fit and predictive relevance

Overall goodness-of-fit indices aim at describing the quality of the measurement and the structural model as a whole. A variety of indices have been developed in the past decades. Although they are all calculated differently and are more or less suitable for alternative SEM procedures, they inherit a common core: summarizing discrepancies between observed and expected values. Due to their ease of understanding, they are very popular in the research community, but still the subject of ongoing disputes in the area of PLS-SEM (Henseler, 2018). Generally, PLS-theorists advise great caution when using and interpreting model fit indices. Dominantly. these have been developed for confirmatory methods and have to date not been sufficiently validated in the context of PLS models. However, some of them may be useful for comparison (Dijkstra and Henseler, 2015a; Hair et al., 2017) and some others have been newly developed especially for PLS and could thus be insightful (Henseler et al., 2016a). However, the latter also require further development, as little is known about their interpretation. Henseler (2018) provides structure to the discussion about goodness-of-fit measures in PLS. Accordingly, the general assessment of the model should match the research goal, which in practice is often a mixture of different research objectives. Model fit indices particularly matter for a confirmatory research goal.[68] In the present case, confirmation of UTAUT2 for smart mobility is a secondary research goal. Thus, the results of model fit are of importance for this scope. In line with the suggestions of Kline (2015) and in order to comply with the challenge of difficult interpretability, I calculated a heterogeneous selection of PLS-suiting indices as shown in Table 2.29. The chosen indices will be briefly highlighted.

Table 2.29 Model fit across use cases

Fit indices	Reference	Target values	SM	FAV	MA	EB
SRMR	(Henseler et al., 2016a)	< 0.08	0.06	0.06	0.06	0.07
χ^2/df	(Wheaton et al., 1977)	< 5.00	3.35	3.51	3.33	3.81
GoF	(Wetzels et al., 2009)	> 0.36	0.62	0.55	0.62	0.63

[68] See the four areas of SEM research in Section 2.3.1.2.

According to literature (Henseler et al., 2016a; Hair et al., 2017) the Standardized Root Mean Square Residual (SRMR) is the best understood model fit index and thus the recommended model fit indicator for PLS-SEM. It is an absolute fit criterion and implies a value of zero, provided that the model is a perfect fit. To calculate the SRMR, the deviations of the elements of the model-implied covariance matrix are summarized using standardized residuals from the empirical covariance matrix. Values of less than 0.08 are considered a good fit.

An older measure is the chi-square to degrees of freedom ratio, χ^2/df, which is the basis of the popular Tucker-Lewis Index (TLI) and the Root Mean Square Error of Approximation (RMSEA). However, these two are not available in PLS, because PLS algorithms do not calculate a covariance reproduced matrix. In order to still approach the informative value of these two overall fit indices, χ^2/df can be employed as a substitute (Kline, 2015). This index is sensitive to small sample sizes, which is fortunately less of a problem in the present case. Values of 5 or less are a common benchmarks (Wheaton et al., 1977; Hu and Bentler, 1999).

To complete the selection, a relatively new PLS-specific indicator was calculated: the so-called global Goodness-of-Fit index (GoF) developped by Tenenhaus et al. (2004). It aims to combine the reconstruction of both, the measurement and the structural model. The index thus yields a measure, which is a compromise between communality[69] and redundancy.[70] However, critics warn against its use. In their view, the GoF is unable to distinguish valid from invalid models. For instance, Henseler and Sarstedt (2013) find evidence that GoF is a measure of goodness and accordingly valuable for model comparison, but not of data fit. Undeterred by these arguments, the GoF is largely applied in PLS research due to its ease and comprehensibility and shall therefore also be listed here. Wetzels et al. (2009) deduce from prior theory (Cohen, 1988) that a GoF > 0.36 indicates a largely better-than-average model performance.

Finally, Chin (1998) suggests the consideration of Q^2 as a global goodness-of-fit criterion due to the forecast-oriented nature of PLS. In assessing predictive out-of-sample power Q^2 is examined to control for predictive relevance. If predictive relevance can be established, this means that the model is suitable to accurately predict data not used in results estimation. Q^2 is obtained through the blindfolding procedure, which delivers values of $Q^2 = .58$ for BI and $Q^2 = .42$ for UB. Thus, with $Q^2 > 0$ the model can deliver meaningful outcomes with data other than the data utilized for parameter estimation. This is supported by concurrent findings in the remaining scenarios.

[69] Equals AVE in PLS.

[70] Equals R^2 in PLS.

All indicators in Table 2.29 state that the model is a good fit of the data. Furthermore, these findings can be proven valid for all four use cases. In a nutshell, the model's overall performance can thus be regarded being very satisfactory in terms of UTAUT2 confirmation. In the novel context of smart mobility, the extended model explains and predicts acceptance comprehensively and precisely, although at most half of the proposed relationships exercise meaningful effects. This and the unclear role of the proposed model's extensions leaves opportunity for supplementary investigations.

2.3.4.4 Additional Results

In seeking to enhance our understanding of the complex interrelationships that constitute the 'black box' (Hair et al., 2018) of smart mobility acceptance, some advanced methods were applied. In contrast to assessing model fit, this is not for confirming a model, but to explore and explain acceptance better through unraveling insights hidden in the data. The methods applied in the following put the results into new comparative structures (e.g., importance-performance mapping, technology comparison), or reflect differences of subgroups which constitute through model context (e.g., multi-group comparison, interactions, and heterogeneity).

Importance-performance mapping

The importance-performance mapping procedure IPMA was developed by Ringle and Sarstedt (2016) to gain more insights from PLS-SEM output. It confronts a dependent variable predecessor's importance with its performance. Importance thereby refers to the total effect size exerted by an independent variable on a target construct. The total effect is the sum of direct and indirect effects. Direct effects are equal to path coefficients while indirect effects are obtained through mediation analysis. Performance, on the other hand, is a simple representation of the variables descriptive mean. The goal of IPMA is to identify those constructs that have a relatively high importance for the target construct, but at the same time, a relatively low performance. This indicates a potential for improvement. All required values for IPMA are provided as standardized and unstandardized values by the PLS algorithm in SmartPLS. In drawing Figure 2.29, unstandardized values were utilized since they better represent the ceteris paribus considerations. I adjusted the model with mean performance and mean importance values as suggested by Ramírez-Correa et al. (2019) to simplify interpretability.

Within the UTAUT stream of acceptance research, *use behavior* is the final construct to be explained by the model. For *use behavior* of smart mobility, Figure 2.29 implies that the largest influences are pursued by directly linking

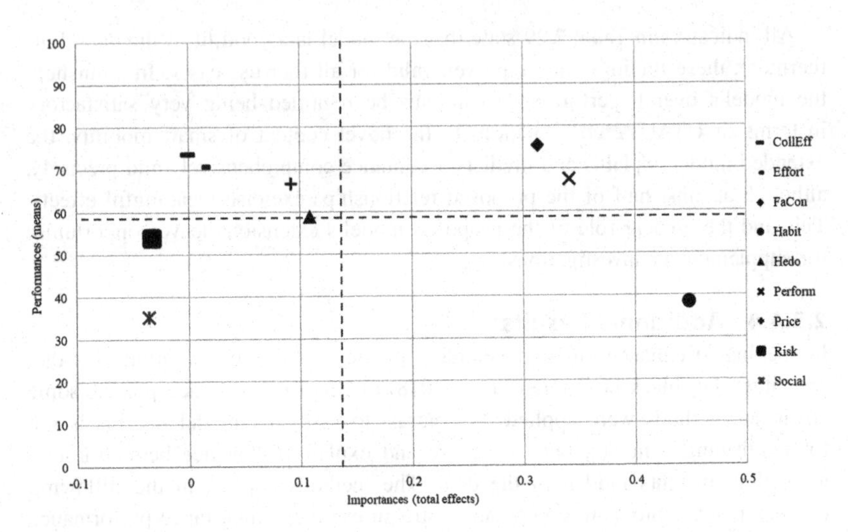

Figure 2.29 IPMA for smart mobility use behavior

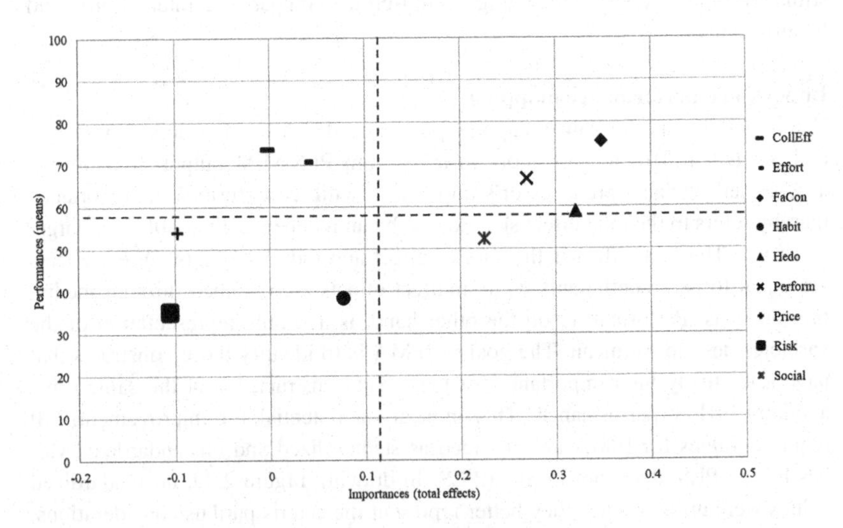

Figure 2.30 IPMA for smart mobility behavioral intention

habit, BI, and *facilitating conditions.* It is striking that the most important construct *habit* has, in the same way, one of the weakest performances. *Facilitating conditions* and BI already perform comparably well and thus leave minor room for improvement. Nevertheless, these results suggest that the three variables mentioned above need to be primarily addressed if one aims to increase use of smart mobility. All other model variables act through BI. Captured indirect effects are comparably small. Thus, a second IPMA on BI, which is commonly the second major target construct of acceptance, can shed light on potential alternate acceptance development strategies.

As illustrated in Figure 2.30, meaningful effects on BI originate from *social influence*, *performance expectancy*, *facilitating conditions*, and *hedonic motivation.* Of these variables, *hedonic motivation*'s performance is average and *social influence* even performs below average. Therefore, it can be stated that increasing the score of *social influence* and *hedonic motivation* would be a sensible strategy recommendation, next to levering habit. Unfortunately, *facilitating conditions* and *habit* are not explained in the model. In this respect, the UTAUT thus limits a further immersion into these two causal factors of acceptance.

When analyzing the other scenarios, I find that *habit* is the most promising lever for increasing usage throughout. This is particularly evident for the eBike with an importance of .44. When going into detail about potentials to increase the BI, the picture becomes disparate. Admittedly, *social influence* continues to be an important but weakly performing factor throughout the four technologies. Additionally, technology-specific strategies can be identified for improving BI. Firstly, it would be appropriate to try to increase the value of *hedonic motivation* if one wants to increase BI for smart mobility. Furthermore, improving weak *habit* performances would lead to enhancing not only *use behavior* but also BI for eBikes and MaaS. Finally, for automated vehicles a supplementary potential lies in improving the perception of *facilitating conditions.* Besides the approach to increase bad performances, another strategy can of course be to strictly address the highest importances as show in Table 2.30.

Multi-group analyses

In order to understand the acceptance of the future smart mobility system better and to detect group-specific ways of addressing different population groups, a series of multi-group analyses (MGA) was performed. MGA is a procedure that examines potential moderating effects of model context in an integrated manner. The core of MGA is a test for significance of differences in the path coefficients of two or more different subsamples. It can thus help to account for observable

Table 2.30 IPMA results comparison

Technology	X → Behavioral Intention				
	Importance*	Rank 1	Rank 2	Rank 3	Rank 4
Smart mobility	Construct	Hedo	FaCon	Perform	Social
	Performance**	0	+	+	–
Automated vehicle	Construct	FaCon	Habit	Hedo	Social
	Performance	0	0	+	–
eBike	Construct	Hedo	Perform	Habit	Social
	Performance	+	+	–	–
Mobility-as-a-Service	Construct	Perform	FaCon	Habit	Social
	Performance	+	+	0	–

* ranking of effect strengths; ** evaluation: (+) above average; (0) average; (–) below average

heterogeneity[71] in the data set using categorical variables. In the present study, model context variables include age, gender, and experience. Table 2.31 outlines contrived grouping.

Prior to applying MGA, some preconditions of measurement model invariance with respect to the subsamples must be met. These three steps build up on each other cumulatively:

- Step 1: Invariance of the configuration: content-related equality of the constructs and their parameterization.
- Step 2: Invariance of composition: constructs correlations between groups A and B should not be significantly lower than 1.
- Step 3: Equality of means and variances: the values of the constructs are not statistically significantly different within the samples.

[71] In contrast, Becker et al. (2013) explain that unobservable heterogeneity refers to potential estimation distortions caused by unknown segmentation patterns in the data. These tend to occur if no observed heterogeneity is found or if little is known about a model's theoretic background. For the analysis of unobservable heterogeneity, FIMIX (finite mixture PLS) or POS (prediction-orientated segmentation) can be applied. However, due to the relatively high number of observable control variables and given our knowledge about UTAUT, other differences in the data structure should be negligible. A review of the FIMIX segmentations and associated entropy values confirmed this assumption. This suggests that the issue of unobservable heterogeneity is not a major issue in present data.

Table 2.31 Data groups for MGA

Grouping variables		Group 1	Group 2
Gender	subgroup	Females ($n = 277$)	Males ($n = 409$)
	condition	Gender = 1	Gender = 2
Age	subgroup	Younger ($n = 207$)	Older ($n = 92$)
	condition	Age < 25	Age > 55
Experience* with smart mobility	subgroup	loExpSM *(n = 245)*	hiExpSM *(n = 106)*
	condition	(0) or (1)	(4) or (5)
Experience with eBikes	subgroup	loExpEB *(n = 170)*	hiExpEB *(n = 196)*
	condition	(0) or (1)	(4) or (5)
Experience with automated driving functions	subgroup	loExpFAV *(n = 133)*	hiExpFAV *(n = 105)*
	condition	(0)	(4) or (5)
Experience with digital mobility services	subgroup	loExpMA *(n = 215)*	hiExpMA *(n = 191)*
	condition	(0)	(4) or (5)

* *scale:* none (1); little (2); some (3); a lot of (4); very much (5)

The purpose of testing for measurement invariance is to control for whether variations in structural relationships among latent variables are due to differences in the meanings that group respondents attribute to the measured phenomena, rather than actual differences in structural relationships. (Hair et al., 2017). The invariance can be controlled for by the so-called measurement invariance of composite models (MICOM) procedure, which performs steps 2 and 3. Its basis is the permutation algorithm.[72] This procedure allows to test whether predefined groups of data show statistically significant differences in their loadings and path coefficients (Edgington and Onghena, 2007). Once Step 3 is fulfilled, it means that the complete measurement variance of the data is given and they can be analyzed pooled, i.e. as one data set. If Step 3 is fulfilled, full measurement variance is established. If Step 3 is not fulfilled, it means that merely partial or no measurement invariance of the data is given. For applying MGA, partial invariance is the desired state of data (Henseler et al., 2016b). In any case, it is important to ensure a sufficiently large sample size in the subgroups (Cohen, 1992).

In present modeling the same configuration is used in comparing groups 1 and 2, and the invariance of the configuration can therefore be taken as given. To assess invariance of composition and test equality of means and variance, a series

[72] Please see Section 2.3.4.1 for details.

of MICOM was performed in SmartPLS. MICOM Step 2 results show that the invariance of the composition can be established after omitting one to three items per grouping scheme from further analysis.[73] Finally, Step 3 analyses indicate that partial invariance exists in all models. This means that either means or variances are equal across groups. Thus, the conduct of MGA appears promising (Henseler et al., 2016b).

The core of MGA is a test for significance of differences in the path coefficients of two or more different subsamples. For this purpose, PLS provides several test outputs which are presented in Table 2.32–2.34. The best-known tests for this are the parametric test and the Welch-Satterwaith test. Both tests inspect the significance for the difference of group-specific PLS-SEM results, but while the parametric test assumes equal variances across groups, the Welch-Satterwaith test presumes unequal deviations (Hair et al., 2018).

As can be seen in the tables, several significant moderation effects were discovered through conducting the multigroup analyses. Pervasive, multiple interaction effects were not detected.[74] However, individual moderation effects were found for all three context variables tested, with varying degrees of strength. The most frequent moderation effect has to be attributed to the different strong influence of risk. Arranged by use case, the results are as follows:

For the smart mobility scenarios, one can notice that *effort expectancy* has a significant influence ($p < 0.05$) on BI for women but not for men. At the same time, smart mobility acceptance of older adults is more heavily influenced by the factors of risk (decreases acceptance) and pleasure (increases acceptance) than for the younger group. Furthermore, the relation of habit and intention is moderated by experience, meaning that for more experienced users, the influence of habit on intention is becoming stronger. Unfortunately, these results can hardly be reprojected in the other three scenarios. This does to a certain extent contradict the generalizability of acceptance in the smart mobility environment. Hence, each technology appears to show its own peculiarities.

Interestingly, however, it is often the same levers within technologies where moderation effects come to light. For example, in fully automated vehicles, the path 'Perform → Intention' is moderated by both gender and experience resulting in a larger effect for men and more experienced individuals. In addition to this

[73] These involved random items across variables. Hence, no pattern of group specific differences in item perception was unmasked.

[74] In preparing for MGA, a series of moderation and multiple interaction tests was conducted as proposed by Venkatesh et al. (2012). However, no meaningful effects were revealed.

outcome, we can observe that older people show a significantly stronger effect of habit on BI for automated vehicles.

In the eBike scenario, both moderations discovered relate to risk perception in such a way that older adults and females are more strongly influenced by the level of risk perceived. No significant moderation effect of experience on eBike acceptance was discovered.

Table 2.32 Moderation effects by gender

Model	Path	β (females)	β (males)	Parametric test	Welch-Satterwaith test
SM	Effort→ Intent	.13	–.06	**	**
EB	Risk→ Intent	–.08	.06	**	**
FAV	Perform→ Intent	.05	.27	**	**

Table 2.33 Moderation effects by age

Model	Path	β (older)	β (younger)	Parametric test	Welch-Satterwaith test
SM	Hedo→ Intent	.49	.16	**	**
SM	Risk→ Intent	–.28	–.11	*	**
MA	CollEff→ Intent	–.11	.16	***	***
EB	Risk→ Intent	–.21	.04	*	*
FAV	Habit→ Intent	.40	.09	**	*

Table 2.34 Moderation effects by experience

Model	Path	β (loExp)	β (hiExp)	Parametric test	Welch-Satterwaith test
SM	Habit→ Intent	.17	.34	**	*
FAV	Perform→ Intent	.11	.37	*	*
MA	Social→ Intent	.25	.09	*	*

*** $p < 0.01$; ** $p < 0.05$; * $p < 0.1$

Finally, for MaaS, three effects were found that are not related to risk, habit, or performance. First and foremost, the path 'CollEff→ Intent' is moderated by age, in such a way that the effect is significantly ($p < 0.01$) stronger for younger adults. *Collective efficacy* herein even causes a negative effect for older adults. The second and third moderation are found in the context of experience, which moderates the influence of *hedonic motivation* and *social influence*. What happens here is that the influence is initially positive for inexperienced individuals and then turns into the opposite or insignificance for experienced ones. So, while people initially approach MaaS with curiosity and social backup, for experienced user the factors are no longer relevant.

Structural comparison of the use cases

In this study, four central use cases are used to examine the acceptance of smart mobility as a whole in a form appropriate to the complexity of technological disruption. For this purpose, a taxonomy of use cases was hypothesized in Section 2.1.1 on the basis of a level of system integration. This assumes a change in the acceptance structure due to growing user experience on the one hand, and existing interconnectedness within the socio-technical regime according to the MLP on the other. The basis for the possible empirical evidence of such an effect can only be provided by rigorous comparison of the use case results. In this context, mean statistics were initially contrasted in Figure 2.26, Section 2.3.3.5. Results indicated a good fit of hypothesized and empirical acceptance levels. However, no definitive evidence of the change in effect according to the possible taxonomy can be demonstrated from these statistical indications. Therefore, in a second step, an examination of the structural models, i.e., a test for significance of differences in the path coefficients, must be performed. Analogous to the previous paragraph, an MGA based on technology types would, in theory, be suitable for this. In practice, however, MGA cannot be applied because Step 1 in the MICOM assessment fails to establish invariance of configuration. This, in turn, is due to the fact that the measurement instruments necessarily exhibit certain deviations between the individual use cases (e.g., conjunctive formulations for future technology such as FAVs). For the same reason, conducting series of moderation analysis as known from regression modeling is also not appropriate (Hotchkiss, 1976). To nevertheless detect differences in variable impacts, I utilized the well-known effect sizes benchmarks to create a ranking (Cohen, 1988).

Table 2.35 hence provides an overview of use case-specific and path-related effect sizes. It becomes evident that from this view of statistical power analysis, solely one pattern can be discovered related to the assumed taxonomy: this is that

Table 2.35 Structural comparison of use-cases using an effect size ranking

Path	FAV	MA	EB (non-owner)	SM	EB (owner)
Perform → Intent	++	++	++	++	++
Social → Intent	++	++	++	++	+++
FaCon → Intent	++	++	n.s.	++	++
Hedo → Intent	++	n.s.	+++	++	++
Habit → Intent	++	++	++	+	n.s.
Intent → Use	n.s.	++	++	++	++
Habit → Use	+	++	++	+++	++
FaCon → Use	++	++	n.s.	n.s.	n.s.

Effect sizes: non-significant (n.s.), only significant (+) small effect (++), medium-strong effect (+++)

as the level of system integration increases, the role of habit changes from influencing intention to affecting *use behavior* directly. At the same time, *facilitating conditions* become less important for using smart mobility technology.

Furthermore, some interesting insights can be drawn from the procedure. Namely, we can extract that the important acceptance factors of smart mobility, namely *performance expectancy*, *social influence*, *hedonic motivation*, *facilitating conditions*, and *habit* are of relevance for almost all use cases while *price value*, *perceived risk*, and *collective efficacy* are not. Within the relevant construct some anomalies can be observed. For example, it is noticeable that BI of MaaS is, in contrast to all other scenarios, not empowered through *hedonic motivation*. At the same time, people value eBikes especially for the fun assumed in using them. Potentially, eBikes are hedonic goods while MaaS is a utilitarian technology (Van der Heijden, 2004). The last thing that stands out in the analysis of the table is that *social influence* is particularly weighty in owners deciding whether to continue using eBikes. This means that a change in social influence perception will cause a strong reaction in acceptance.

2.3.5 Model Modification

A proposed model is seldomly the best-fitting model (Weston and Gore, 2006). Fortunately, the real exploratory strength of PLS-SEM in comparison to other approaches lies especially within the flexibility and ease of its structural modeling (Henseler, 2018). Consequently, the final step in PLS-SEM can be a post-hoc model modification, which has already been recommended by PLS-inventor Wold (1980) and highlighted in subsequent research (Bagozzi and Yi, 1988). Altering the initial model can improve our understanding and resolve inconsistencies that may exist. Further goals may be:

- Discovering a model that explains more variance in dependent variables
- Unraveling new and unknown path connections
- Improving fit and predictive power
- Setting up a more efficient and parsimonious model

Notwithstanding these advantages, many SEM researchers are rather reluctant to respecify their models due to potential methodological errors. Thus, a search for the best or a better model could lead to a model that fits one data set, but loses its general validity when being tested with other data. However, the question remains whether, for the sake of methodological correctness, one is willing to opt out of the search for better explanations. Bollen and Long (1993) have dedicated a larger anthology to this discussion, after the reading of which one has to come to the conclusion that model modification is necessary and feasible, as long as it is guided by certain principles. These are for instance following searching routines or developing the modification in front of rigorous theoretical background (Kline, 2015). Finally, a new solution should be cross-validated to indicate its general applicability.

2.3.5.1 Re-Conceptualization

The model discussed in this section contains issues that may be starting points for modification:

- Of the three variables that predict *use behavior*, solely one, namely BI, is explained well by the model.
- The hypotheses on the effect of *effort expectancy*, *perceived risk*, *price value*, and *collective efficacy* could not be confirmed.
- Instead of taking a moderating function, model context such as *experience* may directly influence acceptance
- The model includes a fairly large number of variables

To address this list, I treaded the path of reconceptualizing according to related theory. On this occasion, Venkatesh et al. (2012) recommend integrating different theoretical streams on the same topic. One source from the literature review accommodated this recommendation in a very accurate manner. Specifically, Sovacool (2017) expands and integrates a theory of mobility (Automobility (Flink, 1975)), and a theory of science and technology (Actor Network Theory (Law, 2009)) with UTAUT to arrive at a new mobility framework. The selection of these models is grounded on 35 interviews with socio-economic experts. As Sovacool (2017) further explains, although the models are methodologically completely incompatible approximations, the theories interlock almost perfectly at crucial points, so that one compensates for the weaknesses of the other. In the framework that the author finally derives from these thoughts around the virtues of theoretical synergism, four constructs are included:

- *Motile pleasure* (MP) combines hedonic and utilitarian attributes. It hence incorporates *performance* and *effort expectancy* as well as *hedonic motivation* from UTAUT to arrive at a construct that refers to the joy of driving rooted in everyday inevitabilities. Motile pleasure, by this design, also includes all sorts of non-economic issues such as safety concerns, travel times, or other personal mobility requirements. After all, these aspects derived from Actor Network Theory's *problematization*. This anxiety-orientated construct can potentially be translated to the present model's *perceived risk*, which I thus attached to MP.
- *Sociomaterial consumerability* (SMC) refers to the overall compatibility of a technology with an individual. Here, manifest infrastructure requirements play just as important a role as the perception of one's own financial scope of action. Furthermore, SMC also addresses systemic frameworks such as the integration of the technology into a network/support system or the connection to a trusted organization (e.g., OEM). Interweaving these aspects, the construct can be approximated through UTAUT's *facilitating conditions* and *price value*.
- *Sociality* (SO) 'blends together social influence, social identification and cultural symbolism' (Sovacool, 2017). The baseline of the construct is the thought that people are more likely to adopt new mobility when they unconsciously believe that doing so will trigger a good feeling in others as it does in themselves. This implicit association may be based on pro-environmental sentiments or other group-related normative beliefs. For the present study, I can best capture SO through *social influence* and *collective efficacy*.
- *Habitual momentum* (HAM) assumes the existence of certain natural conditions within social structures. Following this, every human being lives within a framework of available options that manifest themselves out of the history of one's own consciousness and are, as such, taken for granted (Nye, 2001). Overcoming these habits is very challenging (König and Neumayr, 2017).

One way to create new behavior might thus lie in habit's interaction with experience. Many studies have found evidence that experience can create a momentum to recalibrate behavior.[75] Ergo, I reinforce *habit expectancy* with *experience.*

In further developing the framework, I combined these four elements with the baselines of TAM to arrive at a new structural equation model as outlined in Figure 2.31. This model presents itself to be much more straightforward and economical than the original approach of an extended UTAUT 2. In addition, the role of *habit* is strengthened, as it is now the second pillar of behavior prediction along with BI. The variables SMC, SO, and MP, in turn, are the influencing variables for BI and HAM, creating a funnel-shaped structure. The development of the associated hypothesis network results from key effects of the variables included in the new constructs, which were elaborated in the preceding sections. Finally, the resulting model was described by the author as 'interesting and useful' in correspondence with him (Sovacool, 2020).

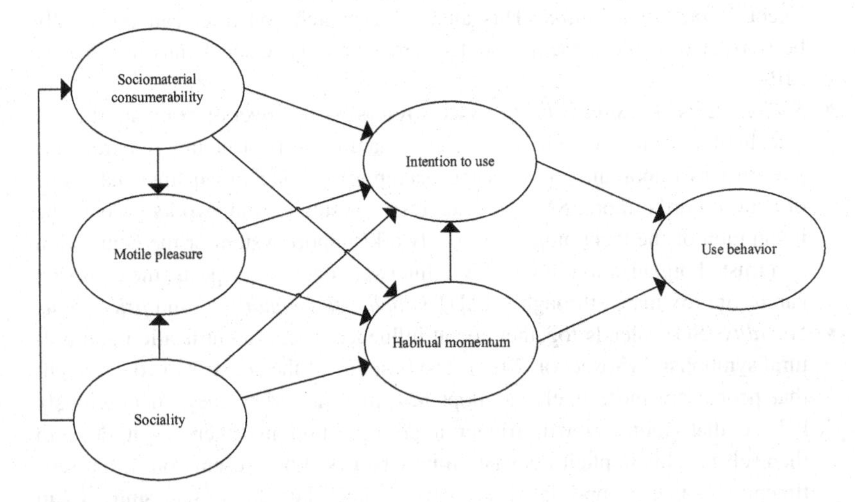

Figure 2.31 Elaborated model

[75] See literature review in Section 2.1.3.

2.3.5.2 Rival Modeling

To test the experimental model, two different methods were applied. In the first elaboration, the related UTAUT constructs replaced the new construct as shown in Figure 2.32. Furthermore, as indicated by the previous findings of this study, *experience* was added as a direct predictor. However, since this does not necessarily satisfy the requirement of model modifying for parsimony, an additional test is to be undertaken. Second-order modeling was therefore implemented in SmartPLS. As Lowry and Gaskin (2014) put it, this is sometimes also called the 'repeated indicator' approach. In this process, the new, as yet unknown, construct are formatively created from the associated variables. The individual items of the variables are reconnected to the construct to be predicted. In our model, for example, *price value* and *facilitating conditions* together form the new construct *sociomaterial consumerability*, which contains the sum of the related items. To allow PLS calculation afterwards, variables obtained by repeated indicator approach cannot be used, however, because their variance is zero. Therefore, researchers have to take the detour via latent variable scores and rebuild the model from these (Hair et al., 2018).

Table 2.36 summarizes the results of rival modeling. Additional to the two alternatives presented, a simplified model with significant hypotheses only was included as proposed by Herrero et al. (2017). In order to compare the models, the structural and measurement models can be confronted with each other individually or the models as a whole can be benchmarked via overarching indices. For this, a model selection criterion, namely the prominent Akaike Information Criteria (AIC), is listed in the table. The AIC is used to compare different model candidates. This is accomplished on the basis of the log-likelihood value, which is larger, the better the model explains the dependent variable. In order not to judge more complex models as consistently better, the number of estimated parameters is included as a penalty term in addition to the log-likelihood. Generally, the smaller (even negative) an AIC is, the better.

Parsimonious modeling demonstrates that this proceeding can deliver almost the same quality of results as the original modeling attempt with only 7 of the 12 variables. At the same time, however, it is a poorer fit to the data, as some informative value is lost.

In contrast, the other two models are detail-enriching, as they explain effects of previously insignificant variables such as *perceived risk* or *effort expectancy.* Furthermore, by adding *experience,* both models explain *use behavior* far better than the original does. However, mapping experience as a second-order construct leads to a weaker effect in Elaboration 2 compared to Elaboration 1. This is also reflected in worse AICs.

Eventually, the goodness-of-fit indices of both models can be considered better than the ones obtained for the original model. This is especially true for the second experiment, which combines the effectiveness of the first elaboration with the efficiency of the parsimonious model. Despite these advantages, there are weaknesses in this model as well. As explained, the two elaborations differ in their structural set up. In case of the second-order approach, the measurement model is invalid, as neither convergent nor discriminant validity can be obtained. The reason for this lies in non-adequate identification due the latent variable scores-based modeling.

2.4 Discussion

Although there is much excitement surrounding the introduction of intelligent transport systems, there is little understanding of the factors influencing the uptake of various smart mobility niche technologies (Madigan et al., 2017).

Hence, the aim of this study was to develop and test a versatile and holistic model for evaluating smart mobility acceptance. Through comprehensively reviewing literature, a model was designed, which extends UTAUT2 with new variables and systemic contexts. Eventually, the postulated hypotheses were carefully tested and revealed a well performing model. In conducting a full PLS-SEM assessment, additional insights were obtained for present vision of smart mobility that require a classification in literature and a translation into practice.

Initially, the general perception of smart mobility is rather positive within the present sample. In benchmarking with similar scales, the mean *intention to use* smart mobility technologies lies above the *intention to use* FCEV (e.g. in Kauschke et al., 2021), which can be classified as a technology in market preparation, but below BEVs (e.g. in Mohamed et al., 2018), which are currently at the border to transition from market uptake to mass market. Overall, individuals consider smart mobility as compatible with their life, easy to use, and well-performing.[76] In line with findings achieved in studying smart homes (Baudier et al., 2018), significant drivers of performance evaluation were expected gains in comfort (1st rank), flexibility (2nd), and safety (3rd). However, in contrast to these positive ratings, merely a 20% minority do currently use smart mobility

[76] M.(FaCon): 5.09; M.(Effort): 5.31; M.(Perform): 4.91.

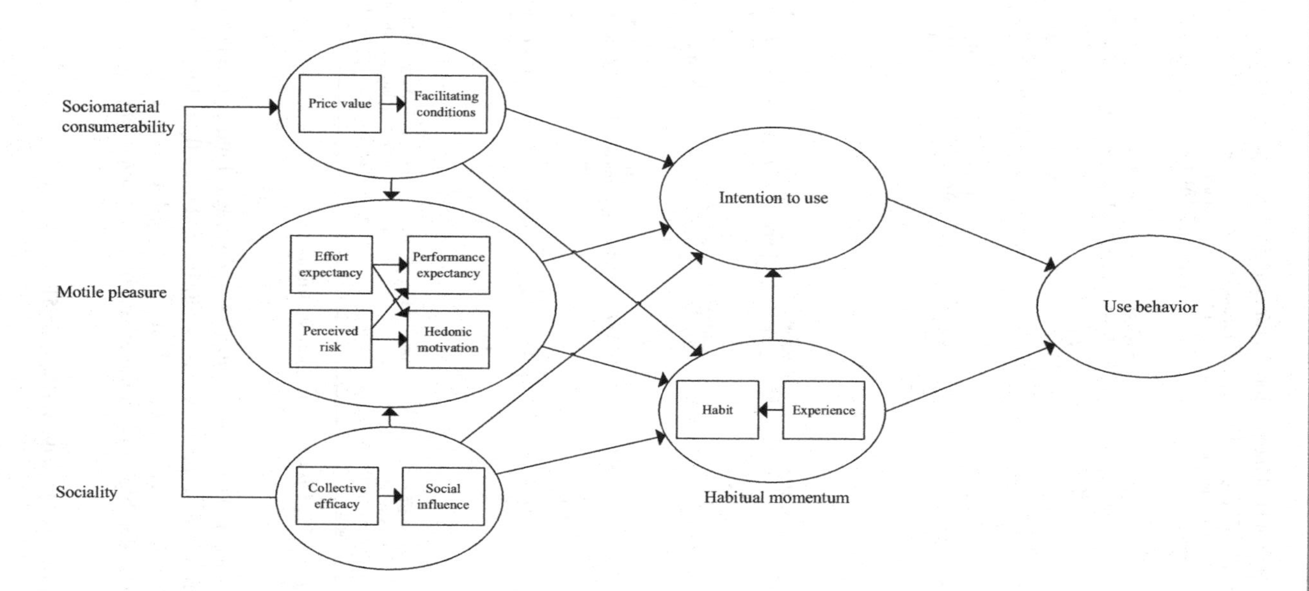

Figure 2.32 Elaborated model with UTAUT and experience constructs attached

Table 2.36 Evaluation of rival models for smart mobility

	Parsimonious model Significant only	**Theoretical model** Extended UTAUT2	**Elaboration 1** Restructured UTAUT2	**Elaboration 2** Second-order approach
Model selection and fit criteria				
AIC (Intent)	-752.001	-765.253	-778.619	-727.699
AIC (Use)	-399.704	-398.344	-726.117	-492.376
SRMR	0.06	0.06	0.08	0.12
Chi Square /dF	7.27	4.75	5.32	0.01
GoF	0.65	0.62	0.53	0.53
$Q^2 > 0$	+	+	+	+
R^2				
Use	0.43	0.43	0.64	0.50
Intent	0.66	0.67	0.67	0.64
Other (mean R^2)	0	0	0.36	0.41
Measurement model				
Reliability	+	+	+	+
Validity	+	+	+	-
Paths				
Perform → Intent	0.22***	0.20***	MP → Intent	0.46***
Effort → Intent		n.s.	MP → Habit	0.51***
Social → Intent	0.20***	0.20***	SO → Intent	0.15***
FaCon → Intent	0.29***	0.25***	SO → Habit	n.s.
Hedo → Intent	0.30***	0.26***	SO → SMC	0.52***
Price → Intent		-0.06*	SO → MP	0.35***
Habit → Intent		0.08*	SMC → Habit	0.19***
Risk → Intent		-0.08*	SMC → Intent	0.15***
CollEff → Intent		n.s.	SMC → MP	0.46***
Intent → Use	0.29***	0.29***	Habit → Intent	0.18***
FaCon → Use	0.12**	0.12**	Habit → Use	0.55***
Habit → Use	0.37***	0.37***	Intent → Use	0.22***
Summary of path analysis	7/7 (100%) 6 x *** 1 x **	10/12 (83,3%) 6 x *** 1 x ** 3 x *	34/50 (68,0%) 24 x *** 4 x ** 6 x *	11/12 (91,6%) 11 x ***
Evaluation	More efficient than the original model, but less informative and poorer fit	Good model	Better results than the original, but very complex	Very good model, but no valid measurement

*** $p < 0.001$; ** $p < 0.01$; * $p < 0.05$; n.s. not significant;

more than once a week. Descriptive statistics in this regard indicate smart mobility could suffer from being perceived as expensive and socially untoward (Fleury et al., 2017). After all, smart mobility is not perceived as something ordinary.[77] This indicates considerably more complex acceptance components.

[77] M.(Price): 3.48; M.(Social): 3.79; M.(Habit): 3.51.

2.4.1 Factors of Acceptance

Based on the insights of my modeling, I can conclude that acceptance of smart mobility primarily relies on five factors: *performance expectancy*, *facilitating conditions*, *social influence*, *habit*, and *hedonic motivation*. In opposition, *collective efficacy*, *price value*, *perceived risk*, and *effort expectancy* do not directly influence acceptance. Hence, the structure of acceptance is determined by the characteristics (or what you aimed at) of several technologies ranging FAVs to EBs. Albeit the acceptance of smart mobility has never been investigated in this way before, essential findings from prior related research (Madigan et al., 2016; Ahmed et al., 2020; Schikofsky et al., 2020) are confirmed. The non-significance of some other hypotheses (Barth et al., 2016; Schlüter and Weyer, 2019; Ye et al., 2020) opens up a space for discussion.

2.4.1.1 Habit

Consistent with Chen and Chao (2011), the most important variable in the model is the often overlooked habit (Tamilmani et al., 2019a) with a total effect of 0.44 on use of smart mobility. Even within rival models, *habit* retains or even expands this prominence. This suggests that when transitioning from automobility to smart mobility, the existing habitual imprints should first be analyzed and then, if necessary, be dissolved. In the same way, future habits, conceptualized as *habit expectancy*, shape future behavior. Therefore, in addition to decoding existing habits, new habits should be built up in congruence with one's needs. IPMA strongly supports this argument, advising to specifically improve *habit* mean scores when wanting to augment smart mobility acceptance as a whole.[78] A successful strategy to promote smart mobility might as a consequence be to design it in a familiar way. This could, for instance, comprise control and design elements, such as having a real, material, and universal key to smart mobility. Further, this may include taking advantage of practiced behavioral patterns; one of the best methods to establish new, positive habits is to tie them to existing habits (Clear, 2018). In personal mobility, this could refer to starting and ending a journey at the usual place, to listening to the same radio or to offering a privacy-protected environment. To stack new habits, fresh behaviors must be repeated in stable and rewarding contexts until automaticity sets in (Duhigg, 2012). Hence, in order to receive a reward for performing positive behavior, elements of gamification can be integrated into the design of a future mobility system. In the end, an iterative introduction of changes is crucial. Establishing new habits will be

[78] See Section 2.3.4.4 for IPMA results.

more successful when behavior smoothly transitions rather than changing radically (Bamberg et al., 2011). Smartness in mobility could consequently first be introduced in the car environment with its well-known interfaces (e.g., routing and information). From there, it can expand more and more (e.g., for payment, multimodality, self-sovereign identities, etc.) up to the point at which the system can detach itself from the car.

2.4.1.2 Performance Expectancy

Besides habit, four other variables make unique contributions to explain UB and BI. The total effects of the constructs *performance expectancy, facilitating conditions, hedonic motivation*, and *social influence* are statistically equally significant and meaningful. To start off with, the important role of PE is not surprising as various related studies (e.g. Zoellick et al., 2019; Ye et al., 2020) produced similar results. To my knowledge, no transport nor mobility-related study failed to find significant results for the effect of PE on BI. Hence, it is evident that mobility is utilitarian due to its simple necessity. Systems will be evaluated strongly by their performance—an effect that is growing as systems diffuse in societies and people gain experience (Kauschke et al., 2021). For this reason, it appears essential to keep the performance of the system high throughout the adoption process. In addition to attributes such as comfort, speed, flexibility, and safety, system performance includes above all reliability and service quality (Ahmed et al., 2020). Currently, the perception of smart mobility performance is satisfactory ($M = 4.91$). However, it is essential to prevent negative experiences and ensure flawless functionality of the smart mobility system, emphasizing its superiority over the current, frequently criticized, public transport services. Therefore, user requirements must be carefully explored and new ideas to solve user problems must be developed. In any case, the system can unfold its full potential only if transport modes and IT systems are integrated to a correspondingly high degree (Lyons et al., 2019). Besides that, a functioning system of smart mobility cannot be imposed top-down on existing mobility, but must rather be tested in protected niches. In such locally limited areas, in turn, a high degree of non-proprietary intelligence must be achieved so that the performance expectation and the performance provision are well-matched. If smart mobility is to be used, then it should be employed in its entirety to realize its power.

2.4.1.3 Hedonic Motivation

Similar to prior research (Plazier et al., 2017; Nordhoff et al., 2018b; Schikofsky et al., 2020), *hedonic motivation* was found to be another strong predictor of acceptance. This means that the use of smart mobility technologies depends on how enjoyable people perceive it. It becomes clear that mobility systems are mostly not purely utilitarian but also hedonic technologies (Van der Heijden, 2004). This is natural, given the notion that driving a car or riding a bicycle can also be a sport. However, taking the train to work might not be as exciting.[79]

Nonetheless, intrinsic motivation can certainly be used to promote key aspects of smart mobility, such as FAVs, EBs, or apps. This is also supported by IPMA, which points out that HM has a strong influence on BI and that there is still a great potential to improve the performance of the value. Generally, mobility is an emotional topic and some people are simply fascinated by future technology. Accordingly, it is, inter alia, the appeal to a person's inner motivation that can convince them to change their behavior. In electric vehicles, Tesla demonstrates how new technology can be successful by activating emotions via B2C-communication of driving pleasure, design, or acceleration. Managers and decision-makers should learn from this to design smart mobility in such a way that it does not simply boost productivity but is also fun to use. For automated vehicles, this can mean marketing aspects beyond the pleasure of self-driving, such as enjoying entertainment or leisure time instead of facing stressful traffic situations. For eBikes, the spirit of health, adventure, or self-determination could fascinate potential new customers. Gamification can be added to the list of ways to use HM as a lever of acceptance. This can play a role in the design of MaaS apps. Gamification can be defined as the use of game design elements in non-game contexts. Yen et al. (2018) continue to explain that it is seldomly employed in transportation settings and therefore bears a promising potential to reward customer loyalty to new mobility options. The reward of using smart mobility could possibly be financial compensation (e.g., 'bonus miles' or 'tax benefits') or non-monetary benefits (e.g., 'social recognition' or 'priority travelling').

2.4.1.4 Facilitating Conditions

The concept of gamification can also highlight to individuals the positive and negative consequences of their mobility choices and thus make an important contribution to perceived self-efficacy (Rachels and Rockinson-Szapkiw, 2018). In recent research on mode choice (Ben-Elia and Avineri, 2015; Lyons et al., 2019), it is often assumed that people are unable to refrain from automobility because they believe that they cannot act in any other way or that they cannot freely and self-determinedly organize their mobility in any other way. This perceived cognitive dependence was therefore identified as a key aspect to hinder adoption of

[79] See Section 2.4.2.1, for details on intrinsic motivation.

alternative mobility offers. It is incorporated in UTAUT through the conceptualization of *facilitating conditions*. The former name of the construct, perceived behavioral control (Ajzen, 1985), sharpens our understanding of what this refers to. In the present study, FC predicts BI as well as UB to be highly significant. The total effect on use lies at 0.32. At the same time, the perception of FC already ranks at a high level, implying the surveyed do not expect to experience great difficulties with smart mobility. Thus, the strategy must be to maintain and develop this perception. Suitable tools to achieve this may be: (1) clear political development goals and communication to reduce uncertainties, (2) the development of a compatible and reliable infrastructure (analog as well as digital), (3) and a design of smart mobility that enables everyone, regardless of location or social status, to participate. Future society needs mobility solutions that do not undermine people's feeling of being free in their choices.

2.4.1.5 Social Influence

The fifth and final variable contributing to explain smart mobility acceptance consistently throughout examined use cases is *social influence*. Its relevance is particularly high compared to prior intelligent transportation acceptance literature (Panagiotopoulos and Dimitrakopoulos, 2018; Wolff and Madlener, 2019; Ahmed et al., 2020), which had achieved similar results. In the present study, its performance is relatively low (M. = 3.79) and its overall importance is large (total effect of 0.25).[80]

These numbers highlight the importance of social norms once smart mobility becomes available on the market. This effect could stem from the growing individual intention to find a substitute for the car's dwindling role as a status symbol. Thus, it might be time to move beyond framing travel behavior as an individualistic decision (Nyblom, 2014). Interpreting IPMA, *social influence* must be considered the second most important starting point, just behind *habit,* if one aims to foster adoption of smart mobility. Hence, one must understand what present results entail. As Barth et al. (2016) explain, the variable *social influence* mainly reflects subjective norms, which describe the perceived expectations of significant others (1. What would people relevant to me think about my actions?). Furthermore, it includes elements of provincial norms (2. What would people in similar situations and life conditions do?), as well as injunctive norms (3. What is approved and disapproved of within a society?). Consequently, targeting all three social norm dimensions may hold a successful strategy for increasing performance to improve *social influence*. For smart mobility, this might imply: (1)

[80] This effect is even larger in rival models.

empower young people with pioneer spirit to make them more resilient to environing opinions, (2) find or develop target group-specific multipliers and social agents that can provoke social learning and imitation, or (3) shape injunctive norms by facilitating a good media reputation for smart mobility and establishing it as a desirable and superior alternative to current mobility. Indeed current mobility might soon face problems, such as bans on private cars in certain areas. To anticipate that, Friis (2020) suggests changing the notions of mobility from providing freedom in terms of flexibility and convenience to life quality measures addressing freedom from pollution, freedom from spending time at gas stations, and freedom from oil producers.

2.4.1.6 Effort Expectancy

In contrast to the original findings of Venkatesh et al. (2012) and several successors (Tamilmani et al., 2020), the hypotheses that *effort expectancy* and *price value* will affect the *intention to use* smart mobility were surprisingly rejected across all four scenarios. Hence, in the current state of smart mobility transition, neither *ease of use* nor pricing appear decisive for smart mobility acceptance. In the case of *effort expectancy*, similar results are, i.e., known from intelligent transport systems (Adell, 2009) or automated road transport systems (Madigan et al., 2017).[81] The authors conclude several reasons, which may also apply for smart mobility. Firstly, people already perceive smart mobility as easy (M. = 5.31). Thus, people might assume using smart mobility does not require learning new skills. Additionally, the concept of smart mobility itself entails increased ease and comfort, making it obsolete to worry about. Secondly, it might be that people simply don't mind learning how to use smart mobility because it appears inevitable in the light of ubiquitous digitalization. The third reason, however, points in another direction: from its theoretic conception, *effort expectancy* includes issues mapped in *facilitating conditions.*[82] Within present elaborated modeling, *effort expectancy* hence proves to be highly correlated with *facilitating conditions* and unfolds a larger predictive power for acceptance through indirect effects. Herrero et al. (2017) had made similar observations in their study on social networks. In this sense, making smart mobility easier will eventually elevate perception of performance and help establish habits, which will then again enhance use and acceptance. Thus, corresponding strategies should incorporate aspects of *ease of use*.

---

[81] See Table 2.8, Section 2.1.5 for prior research findings.

[82] See Section 2.2.1.2 for an explanation.

2.4.1.7 Price Value

In case of *price value*, present literature review elaborates that only a few minor effects of pricing are evident within the smart mobility environment, although in other areas, its influence is quite weighty. These effects mainly refer to material products such as electric vehicles (Fazel, 2013; Park et al., 2018). Thus, the non-significance of *price value* for smart mobility might relate to its immateriality. Supposedly, customers cannot really imagine the system and its services. People might find it hard to estimate an appropriate price in this very early diffusion phase. Another point may be that within digital environments, today's IT services are often free from entry or usage costs. This can result in people finding it even more difficult to confront the future use of smart mobility services with the current use of mobility goods. However, one would normally still assume that prices influence usage and acceptance. A survey conducted in the German Saarland region revealed in this context that people expect smart mobility, above all, to reduce their mobility costs (Kauschke, 2020). Kapser and Abdelrahman (2020) also found that people hardly want to use new automated systems if it does not reduce their costs. However, in contrast to *effort expectancy*, I did also not find indirect effects in alternative modeling. Further research or new theoretic concepts are necessary to untangle such conflicting evidence.

2.4.1.8 Perceived Risk

The non-meaningful influence of *perceived risk* and *collective efficacy* deserves specific justifications, given that I proposed their inclusion in the UTAUT2. First, the well-known *perceived risk* shall be discussed. Its performance is low (M. = 3.19). Thus, one might conclude that smart mobility is not regarded as a risky endeavor. Potentially, individuals relate their expectations to what they experience within their present mobility, which is more or less a safe and routinely practiced exercise. Furthermore, it is possible that people's acceptance is not generally influenced by risk, but rather only by conceptual parts of it. Prior related studies identified, for example, privacy concerns (Lutz et al., 2018), distrust in technical reliability (Ooi and Tan, 2016), or safety anxiety (Prati et al., 2018) as influential negative factors. Some deeper and technology-specific investigation may provide clarification. Overall, *perceived risk* is an unsteady predictor for future mobility technology. Liu et al. (2019a) speculate that respondents' risk beliefs might not relate to available cognitive processes, resulting in a missing link of *perceived risk* to BI. Anyhow, as in the case of *effort expectancy*, the elaborated modeling highlights some alternative paths of effect for *perceived risk*. Namely, in line with Zhang et al. (2019), the variable shows highly significant effects on *hedonic motivation* and *performance expectancy*. Thus, if a strategy to develop acceptance includes improving these two variables, risk perception should be lowered. When considering this association once again with the performances at item level, we can interestingly discover that items on risk, fear, and distrust score comparably low (M. < 3.00). Hence, they do not bear a lot of potential for improvement. In contrast, the two items 'Smart mobility is beyond my control' and 'There are too many open questions about smart mobility' score relatively high (M. < 3.50). This suggests two possible strategies to build acceptance: first, providing people with information and strategic guidance on the big picture of smart mobility, and second, guaranteeing users' safety and control in using the system. In the actual design of smart mobility, this can also mean safeguarding users against technical risks of failure, implementing ratings for service providers, and developing non-intrusive data exchange (Wang et al., 2019; Ahmed et al., 2020).

2.4.1.9 Collective Efficacy

Quite different from risk, the feeling of *collective efficacy* of smart mobility toward the environment displays a larger mean value of 5.27. First of all, this indicates that people are in a positive mood to jointly generate positive environmental benefits through smart mobility. However, the present study was not able to replicate the assumed direct effects of *collective efficacy* on intention as Barth et al. (2016) for any of the examined smart mobility use cases. Maybe, EVs are considered a 'green' technology[83] while IT in general is not (Wang et al., 2018a; Wolff and Madlener, 2019). However, this explanation does not seem sufficient for smart mobility acceptance, whose technological components such as FAVs (Wu et al., 2019) or EBs (Chen, 2016) are on the contrary very much influenced by environmental attitudes. Hence, societies require discussions about smart mobility and sustainability (Jeekel, 2017). Based on the available results, people do not seem to directly associate these two aspects. For this reason, alternative effect paths were also investigated. In respecified modeling, some new insights were obtained that are coherent with what Gimpel et al. (2020) had found about the influence of environmental concern on smart energy acceptance. Personality-trait *collective efficacy* influences most other variables significantly. In particular, *collective efficacy* has significant effects on performance, price, and motivation. Thus, one could conclude that environmentally conscious people tend to enjoy smart mobility more and have greater appreciation for its performance.

[83] In the present study, *collective efficacy* served as a representative of the ecological component, which had been diversely conceptualized in past research (see Table 2.10, Section 2.2.2.2 for an overview).

This is also reflected in an increased willingness to pay. On top of that, *collective efficacy* displays its strongest impact through *social influence*. This implies that when people feel like being able to bring in change through their behavior, they start committing themselves more fully to social norms. Thus, assuming a logical cognitive process, people would seek social validation of their personal norms in order to derive appropriate actions from them. Through all these indirect effects, the environmental variable in the elaborated model finally makes it to a remarkable total effect size of 0.23 on BI. Hence, engaging with the belief in *collective efficacy* must be viewed as a good option to foster smart mobility adoption. One way to address this may be running campaigns that promote collective engagement in pro-environmental actions. Efforts to commit oneself on a personal level might pose a threat to one's sense of personal control since the challenge is so huge. In line with Jugert et al. (2016), I therefore recommend changing communication about smart mobility transitions from being a threat to individual automobility to enriching our transport possibilities while, at the same time, collectively achieving positive effects for sustainability. This delivers public value (Docherty et al., 2018) that should continuously be communicated during the uptake of smart mobility. Raising awareness of environmental impact of daily transport is important (Møller et al., 2018). People must know: How much CO_2 has already been saved by a solution? How have punctuality and customer satisfaction has developed since the introduction of a system? How much do the state and citizens save through efficient management? Good news is welcome. Considering the adoption of smart mobility should be seen as a private way to achieve a collective goal and thereby become effective in the face of an environmental crisis.

2.4.2 Acceptance by Use Cases

The results discussed above hold true for general smart mobility. Regarding the other use cases, this study's modeling includes further insights for the acceptance of eBikes, fully automated vehicles, and mobility-as-a-service. Overall, results are quite similar. Therefore, the strategies proposed above remain predominantly valid. However, cross-comparison of the technologies highlights that smart mobility comprises heterogeneous systems in different innovation phases, whose acceptance promotion can also involve some individual strategies. In the following, the most relevant findings shall be highlighted.

2.4.2.1 Mobility-as-a-Service

MaaS focuses on public transport, sharing services, and apps. In contrast to the other two use cases, MaaS is, as its name may suggest, very service- and little product-oriented.[84]

Within this examination, MaaS achieves consistently positive ratings[85], much higher than for example other future technology such as FAVs. It is the most popular technology of the study. People appear being attracted by the MaaS offering of seamless, flexible, and reliable door-to-door mobility. They can even imagine that using MaaS can become a habit. The fact that MaaS is so closely associated with people could be due to the fact that a comparatively large number of people are already using mobile services today. At the same time, MaaS is potentially the most socially and environmentally sustainable future mobility solution currently being discussed (Jeekel, 2017; Liu et al., 2019b). In recognizing this hybrid potential, there is a great opportunity to further advance R&D policy agendas in the industrialized nations, which is still strongly automobile focused. Instead of building systems proprietarily around vehicles, whole new digital ecosystems should be engineered and tested under real-world circumstances. From an economic perspective, seminal new concepts must be developed that enable innovative forms of transport-related value creation for industry and services. A first approach to this is for example introduced by the Institutional Role Model (IRM) (Schulz et al., 2021b).

The models for MaaS acceptance show three major differences to the general model for smart mobility. The first difference becomes evident within the standard model. There, the present study did not find a significant path HM → BI for MaaS. Unfortunately, this is inconsistent with the findings of Schikofsky et al. (2020), who found a highly significant relationship. I assume this contradiction originates from different MaaS conceptualizations. While Schikofsky et al. (2020) presented in-depth app functionalities, the present study's survey involved only an overview and the emphasis was on public transport. Thus, when marketing MaaS, people should not be confronted with pictures of trains and buses, but a message must be created that involves innovativeness, simplicity, and fun in designing one's own mobility, for instance through an attractive HMI.

The second difference refers to the notion of *performance expectancy*, which emerges as a stronger predictor for MaaS than for all other use cases.[86] Further to the explanation in Section 2.4.1.2, high customer expectations must be met for

[84] See Figure 1.5, Section 1.4.3 for additional information.

[85] See Table 2.18, Section 2.3.3.2 for descriptive statistics.

[86] See Table 2.30, Section 2.3.4.4 for IPMA comparison.

MaaS. When developing systems, priority should satisfy these expectations. This is achieved by providing complete functions and rolling out MaaS not until a high degree of system integration is realized (Lyons et al., 2019). Polydoropoulou et al. (2018) specified this, recommending offering a reliable, flexible, real-time system that includes all regional means of transport under one umbrella. Ye et al. (2020) add that MaaS must be capable of solving various processes simultaneously such as querying information, booking, finding a site, waiting for a vehicle, transferring, authorizing at a vehicle, checking tickets together with payment. From an acceptance modeling perspective, it would be interesting to explore further factors that influence *performance expectancy*. Schikofsky et al. (2020), in this regard, explored socio-psychological mechanisms and discovered that *autonomy* and *relatedness* predict PE. However, future research might start decoding other influences outside the UTAUT universe, such as time-savings or economic or ecological benefits.

Thirdly, present analysis in a sense suggests that MaaS is favored by younger urban people and women. This is because ANOVA revealed that women have significantly stronger faith in the environmental effectiveness associated with MaaS and on average display higher acceptance performance (although non-significant). Furthermore, residents of rural or suburban areas possess significantly lower scores of MaaS BI. The same applies to older people compared to younger people. The findings are overall consistent with a preceding study (Kauschke and Maringer, 2019). By performing MGA, another exciting connection was found: a meaningful moderation effect of age on the relationship between *collective efficacy* and BI. This indicates that the actions and intentions of younger people toward MaaS are indeed influenced by collective environmental attitudes, but the ones of older adults are not. It is vital to take this into account when developing suitable marketing tools such as targeted information.

2.4.2.2 The EBike

Bicycles in general are not necessarily smart. However, electric bikes offer more potential for implementing smart systems (Prati et al., 2018) or connecting them to sharing environments (Nikitas, 2018). Moreover, their use can also be a door opener for switching to new mobility. In a similar vein, Schlüter and Weyer (2019) were able to show that car sharing users have a higher willingness to try electric cars. Why should it be any different with eBikes and new mobility? Especially in suburban areas, they offer the potential to substitute vehicles. Compared to other smart mobility technologies, eBikes also already appear to be widespread and well accepted (Hung and Lim, 2020). At this point, however, one must differentiate. While smart mobility addresses all people, eBikes address a much smaller target group. Many people cannot or do not want to ride a bicycle due to internal and external factors. Popular cycling capitals such as Amsterdam or Copenhagen exemplify that up to 30–40% of urban transport can be achieved through cycling (Urban Audit, 2013). EBikes could help to significantly increase this share. For the present study, however, it must be seen that by considering a population and not only potential cyclists, but some bias is also present in the data that requires clarification. The relatively low mean eBike acceptance score of M. = 3.92 and even lower usage, with 57% of survey respondents reporting they have never used an eBike, signal the need for differentiation. Therefore, data on eBike ownership were collected. The following analyses could thus always be performed for owners and non-owners. It was expected that the innovation would be in different acceptance phases for the people concerned. Refined indications might on the one hand explain how user acceptance can be maintained (continuous acceptance), and on the other hand demonstrate how more acceptance can be created for those not yet involved (initial acceptance).[87]

As outlined in Section 2.3.3, the eBike is viewed by individuals as particularly easy and low risk. Conclusively, *perceived risk* does not influence acceptance in modeling either. There may have been a change in perception here or Germans may assess the risks differently, as prior studies in Greece (Galanis et al., 2014; Nikitas, 2018) found *perceived risk* to seriously deter travelers from biking. Perhaps an eBike also induces a feeling of safety over a conventional bicycle. A group whose acceptance is indeed influenced by the risk of using an eBike are the elderly and women.[88] Thus, although effect sizes are small, marketing might need to come up with strategies to make electric bicycles more appealing to these groups. In line with van Cauwenberg et al. (2019) present data exhibits that older generations are especially attracted to eBikes. Given this group's general reluctance to change (Kunze et al., 2013), it might be a good idea to engage this group in the smart mobility transition via eBikes. On the one hand, the user potential should be further exploited and, on the other hand, existing users of eBikes should be further inspired for digital services. This can lead, for example, to eBike users being able to link up more easily with public transport by means of integrated multimodal routing or the eBike being carried free of charge. Finally, no clear indication can be given as to whether it is urban or rural residents who are more accepting of them. Potentially, there seems to be application potential

[87] See Table 2.35, Section 2.3.4.4 for effect comparison; See Figure 2.25, Section 2.3.4.2 for levels of acceptance according to taxonomy.

[88] See Table 2.32 and Table 2.33, Section 2.3.4.4 for moderation effects by age and gender.

in both spaces. Two very conceivable scenarios are biking for fun in nature or commuting to work.

To deepen my insights on which strategies could be applied for eBike promotion, I ran an IPMA on the eBike sample and refined it by running the procedure again for users and non-users.[89] To my knowledge, UTAUT2 has never been applied to eBikes. So, acceptance has not been modeled and related to one another in this way before. The interpretation of the results[90] consequently revealed some interesting points:

(1) The use of eBikes is heavily affected by the underperforming constructs *habit* (0.45), *social influence* (0.31), and BI (0.20). The latter is again, i.e., predicted by low performing *social influence* (0.58).
(2) Hence, the initial strategy for eBikes, as for smart mobility, should be improving *habit*. Currently, also in comparison to other smart mobility, the by far least habituated behavior is associated with eBikes. This is potentially due to the aforementioned issue with limited target groups. Hence, no clear recommendation for action can be derived, but to care for continuous acceptance in users. IPMA for users in this regard points at strategies that improve prices for customers, develop the infrastructural surroundings, and, again, alter norms toward a preferable and desirable social biking environment (Wang et al., 2018b).
(3) As a consequence, decision makers and managers should primarily focus on enhancing *social influence*. The performance of this construct for BI is relatively weak among both owners (4.07) and non-owners (3.09).[91] Thus, in line with Chen (2016)'s results, the potential for improvement at this point is exceptionally high. Once again, it is worthwhile controlling at item level to ascertain where to start. Thereby, it becomes apparent that the items for injunctive and provincial norms do not contribute to *social influence* for eBikes. On closer analysis, they are therefore subjective norms[92] that need to be addressed. This is consistent with the findings of Le Bris (2015), who studied mobility careers and found that subjective social scripts most importantly must be changed to help diffuse eBikes. This could be done, for example, through mobility-related educational work at an early age. Another possibility

[89] The elaborated model was employed.
[90] The interpretation was conducted analogously to the methodology in Section 2.3.4.4.
[91] A reminder: on a seven-point Likert scale (with value 4 as its median).
[92] These refer to what people relevant to an individual may think about their actions.

would be integrating eBikes to corporate mobility management, thus developing a certain compliance toward eBikes. Furthermore, marketing should take place in acknowledgement of people's everyday practices. For it is within this subjective resonance space that behavior is significantly formed (Weichhart, 2003). Why not advertise eBikes on steep mountains or at well-known traffic jams on the way into town? Chaney et al. (2019) brings in another perspective, explaining that certain social fears and prejudices surround eBikes. People often feel the need to socially justify for using an eBike, because it might be considered 'cheating', implying rapid travel without paying the price of physical training. Normal bikers might as well envy electric riders for being able to afford an eBike and so on and so forth. To calm this conflict, measures should be taken to create a greater sense of collegiality amongst cyclists. These could involve public campaigns, legislation issues, or infrastructural projects. In particular, when cyclists realize that their collective presence (e.g., in associations and interest groups) enables them to be more effective (e.g., in urban and regional planning), the collective awareness to being one could be sharpened.

(4) Different from bike sharing (Chen, 2016), green attitudes do currently not seem to affect eBike adoption. So, it might be a good idea to reconnect eBikes and sustainability in the public eye. Of course, this can only be plausibly conveyed if the eBike replaces a car. Thinking outside the box: why not offer a scrapping bonus for the elimination of the car when buying an eBike?

(5) The second two pillars besides social norms that influence BI toward eBikes are *performance expectancy* and *hedonic motivation*. Their influence is powerful for non-owners and owners alike. However, the mean values are average for non-users only. Hence, increasing perception of fun and performance indicators in this group bears a potential to sell more bikes. Although marketing efforts in this regard have arguably largely fallen asleep, the cycling industry is currently booming (Reid, 2017). It is therefore worth investigating which factors can contribute to promoting performance and fun. A recent Dutch study by Plazier et al. (2017), for instance, proposes spotlighting enjoyment of speed, reduced physical effort, and associated higher ranges manageable on a regular basis. Continuing this thinking, the nature immersion experience or doing something healthy could be attractive intrinsic features for users. Furthermore, in order to increase performance, some current issues of eBikers may be addressed. For example, users often want more convenient parking and securing options, better batteries, and generally lower-maintenance technology. Young people are one of the largest user groups of bicycles today. Having said that, they can still rarely afford an eBike (Jones et al., 2016).

For this reason, it will be important to establish a trustworthy second-hand market for electric bikes and to recondition batteries accordingly.

2.4.2.3 Fully Automated Vehicles

The third application example of smart mobility was the vision of the fully automated vehicle. As can be read in the literature review, numerous studies have already been conducted in this area. Nevertheless, present modeling can enrich our knowledge with some interesting aspects. This is especially due to the fact that, in contrast to previous Level 4 studies (Alessandrini et al., 2014; Madigan et al., 2017; Nordhoff et al., 2018b), the vision of Level 5 vehicles has been investigated rather seldomly (Payre et al., 2014). As a result of examining Level 5 vehicles, the data do not allow a reliable interpretation of current usage or prior experience. However, statements can be made about the factors influencing acceptability.

The first remarkable finding that emerged from our data became evident in descriptive evaluations.[93] Accordingly, it is primarily men who are fascinated by FAVs. This is consistent with observations made by Kyriakidis et al. (2015) and Hohenberger et al. (2016). Present data illustrate this outcome again particularly impressively through cross-comparison with other technologies. With these, there is almost no gender difference. This attraction that FAVs exert on men has several plausible causes. First, it is generally assumed that men have more affection for new technology (He and Freeman, 2010). Second, in comparison to the other use cases, fantasies about FAVs include a car, which again seems to be a male technology (McKelvie et al., 1993). Third, it is predominantly men who drive cars, whether alone or in passenger situations (Morris, 2009). Thus, it might also be men who would most appreciate benefiting from Level 5's new free time offering. Fourth, studies have shown that at the current stage of technology diffusion, the evaluation of FAVs depends heavily on the factors of trust and risk (e.g. in Zhang et al., 2019). Thus, men's higher willingness to take risks (Abay and Mannering, 2016) may be another factor that helps explain this extraordinary favor.

Investigating the general acceptance scores, one can see that acceptance of FAVs is lower than acceptance of electric vehicles or connected services (Kauschke and Schulz, 2017). It ranks at about the same level as FCEVs (Kauschke et al., 2021). Therefore, I assume that the low value is primarily due to the fact that there are quite few concrete notions of the technology in public.

[93] See Figure 2.23, Section 2.3.3.4 for gender effects across use cases.

Currently, the cognitive effort to imagine a system of fully automated vehicles and its socio-economic implications is large, while the possibility of gathering information and experience is small. However, it is by no means the case that FAVs are not accepted per se. In fact, the people in the present study expect Level 5 vehicles to deliver the highest benefit in terms of performance and usefulness. They also believe that they will use FAVs quite routinely in the future. In the here and now, of course, the vehicles are still considered too expensive and not necessarily socially desirable, given that the technology appears overall technically complicated and risky. As Nordhoff et al. (2020) have put it, consumers seem to lack trust in the machine. Conceivably, people are taking notice of the discussions around ethics and liability. This may leave them in a state of uncertainty. Literature has extensively focused on these issues and identified that, in essence, legal amendments must be based on democratic and honest ethical discussions. In the end, this may have to lead to legal equality of human drivers and computers (Adnan et al., 2018).

The present SEM outcomes concur with our literature screening[94] and other state-of-the-art reviews (Jing et al., 2020; Kaye et al., 2021) on the topic. Moreover, from an UTAUT perspective, not a great deal of difference can be seen between the acceptance of FAV Level 5 compared to previously examined Level 3 and Level 4 vehicles. Apparently, people find it cognitively difficult to differentiate accordingly at this point. It would be the task of politics, media, and science to better communicate the SAE levels and to outline the road toward Level 5 vehicles as a gradual and plausible process. In connection with automated driving functions, one should refrain from speaking exclusively of the final result, frequently referred to as 'autonomous' driving. Furthermore, future studies should not only be designed longitudinally, as demanded by Jing et al. (2020) Jing, for example, but should specifically investigate and elaborate the likely acceptance barriers at SAE level on a comparative basis.

To return to results, additional to trust and reliability, the dimensions of *performance expectancy* and *social influence* contribute to explaining FAV acceptance in the present study as well as in many others. By including the IPMA output, however, one learns that also for FAVs, as well as for eBikes, *social influence* offers the greater potential for improvement due to its still rather weak performance. So how can society create a framework in which FAV becomes more socially desirable? The answer to this would in principle be similar to what has already been described in the sections on smart mobility and eBikes above (e.g., addressing different types of social norms and increasing landscape pressures). Taking

[94] See Table 2.6, Section 2.1.3.4 for UTAUT-based AV studies.

a different avenue here, I have calculated a regression in which *social influence* is the one and only dependent variable. Results display that only one variable of the ten in the FAV model had a statistically significant effect on *social influence*, namely *facilitating conditions* ($f^2 = 0.03$). The strongest items were FC3: 'An FAV is compatible with other systems that I currently use' and FC2 'I have the necessary knowledge to use a FAV'. As a conclusion, I may posit that by providing knowledge about the usage of FAVs, one can influence social norms and ultimately improve acceptance. The issue of compatibility is probably comparable. People may feel like needing to ensure that the technology will fit their lifestyles, daily used infrastructures, and technologies. Thus, advertisements and campaigns about FAVs might focus on daily habitual system interactions instead of promoting pure technological visions. In a video spot about FAVs, I could for instance imagine a person walking toward their personalized service FAV, which is inconspicuously parked outside in the street. The vehicle then unlocks and opens automatically, so that belongings or further passengers can comfortably enter. The power of automation lies in the restraint of technology. Therefore, low user involvement in the route selection and starting process of the FAV are ensured due to minimalistic design elements. The passengers can in the meantime concentrate on the essentials, such as preparing for work or spending quality time with their children on the way to school. The focus must be on the story and everyday life. The vehicle is so simple and compatible that it absolutely fades into the background of the story. The course of events itself remains, shimmering along existing habit, utterly disturbing elements such as traffic jams or technical problems are no longer important. A story like this might set up for a social desire. This evaluation is supported by the fact that the present study furthermore found *habit expectancy* to be a significant predictor for FAV. No previous study had taken this into account.

With respect to *effort expectancy* and *facilitating conditions* results are somewhat more disjointed. Potentially, as in Madigan et al. (2017), *effort expectancy* is mediated through other constructs such as *facilitating conditions*, once the latter emerges as a direct predictor.[95] Thus, the *ease of use* of an FAV can be seen as a kind of precondition for perceiving behavioral control over the technology. Overall, *facilitating conditions* are hence an important predictor of acceptance, which underperform in comparison to the other use cases. To improve its score, designers and developers should consider supplying an appropriate material and immaterial ecosystem. On the road infrastructure side, for example, sections could be defined and signposted that are permitted and certified for Level 5 automated

[95] See Section 2.2.1.2, for a theoretic excurse on the issue.

driving. On the market side, in turn, there is a need for greater supply and adoption of automated and assistance systems at levels 2–3. Kyriakidis et al. (2015) found that people who use driver assistance systems are more likely to accept FAVs. However, in comparing different levels of autonomy Rödel et al. (2014) showed that perceived control continuously decreases with higher autonomy. Hence, I conclude that a bottom-up approach in introducing automated systems stepwise is more promising in terms of acceptance than fast technological disruption. Regarding data infrastructure, functioning safety instrumented systems / fail-safe systems must be built while also ensuring ubiquitous connectivity to other road users and systems. Furthermore, service systems for repair and maintenance must be established to give people a sense of controllability. In order to enhance the likelihood of acceptance, it could make sense from a design psychology point of view to give people in Level 5 vehicles the possibility to authorize themselves as drivers in certain situations since people value and welcome emergency systems (Payre and Diels, 2020).

Finally, as in almost all vehicle-related acceptance papers but few AV studies (e.g. Nordhoff et al., 2020), our study finds a highly significant association of *hedonic motivation* and BI to use FAVs. Even though many people actually appreciate driving themselves (Rödel et al., 2014), FAV seems to be a fun, innovative, and exciting technology. Thus, *hedonic motivation* can potentially serve as a significant proxy for people beginning to engage with the technology and wanting to gain experience. However, this factor should also be addressed in more depth as soon as the vehicles become more common sight and actually start to compete with today's automobility. R&D needs to reflect on how to render automated driving enjoyable, how to convey the joy of driving, or what alternative ways of addressing people can be found to get them intrinsically excited about continued use. Berliner et al. (2019) give some first insights for what Californian early adopters enjoy about AV: reduced driver fatigue, comfort, and a gain in free-time.

2.4.3 Context of Acceptance

2.4.3.1 Experience

The experience an individual holds toward a technology often positively influences its acceptance. However, the proxies, operationalizations, and role of the variable vary from study to study, and from technology to technology. Literature reveals significant heterogeneity on how experiences translate into acceptance.[96] As anticipated in hypothesis H14, the data thereby suggest that, also for smart mobility, the multiple moderator effects originally proposed by Venkatesh et al. (2012) cannot be replicated. Hence, experience neither influences the effect of *intention* on *use*, nor the effect of *facilitating conditions* or *hedonic motivation.* This is in line with what other authors have recently concluded (Nordhoff et al., 2020; Tamilmani et al., 2020). So, does smart mobility experience not have any effect on adoption? It is doubtful, as outlined in Figure 2.25; descriptive analysis bears significant variations in level of experience. Thus, I deduce that the applied operationalization must be effective and appropriate. Nevertheless, the lack of evidence for the moderator effect could have its reason in another theoretical aspect: the use case-based transversity of data collection. In contrast, Venkatesh et al. (2012) studied longitudinal data, which hardly any subsequent study has done. This must be seen as a possible limitation of existing as well as numerous comparable studies. Measuring acceptance as a process and identifying interaction effects such as moderation is likely to be most successful when individuals are interviewed before and after exposure to a certain technology. In terms of our taxonomy,[97] this corresponds to the border from acceptability to acceptance at which it becomes particularly exciting. However, longitudinal field trial-based acceptance analysis for new mobility (Madigan et al., 2017; Kauschke et al., 2021) also failed to find any moderation effect.

This, in turn, allows it to appear conclusive that present data display a different path of action. In the restructured UTAUT,[98] I included experience as a direct predictor of the core constructs *use behavior*, BI, and *habit*. In principle, this connection had been a recurring theme in early research (Kim et al., 2005; Limayem et al., 2007), but was largely sidelined after the establishment of the UTAUT. Eventually, the inclusion of *experience* as a direct predictor substantially increases the coefficient of determination R^2 of the usage behavior from 0.43 to 0.64. Acceptance can accordingly be much better explained by such a model. Further, however, *experience* with smart mobility shows little direct or indirect effect on BI. I conclude that for the present smart mobility use cases, *experience* significantly increases acceptance. It does this through the fact that unconscious and more or less automated processes shape our habits based on prior exposition to a technology. Moreover, it becomes the strongest direct predictor of usage. In this process, it bypasses the effects created by habit automaticity (Hornbæk

[96] See Section 2.2.3.2 for related theory.

[97] See Figure 2.2, Section 2.1.1.3 for the taxonomy of acceptance.

[98] See Figure 2.32, Section 2.3.4.4 for associated modeling.

and Hertzum, 2017) and causes *facilitating conditions* to become insignificant. Thus, from the moment a technology is available and real experience can be gained, the dynamics of acceptance of smart mobility drastically change. Managers and decision-makers should thus enable people to experience new smart mobility technologies directly and easily. One possibility at this point would be to make new technologies initially available free of charge for testing purposes, as known from the app market, for example. In the future, value creation will be generated more via downstream services.[99] Against this backdrop, Tsou et al. (2019) introduce the variable *service experience*, which, in turn, elicits a value-oriented response from individuals. In any case, the aim must be to create functional experience spaces and real laboratories for the overall smart mobility system. Since only when services interconnect seamlessly, will people collect positive experiences and start appreciating the new smart mobility offer.

Further research must aim at clarifying conceptual overlaps between *habit*, *use behavior*, *experience*, and *perceived behavioral control*. In addition, *experience* should possibly be dissected in more detail. Exemplary questions to be answered, which are located more in basic research, would be which types of experience[100] have which impact on the acceptance. Combining TAM-based research with UX-based research[101] might also be promising. An integrated understanding could illustrate how future experiences need to be designed to foster acceptance.

2.4.3.2 Level of System Integration

By linking smart mobility with its technology types, I proposed a taxonomy that allowed for classification of use cases. The basic idea was to allow for a transversal comparison between use cases that integrates collective stages of cognitive effort. Depending on this level, one finds more experienced people in a group. The levels are based on Stradling (2006)'s observation that mobility requires physical, cognitive, and affective (i.e., emotional) efforts. According to the taxonomy, acceptance consequently varies not only because the technologies have different characteristics, but also because they are, ceteris paribus, at different thresholds of integration into the socio-economic regime. Conceptually, this fits into the framing of hyperbolic discounting. According to this behavioral economic phenomenon, people prefer to be rewarded for their decision today rather than in the future, even if the future decision potentially contained the

[99] E.g., gaia-x.eu for prospective service directories.

[100] E.g., knowledge, seeing, feeling, time and intensity of exposure as well as related technology.

[101] See Section 2.1.2.5 for an introduction.

higher reward (Ainslie and Haslam, 1992). In a way, it is also consistent with Rogers (1995)' stages of technology diffusion. However, it must be stated that this conception is still at a very experimental stage and consequently requires further examination. In this context, it seems to be useful, for example, to map already existing studies via the procedure. In the sense of such a review, however, the time component should also be covered in order to enable longitudinal conclusions and side-by-side comparisons. It is vital to understand the point a technology is at when trying to learn from other diffusion processes.

Taking the taxonomy perspective, acceptance is assumed to increase with the level of system integration and exhibits more utilitarian traits. Collective experience values may then again provide certainty about the coming technopolitical development. In the broad outlines of the data available, these assumptions can be confirmed. Results[102] show that the more integrated and available a technology is, the higher experience values are. However, it is impossible to determine what is cause and what is effect here. If one still wants to follow the idea of the taxonomy, one could, for example, extract the following lessons:

(1) At level 0, FAVs face issues related to a-priori acceptability. They are not at all integrated into our system. Hence, people initially need information to develop trust and a meaningful opinion.
(2) With the ubiquity of apps and smartphones, people have gained an initial idea of MaaS. Collectively, many people could be looking for experience at this point.
(3) Level 2 includes automated driving functions with which some people are already more intensively in contact due to their readiness for series production. The question of use and acceptance could revolve around whether the framework conditions are right and whether one can reconcile such systems with one's personal norms.
(4) Digital mobility services are found at Level 3. Most people already use them. They may serve as proxies for bringing in smart mobility. To achieve this, however, they must be easy, compatible, and socially desired.
(5) Next, eBikes range at Level 4. They represent a smart mobility technology which is already widely diffused. In order to gain further market shares, they should add value compared to existing mobility (e.g., in terms of prices, fun, and performance).
(6) No technology examined ranges at Level 5, which describes habitual acceptance. The collective potentially inherits such acceptance pattern toward the

[102] See Figure 2.26, Section 2.3.3.5 for allocation of use cases within the taxonomy.

automobile. In line with the present research above, familiarity and predictability of technologies may be decisive in elevating a technology to this stage.

Finally, the possible relationships described here could not be confirmed in the structural modeling of this study. H15 is rejected. Further research on the temporal change of technology acceptance is advised. Additionally, a construct such as *perceived system integration* could be investigated as a predictor of acceptance to shed more light on the issue of how different emerging smart mobility technologies can lead to change in the socio-technical regime.

2.4.3.3 Excursus: The Thomas Principle

It is common practice in acceptance research to control for demographic effects. In analyzing present data, some interesting aspects came to light. As expected, men overall displayed higher acceptance of smart mobility. This was particularly evident in the case of FAV. Hence, if automated technologies are to be brought to the streets quickly, it is reasonable to consider men as a special target group. This segment, in turn, shows a significantly stronger influence of the expected performance on technology acceptance. Promoting the technical capabilities appears a good strategy in this regard. I furthermore find that women are affected more strongly through risks and that older people's acceptance is much more strongly influenced by hedonic motives. The present study thus might have detected a phenomenon that Jeekel in his 2017 study refers to as 'toys for rich boys'. His point is that smart mobility runs the risk of bypassing the majority of people while serving solely a privileged, wealthy class with an affinity for technology. This, in turn, would contradict the goals that transport transformation should actually have: efficient and socially just resource preservation.

Motivation

Altogether, these aspects encouraged me to discuss the role of demography in a more provocative way, referring to a topic that has been repeatedly discussed in recent smart mobility publications. The issue revolves around a regime-inherent gender-bias in smart mobility. Authors refer to it, inter alia, as the 'gender gap in urban mobility' (Singh, 2019). Woodcock et al. (2020) further elaborate that differences in the perception of smart mobility do not stem from biology only but from structural disadvantages. In planning new digitalized transport options, the needs of women in mobility, some of which are fundamentally different from those of men, are hardly taken into account at all. Too little emphasis is placed on improving access, safety, and comfort for women. Women do not focus their mobility on commuting to work as men do, but travel for diverse reasons such as household, visits, and family issues. Hence, women value flexibility over time-efficiency (Singh, 2019). At the same time, women drive significantly fewer miles in cars, but walk more or use public transport (CIVITAS, 2015). Contrary to popular belief, these mobility patterns are found evident both in the global south and in the global north. It is assumed that the different everyday mobility behavior is primarily derived from the individual roles in the labor market and related family care-taking duties (Lenz et al., 2019).

The authors above outline in their studies that designers and decision-makers of smart mobility are predominantly men, who then again shape mobility for men. This idea, in turn, has recently become popular in the context of unequal staffing in management positions. It is also referred to as the 'Thomas Principle'. In 2018, the German-Swedish AllBright Foundation published an attention-grabbing study titled: the power of monoculture. In this study, German boardrooms were examined as to their gender distribution and demographics. It became evident that most people who hold such board positions are very similar in almost all respects. Nearly 93% of German board members have identical résumés, including similar age, origin, gender, education, and even names. Thomas or Michael appear in 49% of all German board member lists, while there are as little as 43% women in total. On average, executive positions are given to 53-year-old, white men born in Germany (AllBright, 2018).

According to the authors, the main disadvantage of the Thomas Principle in Figure 2.33 is its uniformity. With almost identical people, there is little chance for innovative and unconventional ideas. In addition, it has been proven that in the recruitment process, Thomas tends to reassign a new leadership position to the next Thomas. Further, leading roles have an important influence on products and services, as they decide whether an idea is good or bad. Hence, designs will be based on a limited set of personal wishes and experiences, neglecting the needs of groups outside the Thomas bubble of thought. In mobility this, inter alia, provokes vehicles being tested with male dummies only, women voice in car navigation, or premium smartphones simply becoming too bulky for female hands. In public transportation, in turn, getting on and off the bus in time with small children is a challenge. And here we arrive again at the gender gap. As a deduction, it can be hypothesized that the typical Thomas character will also lead to services of smart mobility being tailored to (older) men, leading to a very limited target group. To explore whether present data also contain evidence of the presence of a Thomas effect, I tested three hypotheses. Associated procedures and results are highlighted below.

Figure 2.33 The Thomas Principle in mobility

1. **Hypothesis: Thomas displays higher acceptance of smart mobility**

First, I defined a group as Thomas: male, older than 50, with an above-average income (>€2000 net / month).[103] This produced a group of 88 Thomases. Second, I analyzed descriptive group differences for all use cases combined using ANOVA in SPSS. The mean value for the Thomas group was M. = 4.67 on the seven-point Likert scale, while the No-Thomas scored at M. = 4.43. This difference, however, is not significant. Hence, I analyzed the individual characteristics of income, gender, and age. It became apparent that the influence of age on acceptance is hardly relevant. For example, the group of people under 25 has a mean acceptance value of M. = 4.42, while those over 55 have a value of M. = 4.39. High incomes, in turn, show significantly higher values. Those who earn more than €3000 per month have an acceptance value on the level of M. = 4.99.

[103] Due to many missing data, income was not considered in the initial evaluations.

If I consequently leave age out of the definition of Thomas and raise income threshold to €3000 instead, I obtain a group of 114 Thomases. This group now displays significantly ($F(668) = 11.55$; $p < 0.001$) larger acceptance values than the No-Thomas group of 555 individuals. From the variables involved, *facilitating conditions* are perceived especially better by Thomas.

From these insights, two conclusions are drawn. First, the hypothesis that Thomas accepts smart mobility better than others is true. However, it must be noted that no causal relationship can be deduced whether this is due to a design bias of smart mobility or simply due to differing perceptions among kind of diverse population cohorts. Secondly, it can be assumed that rather than Thomas' age, it is his gender and role of power within society that influences his perception. This also fits the picture that AllBright paints in a preceding study (AllBright, 2021). Herein, the foundation investigated board members of stock exchange newcomers and start-ups. The authors conclude that Thomas' new name is Christian and he is 48 years old. Yet still, there are very few women.

2. **Hypothesis: Thomas favors car-related technology over public transport**

'*The man is in the middle and therewith in the way.*' This famous quotation of poet Pablo Neruda illustrates the mindset with which critics of the smart mobility gender gap regard men such as Thomas. They believe Thomas is obsessed with cars, whereby his mental focus hinders the development of alternatives. It seemed appropriate to me, therefore, to test this assumption using the example of data from automated vehicles and mobility-as-a-service. The first technology is the representative of the car-centric future and the other represents the sustainable community solution. Looking at the corresponding MANOVA results, the differences for FAVs are significant for the comparison Thomas to non-Thomas as well as for the comparison of men to women. However, it turns out that wealthy men find FAVs even better than men alone. Nevertheless, the mean value is on average still behind the assessment of MaaS, where no significant differences came to light (Table 2.37).

Table 2.37 Mean value comparison between men, women, and Thomas

Use-Case	Women	No-Thomas	Men	Thomas
FAV	3.40	3.84	4.37***	4.63***
MaaS	5.00	5.07	5.24	5.35

*** significantly larger than women and No-Thomas group

With regard to the hypothesis, it can hence be concluded that it has to be rejected. However, the data give indications for the assumption that the Thomas type somewhat prefers automotive technologies. However, no automatic devaluation of public transport can be derived from this.

3. **Hypothesis: Thomas sets social norms**

Essential to the Thomas Principle is that it does not purely identify Thomas and his preferences but aims to have Thomas pass on his ideals and values. Thus, socially-normatively, Thomas produces the next generation of Thomases. Consequently, one question arises for the present study. Namely, whether it can be proven that Thomas is influenced in a special way by social norms or not. If so, one could assume that he himself has a great interest in influencing norms in smart mobility. To understand how the acceptance process of Thomas might work different from No-Thomas, I performed a series of PLS runs and multigroup analyses in SmartPLS.

However, the results were different than expected. The path coefficients and effect sizes of Social→Intent tended to be stronger in the No-Thomas group. After all, no significant difference concerning *social influence* was confirmed in MGA. Instead, we saw that Thomas' adoption was significantly more driven by *performance expectancy* (FAV use case) and *facilitating conditions* (SM and MaaS use cases). So, polemically speaking, Thomas values mobility that is empowering and inherits a sense of controllability. No evidence is found that social feedback is of a special relevance for him.[104] Hence, the hypothesis cannot be accepted with available data. For a serious answer to the question, a targeted research design is needed.

Interpretation

The Thomas Principle draws attention to the fact that diversity at the managerial level is rather inadequate. Systemic managerial perspectives, including smart mobility, are therefore likely to be skewed and male dominated. However, technology must / should not only be made for wealthy men but rather, considering a more social perspective, include the perspectives and needs of other groups such

[104] I ran additional regressions with Thomas being a direct predictor of different kinds of norms (injunctive, subjective, etc.). Across all four use cases, no significant results were obtained.

as women, the poor, or the underprivileged. Jeekel (2017) notes that smart mobility will only be a breakthrough for humankind if social sustainability becomes a target component.

In the present excursus, I explored whether smart mobility does not only include a gender gap (Singh, 2019), but also a Thomas Principle (AllBright, 2018). However, this can merely be detected superficially. Functional or even causal relationships cannot be uncovered. The data and scope available do not allow for a deeper investigation. It should be worthwhile to conduct further investigations at this point. As the social-normative network of an individual is one of the most important levers of acceptance in smart mobility, Thomas could have a leverage and multiplier function. If so, how should it be dealt with? In general, this excursion has made another intriguing observation; if smart mobility is better accepted by Thomas and he values performance and control, does that mean mobility has thus far been designed for him? Is it possibly no coincidence that Thomas and the positively connoted early adopter are so similar? Do individuals find themselves in a closed design cycle? If so, how can one break free? These and other questions need to be addressed in future research.

2.4.4 Implications for Theory

The purpose of this research was to develop and empirically validate a model to explain the acceptance factors of smart mobility from a holistic user perspective, which has not been addressed in previous literature. For this reason, a content-related and methodological extension of the UTAUT2 was implemented. On a meta level, a reference to the Geels and Schot (2007)'s MLP was established. After conducting the study, there are not only substantive findings, but also some critical implications for the theoretical underpinnings of this thesis, which will be presented together with several future directions of research.

2.4.4.1 Tooling

First of all, UTAUT2 was successfully applied to smart mobility. It proved to be a robust and suitable tool for present analysis. With $R^2 = 0.66$ in the basic model, it displays high explanatory power (Chin, 1998). With 83% of hypotheses being true, acceptance can be described in a well-differentiated way. In cross-comparison with other UTAUT2-based work, these results are considered above-average. This applies both inside (Adell, 2009; Madigan et al., 2016; Ye et al., 2020) and outside the mobility (Raman and Don, 2013; Kraljic and Pestek, 2016) domain. Against this background, this research significantly contributes to evolving UTAUT2 from its sometimes-strong focus on the study of ICT to a universal, cross-technology acceptance model.

2.4.4.2 Constructs

Some of the present findings differ severely from what original theory (Venkatesh et al., 2012) and following adoptions posit. From my point of view, the conception of variables should initially be renewed. Or as Dwivedi et al. (2019) put it: acceptance theories must be tailored to their underlying context. Three examples are mentioned.

First, it was found that the role of *habit* was much stronger than the role of most cognitive beliefs. This corresponds to the fact, which is well known in science, that the behavior of a human is hard to change due to their practiced automatisms of action (König and Neumayr, 2017). In the field of acceptability,[105] *habit* had generally hardly been used so far, (Tamilmani et al., 2019a). However, it is primarily future technologies that are to be investigated by means of acceptance modeling. Thus, UTAUT and most alternative modeling in use (e.g. Davis, 1989; Ajzen, 1991) in its current form, albeit claiming to be holistic and universal, frequently overlook this important acceptance factor. The reason for this lies in the contemporary reference of *habit*'s operationalization. In my view, we need to rethink this. Different solutions exist. Schikofsky et al. (2020)'s *habit scheme congruency* and presented *habit expectancy* construct are two examples of how *habit* may be alternatively conceptualized. A third option might be measuring habits tied to competing technology, such as car ownership in smart mobility, for example, and to incorporate this parameter as a (negative) determinant of acceptance into the model (e.g. Jia et al., 2014). Furthermore, besides the conception of the construct, I would like to encourage future research to dive deeper into the factors that constitute *habit*. In this work, *habit* is indicated to have a potentially stronger effect on *use behavior* than BI. In contrast to this key construct, however, research knows too little about the determinants of *habit*.

Also, for the construct *price value*, this work can identify needs for adjustment. As described in Section 2.2.1.3, conceptual problems with the construct are not new. Since the financial aspect, which is actually very important in mobility (Kauschke, 2020), has a negligible influence in this model, I conclude that it is possible that people find it difficult to assign a price to the use of smart mobility services. Consequently, replacing *price value* with the common *perceived costs* construct is not likely to help, unless people are previously provided with information about current and future costs. These, in turn, cannot be estimated, even

[105] Acceptance before first exposition to a technology (see Section 2.1.1.3 for a definition).

by experts. Apart from that, it is assumed that smart transportation will bring significant efficiency gains. However, it remains unclear whether environmental policy tax elements or future value-added business models will offset these efficiency gains or not. A solution to this conceptual problem might be transforming the financial aspect into a personal attitude such as *price sensitivity* that Goldsmith et al. (2005) established in a study about fashion. Kapser and Abdelrahman (2020) transferred this construct to a UTAUT study about autonomous delivery vehicles and delivered impressive results. Suddenly, in fact, the perception of price was the most important acceptance factor of all.

Thirdly, *effort expectancy* and *facilitating conditions* display strong interactions. It seems that both cannot be significant at the same time in UTAUT.[106] Arguably, this is due to two things: (1) the mixed conception of the construct *facilitating conditions* in which self-efficacy, perceived behavioral control, compatibility, as well as support conditions are merged (Venkatesh et al., 2003) and (2) the IT and computer background of the original items. The feedback that I received most frequently via the survey review function during the survey was that items FC1 ('I have the necessary resources to use smart mobility') and FC4 ('In case of difficulties with smart mobility, I can get help') were unclear. I therefore believe that *facilitating conditions* might have to be readjusted for use in mobility. For this, a focus group-based approach could be promising. However, one should not confuse *facilitating conditions* with framework conditions, but should clearly identify which action control variables determine the environment of mobility and smart services. An interesting paper by Ajzen (2002) outlines how the perceived behavioral control is constituted from a basic psychological perspective and can thereupon be used as a reference for redevelopment. He notes that controllability can depend on internal or external factors and is eventually an empirical question which varies from subject to subject. In Ajzen's view, many studies put themselves on the shoulders of giants and adopt scales that are validated within unrelated areas of investigation.

2.4.4.3 Comparative Methodology

Acceptance modeling is frequently leveraged to draw comparisons. In this way, for instance, demographic, cultural, or temporal changes can be better understood. Overall, acceptance is somewhat relatively sized, best conceived by contrasts. Differences can exist within importance as well as performances of items and variables.[107] Group comparisons are commonly performed in software such as the

[106] See Section 2.2.1.2 for related theory.

[107] See Section 2.3.4.4 for IPMA terminology.

utilized SPSS or SmartPLS. Within the paradigm of UTAUT2, grouping variables are often modeled outside the functional relationships. In a very recent review of 147 applications, Tamilmani et al. (2021) identified eight classes of context variables. One of the classes refers to technology attributes. However, neither this class nor any publication mentioned in the review ever used UTAUT2 for comparing characteristics of different, yet related technologies. The solitary research that had investigated something similar, but in the form of a meta-analysis with access to over 40 TAM data sets, was the e-learning study by Šumak et al. (2011). At this point, the present work thus takes a pioneering role. Accordingly, no proven methodology could be relied upon.

The theory contribution of this work consequently also lies in its methodology of comparing different technologies with the same acceptance model. For this, descriptive facts were compared and an effect size-based benchmarking was undertaken.[108] However, a targeted MGA could not be performed due to a failed MICOM assessment. Since each questionnaire must necessarily contain different technology terms, no invariance of the configuration can be established. Future research in SEM statistics must work on establishing a standardized procedure to compare acceptance models and their parameter estimates with each other.

2.4.4.4 Directions for Mobility Modeling

One goal of this thesis was to arrive at a more general acceptance model for new mobility by means of validation across differentiated use cases. In Section 2.1.2, it was highlighted that UTAUT2 is currently one of the best universal acceptance models. Its strength lies in its multidimensional proceeding and a strong, interdisciplinary grounding. Thus, it should be tested for smart mobility. This project can be considered successful, since the statistical validation as well as the added value in terms of content are undoubtedly given. Nevertheless, from the view of present research, there are still exciting potentials for further development. To improve subsequent research, three directions are given:

Extended modeling

A model is the attempt to render complex realities analyzable. Despite its claim to universality, UTAUT2 often reaches its limits in this respect. Thus, criticism with the model as a whole is growing (Tamilmani et al., 2020). What's more, this becomes apparent within the present work when for instance social norms can only be treated superficially in a summarized variable or when model context was limited to age, gender, and experience. UTAUT may have reached the stage

[108] Please see Table 2.35, Section 2.3.4.4 for a structural comparison of the use cases.

of model elaboration. Hence, a current stream of research exists, which seeks to develop the UTAUT2 into what some call a UTAUT3 (e.g. Gunasinghe et al., 2020). However, a general, even more finely nuanced model has not been able to be established to date. Due to the potential over-complexity of such a model, mostly large frameworks have been introduced so far (Nordhoff et al., 2019; Tamilmani et al., 2021). In the future, these models should be validated. However, it is unclear whether SEM is still the right approach in this scenario. It is possible that AI or neural networks can achieve better results at this level of complexity.

Selective modeling
Another option might be using these frameworks as a construction kit for target and context-specific model extensions. This is kind of what Venkatesh et al. (2016) had already suggested and Herrero et al. (2017) notes as flexibilization of the model. Furthermore, I encourage future researchers to dive deeper into conceptual parts of UTAUT2. From present findings, these can involve, inter alia, social norms, framework conditions, or habits. In smart mobility, research finds these variables to be highly influential, partly even more influential than intentions, although we know little about how they shape and constitute themselves. In this connection, both a readjustment of the constructs themselves and the identification of reflective or formative independent variables can be purposeful.

Alternative modeling
A third avenue would be to fully recalibrate our theoretical frameworks for explaining mobility acceptance. The goal may be to obtain a more convenient model, which on the one hand is just as versatile, but on the other hand is more efficient in the use of variables. This squaring of the circle can succeed if the model is not universal but domain specific. As highlighted in Section 2.3.5, Sovacool (2017)'s model might be promising. The next step in its development would be to establish operationalizations and hypotheses. As a second option, further research may also take a completely exploratory approach. Expert interviews and focus groups offered the opportunity to develop a comprehensive set of appropriate questions. The results of a survey might then be fed into an exploratory factor analysis or MANOVA. Even if this bears the risk of failure, it could also lead to a completely new approach.

2.4.4.5 Interlinking Acceptance and Innovation Research

Lastly, a comment should be made on the usefulness and further development perspective of the use of the multi-level perspective. The MLP sustains the concept of technology growing in niches, while bearing the potential to radically transform the current socio-technical regime once a certain level of saturation is reached (Schlüter and Weyer, 2019). In this work, I proposed that the experimental technology and business model development that is taking place around FAV, MaaS, or eBikes may facilitate a transition toward smart mobility. The point at which they are on a fictional scale to breakthrough to regime change could be qualified as acceptance. To better understand these processes, the development of a regime change readiness indicator for niche technologies might be useful. To this end, I recommend applying my mixed MLP acceptance approach to historical diffusion events. At this point, one could also build on the pilot of a taxonomy of acceptance that I introduced in this dissertation. A clearer distinction between personal and collective experience, level of integration, and influence on the regime could prove a crucial starting point.

2.4.5 Implications for Practice

The present discussion includes several implications for practice. At this point, the main consequences of present acceptance study shall be summarized into five straightforward recommendations for action. These are intended to help decision-makers in politics and business to address fields of action for the smart mobility transition. As Figure 2.34 illustrates, the implementation of measures in all five topic areas can be initiated as early as today. In the short to medium term, viable solutions should be developed and integrated into the surrounding infrastructure and data systems. This is to empower technological niches. At the same time, smart mobility should be cleverly promoted and long-lasting spaces for gathering experience should be created. This will likely increase the landscape pressure and install a powerful social-normative framework. Finally, the transition process must be consistently monitored to enable adaptive governance (Docherty et al., 2018).

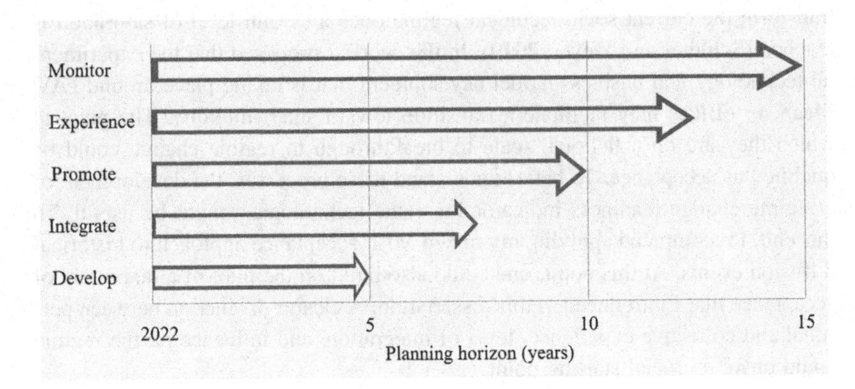

Figure 2.34 A roadmap of smart mobility development

2.4.5.1 Develop

In the short to medium term, the development of smart mobility systems must be driven forward vehemently so that new technologies and business models become practically available. The acceptance factors highlighted and discussed in this work (e.g., *performance expectancy*, *habit*, *hedonic motivation*, etc.) may therefore be included as new requirements in the design process.

While it is certain that smart mobility will transform transportation, the question is whether economies can develop solutions that respond to this. New systems must function effectively, must be psychologically acceptable, and meet the needs of sustainability. Authors such as Schneidewind (2018), therefore, demand a bottom-up approach of development, orientated exhaustively at collective needs. The user must be at the heart (Lyons et al., 2019).

New forms of public-private-social cooperation are needed to meet these development requirements. Sulz (2018) concludes in his in-depth analysis on the topic that in the future, cooperation will be at least a sufficient, possibly even a necessary, condition for the success of industry, mobility service providers, and mobility platforms. Smart mobility comes as a form of disruption to the incumbent mobility regime. Thus, to be successful, cooperation is also needed to overcome system inertia to change.

2.4.5.2 Integrate

This dissertation elaborates that the level of system integration of a smart mobility technology can be crucial for its acceptance. In the medium term, actors and technologies should therefore be intertwined. Lyons et al. (2019) highlight that the pursuit of more convenient and seamless door-to-door travel is an evolving and longstanding process. Integration will first be completed when a comprehensive operational, informational, and transactional implementation has taken place. The goal must be to have fully featured and fully functional solutions that can rival today's automobility in its extensiveness.

Therefore, I want to encourage decision-makers to invest in integrating structures and superior data spaces. Every application of smart mobility is initially based on the availability of suitable data and interfaces. Here, the organizations and companies involved frequently lack information technology and data analysis skills. Together with the lack of prospects for profitable system operation and unclear distribution of roles, this, in turn, often leads to the topic not being addressed at all (Karlsson et al., 2020). Thus, an external framework must be offered that harmonizes the various data ecosystems and enables non-discriminatory access for companies and authorities (e.g. Gaia-X, 2022).

2.4.5.3 Promote

Results of the present acceptance research translate well into marketing tools. Within the present discussion, numerous approaches such as nudging social norms or positive communication of new mobility were therefore examined. If one scales such proposals to the level of regime change, one comes to the point where one has to look at the grand narratives of smart mobility. How are politicians and decision-makers supposed to promote a technology that offers collective advantages but whose individual benefits are not (yet) tangible? Holden et al. (2020) did a conceptual review on 30 years of sustainable mobility communications. They finally distilled their findings into three narratives, which together have the power to whet people's appetites for smart mobility:

(1) *Electromobility*: The rise of electrification can hardly be stopped. Virtually every major vehicle manufacturer in the world has embraced the trend. However, challenges (e.g., range, battery life cycle, infrastructure, etc.) remain to be solved. In any case, smart mobility as a whole can benefit from the tailwind of the behavioral change induced by electromobility by linking sharing or mobility-as-a-service concepts to successful electromobile applications.
(2) *Collective transport 2.0*: Today, more and more people discern great benefits for themselves and the environment in switching to public transport. However, they demand it to be modern, safe, fast, and convenient. And it is often not due to outdated and proprietary organization and infrastructure. Hence, the story to be told must be that of significant investments in the restructuring

and expansion of public transport. Most people are likely to welcome such initiatives and think positively about smart mobility.

(3) *Low-mobility societies*: A narrative that may be unpopular but is essential from a scientific perspective, must also be to reduce mobility and the effort required to achieve it. In this context, communication must be positive if smart mobility is to benefit. This can entail, for example, communicating the ban on cars from cities as a way to improve quality of life. For people in rural areas, homeworking regulations, in particular, can have an enormously positive effect. The smartest mobility might be the one that does not occur at all.

The challenge of these narratives does not lie in telling them, but in making agents believe them. According to Holden et al. (2020), artists, social institutions, and, not least of all, charismatic leaders are also called upon to stand up for new mobility. This is a process that may begin today but will extend over the medium and long term.

In addition to creating narrative meaning, people should also be given access to simple information about smart mobility. The connection of new mobility concepts with overarching objectives must be better understood. The mobility of the future is a systemic innovation that affects everyone: companies, citizens, research, and public institutions. It is fully networked and digitized. In this form, smart mobility offers enormous environmental potentials. It is consequently important to create a common understanding of what smart mobility is and why a transformation of the transport system is necessary. The overall societal benefit and public value of smart mobility must be placed at the center of communication.

Finally, to promote smart mobility solutions, I want to propose a new approach to customer segmentation. In the future, target groups could no longer be clustered according to demographic aspects, but according to acceptance characteristics from psychology. Present study shows, for example, the major role of emotions in mobility. Acceptance research can thus help to portray a mindset of prospective FAV buyers, eBikers, or MaaS users. Furthermore, it may be able to map tastes of current users of trucks, vans, or sport vehicles to detect individual entry points to preference-based change and individually tailored smart mobility services (Mohamed et al., 2018).

2.4.5.4 Experience

Empirical evidence proves that BI predicts *use behavior* along with habits and past experience. These experiences, in turn, guide and reinforce habits. Thus, the level of experience is one of the major key components in creating acceptance for smart mobility. Almost all acceptance factors can be positively influenced by the level of exposition and familiarity with smart mobility. However, as described in Section 2.4.3.1 the experience must be positive. Either way, creating spaces for gathering favorable experiences with smart mobility must be on the political agenda. In this context, it is just as relevant to create short-term experiential opportunities (e.g. Nordhoff et al., 2018b; Kauschke et al., 2021) as it is to build entire local ecosystems (Flügge, 2016; Giesecke et al., 2016) for the long term. Three possibilities are suggested:

(1) *Vision spaces*: Providing experience and information about future technology can be done in special spaces. These could include museums, trade fairs, or universities where permanent places of experience could be established. There, virtual reality, media, or prototypes could bring smart mobility to life and open room for dialogues. A good best practice example for other cities would be the Futurium[109] in the city of Berlin, where interactive exhibitions on important future topics such as climate, living, nutrition, and technology take place on a regular basis.
(2) *Field trials*: The moment smart technologies become available, they still tend to remain solely accessible to a small circle of researchers and enthusiasts. The confrontation with reality is frequently delayed. In the field of mobility, it has therefore become established that new electric or automated vehicles, for example, can be experienced with scientific support in field trials. The same could also be done with digital technologies, for example by releasing a (fully functional) app to a group of test subjects in a closed geographic area.
(3) *Real labs*: Finally, it appears promising to create model regions. Smart mobility is comprehensive and has many interactions. Thus, it makes sense to concentrate transformative pilot efforts in certain areas and data spaces. It is also desirable to integrate new solutions to existing technological testing areas. In such a laboratory, complex systems could be tested under real conditions. This would also require the involvement of stakeholders from business, society, and administration. Consequently, interdisciplinary researchers would have the opportunity to test interventions in a full societal context.

[109] https://futurium.de/

2.4.5.5 Monitor

Evidence from this work suggests that dynamics of acceptance factors may change over time and new factors may play a role for the use cases of smart mobility. Therefore, it seems useful to accompany the transition to smart mobility longitudinally with a suitable set of measurement instruments. Recently, for instance, the German Federal Ministry of Economics has added indicators on the external environmental costs of growth to its reports (Bundestag, 2022). Hence, why not establish a further reporting system that controls the progress of the transformation of mobility on the basis of exemplary acceptance measures?

These and other indicators could be relied upon as the basis for what Docherty et al. (2018) calls 'the governance of smart mobility'. The authors describe that technological change is outpacing the capacity of governance structures to respond to mobility challenges. Thus, the administration itself must be reformed and adapted. The authors go on to note that economies are facing critical junctures where the right decisions need to be made urgently. This refers to facing innovations with optimism without overlooking risks that are known from past transitions, for example. Short-term interventions and adaptive long-term strategies are advised if the government wants to maintain its position vis-à-vis private providers. In Europe and Asia, smart governance is still possible today, but in large parts of North America, the state is already too weak to manage the smart mobility transition. The study advises against getting caught in this lock-in.

In this context, I would lastly like to propose the deployment of integrative socio-political support tools. This is to be substantiated: smart mobility requires bottom-up (user requirements), top-down (governance), combined with transversal (integration and cooperation) management. Therefore, effective, uniform, and permanently established organizational units are needed. This unit could accompany the implementation of strategies, coordinate and support applications, and provide a single point of contact for users, industry, and administration. I hence advocate establishing national transformation hubs for smart mobility. Such projects or agencies could monitor, harmonize, and support the transition process. The example of Germany currently underlines that although cautious development means that no one in the industry is left behind, it also leads to development goals in mobility being achieved more slowly (e.g., compared to the UK) or being missed (Mazur et al., 2015). Thus, another task of such an undertaking would be to speed up transitioning.[110]

---

[110] In Germany, the National Platform for Electromobility has been further developed since 2019 into the National Platform for the Future of Mobility.

2.4.6 Limitations

The findings of the present study have to be interpreted with regards to some limitations, which should not be disregarded.

2.4.6.1 Measurement under Uncertainty

First, it is generally challenging to elicit opinions from individuals when no knowledge is available. If the surveyed do not have meaningful experiences with smart mobility, how would they be able to decide what to think about it? Ricci et al. (2008) call it a '*recurrent assumption*' that people generally have '*a-priori*' opinions. Klöckner (2014) notes that hypothetical decisions are not real and thus need further validation. In future technology surveys, this is commonly reflected in larger proportions of respondents who opt for 'do not know' responses. Therefore, when collecting the data for this study, I deliberately did not allow subjects to choose such an alternate option. Instead, I aimed to encourage respondents to take a moment to think about their opinion. Additionally, the large sample size may have helped to reduce subjectively skewed perceptions of occurrence probabilities regarding the future of mobility.

2.4.6.2 Single-dimensional Study Design

Another limitation of this study might lie in its single-dimensional design. Thus, the data were all obtained at the same time from a similar group of respondents (e.g., Germans). However, my taxonomy elaborates that technology acceptance is temporary and dynamic, as is the relationship between measurement tool and user. This may be the primary reason why there is such wide variability in literature insights. I therefore agree with the request of Zoellick et al. (2019) to establish a clear conceptual framework for employing SEM-based acceptance studies. While the present study was cross-sectional in nature by considering multiple use cases, future research should validate the findings with longitudinal studies and monitor the progression of adoption across societies and population groups.[111]

2.4.6.3 Social Desirability

The answers in this study might be biased by social desirability. This phenomenon describes the reply tendency of respondents to prefer expressing a socially liked evaluation instead of an unpopular opinion (Messick, 1960). This could be the

[111] An interesting option for this purpose would be to employ target group information systems such as the SINUS milieus in Barth et al. (2018).

case in the smart mobility environment, as people like to be regarded as innovative and environmentally aware (Miller, 2011). However, since respondents were not observed during the survey and participation was anonymous, the effect could be weaker. In line with this, my analyses detected neither CMV nor unobserved heterogeneity. Nevertheless, future research might consider implementing social desirability correction scales (e.g. Crowne and Marlowe, 1960).

2.4.6.4 Just Psychology

Lastly, the present modeling is limited to psychological factors. Just as the MLP suggests, factors from alternative disciplines could also play a role. In the present thesis' introduction,[112] environmental economic factors (CO_2 taxes, oil prices, etc.), new business models (sharing instead of owning), or consumer renunciation (linking the smart mobility transition to a social-ecological transformation) were brought into play. These factors may likely affect acceptance through existing variables or moderate relationships. The established model could therefore be used to identify new effects. Furthermore, such an approach would strengthen model validity thanks to the integration of new data sources. Another example would be the use of factual usage data versus self-reported usage, which would help build a bridge from psychology to reality (Chen and Chao, 2011).

[112] See Chapter 1.

The Institutional Perspective

3

The basic assumption of this work is that linear mobility is essentially developing into a service-oriented fabric of interactions. Such an evolution will likely enforce the alteration of roles and actors in the automobile regime. Such processes offer space for pioneering and realigning previous strategic, functional, and task-related positions. Institutions (actors) will increasingly ask themselves what position they can take within a new framework of mobility provision. Untapped market potentials open up when agility is reflected in technical solutions as well as in the economic environment (Schulz et al., 2014).

In this context, role models can offer guidance (Hillemacher et al., 2013). Role models originate from sociology. They describe and explain role expectations and definitions on the one hand, and which scope for interplay and action is open to a role on the other. Originally, this concept was developed to analyze social action systems, but its core ideas might also account for economic and institutional research. To specify the area of research, I present three state-of-the-art examples of possible ecosystem modeling for smart mobility:

(1) Actor-Network Theory ANT (Law, 2009) is a vast approach that proposes how people and devices combine into network nodes. These alliances are then the nuclei for an analysis that aims to identify success factors for technology within entire socio-economic ecosystems. For this purpose, ANT also uses tools such as roles and network assemblage mechanisms.
(2) In contrast, Flügge (2016)'s Internet-of-Services IoS model is a straightforward technology-centric approach that defines seven simple service roles. These are derived from the layered architecture of the Internet. However, it

L. Kauschke, *The Transition to Smart Mobility*, Mobilität – Innovation – Transformation, https://doi.org/10.1007/978-3-658-43001-6_3

remains unclear which of these roles are merely technically or also economically feasible. Moreover, the roles are not linked to the actual actors and institutions.

(3) The Ecosystem Pie Model EPM (Talmar et al., 2020) develops classical stakeholder modeling toward a graphical representation that is easy to understand and yet contains added informational value. This allows complex systems to be represented with a robust practical reference. In return, abstraction to a more generic role level is not possible.

This paper applies the Institutional Role Model (IRM) (Schulz et al., 2019). It features certain strengths of the presented approaches (e.g., holistic view of the ANT, conceptual parsimony of the IoS, or the EPM's graphical interface) while disregarding some of their weaknesses (e.g., the IoS's technocentricity, the ANT's vastness or the EPM's non-scalability). It bears a new understanding of services through its consideration of technical and economic roles and institutions. This is the uniqueness of the IRM. By including empirical elements, it is acknowledged that an external perspective is needed, quasi from the user's perspective and from the perspective of other market participants, in order to depict an ecosystem correctly.

This thesis's institutional perspective on mobility regime change is designed as a companion study. It aims to enrich our understanding of acceptance in the context of the MLP by providing a visionary setup of cooperation in the smart mobility ecosystem. Therefore, this section offers an introduction to the theory and application of the IRM. For this purpose, previous applications of IRM are reviewed. Thus, a new model arises through feedback interviews with experts from a broader range of disciplines (N = 8). The result is a list of technical and economic roles matrixed with potential smart mobility institutions. After all, the IRM signifies an alternative or complementary form to conceptualize and guide smart mobility transition.

3.1 The Institutional Role Model

The IRM is a system model with two goals: reduce uncertainty for stakeholders and develop a relationship of trust. It thereby establishes a cooperative dynamic and creates a basis for pursuing of common interests (Schulz et al., 2021a). The basic components of the model include roles, rules, and institutions. The application's core is the IRM matrix. Thus, an effective, efficient, and non-discriminatory system of smart mobility can be created.

3.1.1 Theory

IRM is inspired by three different disciplines: institutional economics (Schneider, 1995), systems theory (Luhmann, 1997) together with system dynamics (Schulz, 2005). Despite the inherent complexity of such a merger, IRM remains manageable through a cleanly defined yet flexible procedure. The individual terms and components of the model are explained below.

First, institutions in IRM are companies, authorities, associations, and courts. But in economic terms, they are subsystems that regulate behavior, communication, and actions with rights and obligations on how to act. Depending on the objective of the modeling, different typologies of institutions can be deployed. A common theoretical basis is provided, for instance, by the model of Hoddeson and Daitch (2002) in Figure 3.1. This pentagonal approach states that there is no hierarchy between the institutions of science, marketing, technology, production, and research and development. This creates a relevant institutional space for the emergence of innovations.

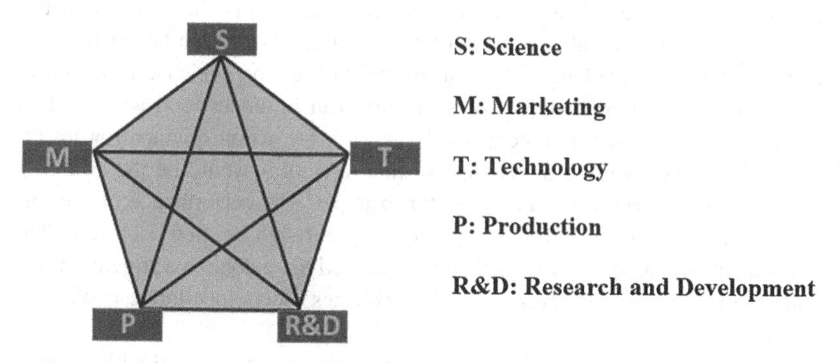

Figure 3.1 Pentagonal model of institutions (Hoddeson and Daitch, 2002)

The IRM can be classified one level below institutions. It describes hierarchical cooperation in which the roles are defined but is designed based on the exchange and cooperation of the respective hierarchies (Herb, 2013). This leads to the question of what is meant by roles.

Second, in IRM, two main types of roles have become established: technical and economic. The former can be derived by engineers from communication architectures or technical system design. They usually represent the central interfaces, services, and management entities of innovations. Each role contains self-descriptions and action characteristics. Furthermore, it is invariably possible to view a role as a meta role and to assign associated sub-roles to it. The examination level (meso, meta, or micro) can always be arbitrarily selected (Schulz et al., 2021c). Economic roles include the key mechanisms to establishing and operating an innovation in a market. In business economics (Schneider, 1995), these are, inter alia, investment, service offering, accounting, corporate governance, etc. However, this typical scheme may also be modified in accordance with the market scenario. In the 5GNetMobil project (Sanmateu, 2018), for example, the overall market for new real-time services was examined at the economics level rather than business economics.

The inherent logic and mechanics of the IRM can be inferred from Figure 3.2. The basic components of the model include the elements arranged in the illustration below: 'roles', 'actors', and rule-based 'actions'. The motivation of an institution or an individual to become an actor stems from the expectation that benefits can be generated as a result of participation. Actors can be existing system stakeholders or belong to the supergroup institutions, which can be system actors or system third parties. Roles, in turn, can be understood and filled by actors. Taking on the role produces a framework for action oriented toward the overall system's goals. In the context of this work, this would be the establishment of a smart mobility data space, for example. The performance of the role can, in turn, be influenced by third parties (e.g., through existing contracts). The assignment of actions to a role is therefore advised to be done in a structured way under non-contradiction. Overall, IRM coordinates the various interests within a system.

Thus, the first step in the IRM process is to set up a role model with an operational and a time dimension. The time dimension is commonly subdivided into market phases (development, growth, loss of maturity, stagnation, and revival). Stakeholders or experts then evaluate the operational model elements on the basis of action intensity levels: low, medium, and high. The results are subjected to an algorithmic calculation. This represents the second model step. The outcome is an optimal basis for organizing a system with maximal interoperability and minimal transaction costs (Schulz et al., 2019). Ultimately, the overall objective

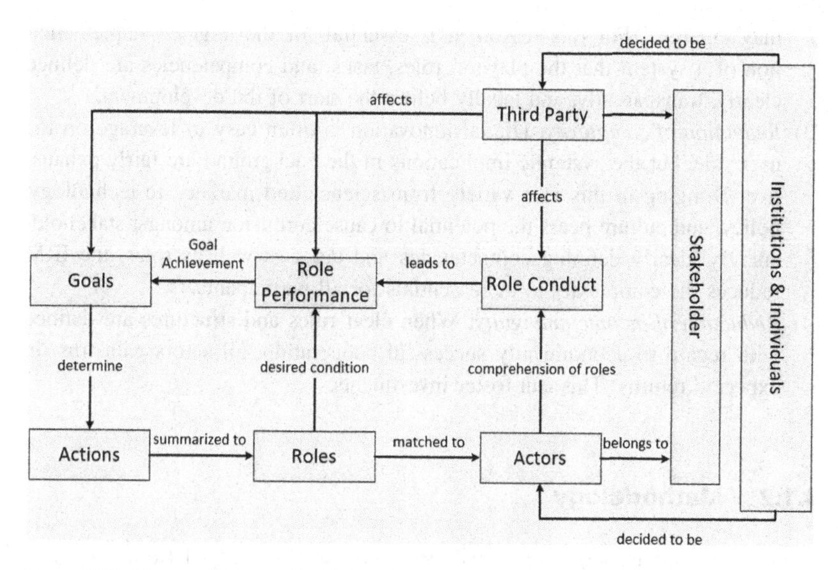

Figure 3.2 How the IRM works (Schulz et al., 2019)

of the IRM is to overcome the weaknesses of the operation model of a given technological system.[1] Additionally, four subgoals are described in the literature:

(1) *System governance*: The IRM serves as an inter-organizational control instrument to ensure the cooperation structure over time, under changing conditions, and everlasting data complexity (Al-Ruithe et al., 2019). The roles support the derivation of adaptive data governance. This refers to a collection of processes, standards, and metrics that enable the effective use of information and help organizations participate per the General Data Protection Regulation (GDPR).

(2) *Reduction of uncertainty*: The establishment of shared data ecosystems is fraught with uncertainty, as there is a virulent risk that other market participants will exploit one's own data against one's own interests (Matheus et al., 2021). Possible further market barriers that cannot be foreseen at this stage

[1] IRM, with its division into technical and economic roles, focuses on ensuring both technical operability and, above all, economic marketability. With regard to the combination of these two aspects, IRM exceeds classical operator models.

may emerge.[2] For this reason, it is essential for the targeted implementation of a system that the players' roles, tasks, and competencies are defined clearly, transparently, and legally before the start of the development.

(3) *Reduction of complexity*: Digital innovation is often easy to leverage on the user side, but the systemic implications in the background are fairly exhaustive. Bringing in this vast variety from science and markets, to technology, policy, and culture bears the potential to cause confusion amongst stakeholders. By clearly defining competencies and tasks as well as roles, the IRM reduces the complexity to the essentials for all participants.

(4) *Reduction of income ambiguity*: When clear rules and structures are defined with regard to economically successful cooperation, all actors gain trust in expected returns. This can foster investments.

3.1.2 Methodology

In the course of its application, the IRM process has evolved from a specific implementation in the field of cooperative driving (e.g. Fünfrocken et al., 2021) to a standardized and transferable process. Recent areas of application hence reach from mobile networks (Sanmateu, 2018) over organizational transformation (Schulz et al., 2021b) to sustainability (Schulz et al., 2021c).

Within the current discourse, two phases and five process steps are recognized. The two phases are the development and application of IRM. During development, defining the object of investigation and the corresponding level of abstraction is crucial[3]. In the application, on the other hand, IRM offers some flexibility so that both qualitatively the development of policy and governance rules and quantitatively the optimal fulfillment of the model roles may be the priority. Phases and process steps are systematized in Figure 3.3.

The core step of the process is the IRM matrix. It serves not only as a graphical illustration of the ecosystem but also as an empirical tool to start the application and further investigate. It is also the starting point for reconfiguring the model if necessary. It thus recognizes Luhmann (1995)'s thoughts that systems should

[2] See Section 3.2.1.3 for potential barriers.

[3] Levels include meso (e.g., economic level), meta (e.g., sectoral level) or micro (e.g., business level) observations.

be self-learning and self-sustaining.[4] The IRM matrix ultimately needs two input vectors: the economic and the technical roles. In the economic roles, the traditional roles are essentially defined and iteratively replaced by the specific roles that characterize the value chain. Technical role models are regularly used in the development of system architectures. For a long time, however, there was no corresponding procedure on the economic side to make it possible to automatically transfer a technical role model into an economic business model. This gap in development methodology can be closed with the assistance of the IRM approach.

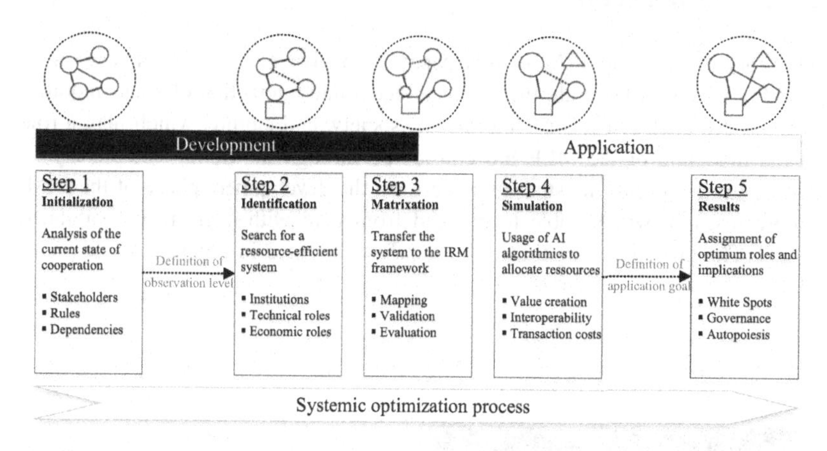

Figure 3.3 The five steps of IRM

In the upstream development, two steps must be performed. The first step is initialization. This stage involves a status analysis of the socio-economic system. Usually, a stakeholder analysis is performed, and the cooperation framework is described. Based on this knowledge, roles and institutions are identified in the following. Literature reviews or expert interviews are common tools (Bogner et al., 2014). Step 2 is called identification. Eventually, a finished IRM matrix can be applied empirically. Workshops or upscaling internet-based formats are suitable for this purpose. The research objective defines the method. In any case, the data can then be fed into the simulation in Step 4. An algorithm determines

[4] In this relation, Autopoiesis paraphrases the emergence of a living and adaptable system from the network of elements that compose it. Our present ideas of the future are to be distinguished from the ones of tomorrow. Analogously, today's optimal solution is likely not to last forever.

a solution based on specifications (e.g., optimization with respect to transaction costs). This solution can draw conclusions for the system in Step 5, allowing so-called white spots to be identified,, for instance (Heinrich and Kauschke, 2017). These roles are not occupied by institutions and thus block the emergence of a market institutionally. Furthermore, design and ordering principles can be derived from the solution of the model, which enables the solution to adapt to the real world.

3.1.3 Application

The goal of applying IRM in this paper is to give a vision for a new smart mobility regime. Above all, the technical-economic interaction of smart mobility is to be depicted. Aspects such as policy, society, and culture, which play a role in the modeling of the MLP, are considered through the initial status analysis. Moreover, the aim of the study is to perform the development phase of the IRM. Future research will be able to proceed from here with a given and validated model. The procedure constitutes the steps as presented in Figure 3.4.

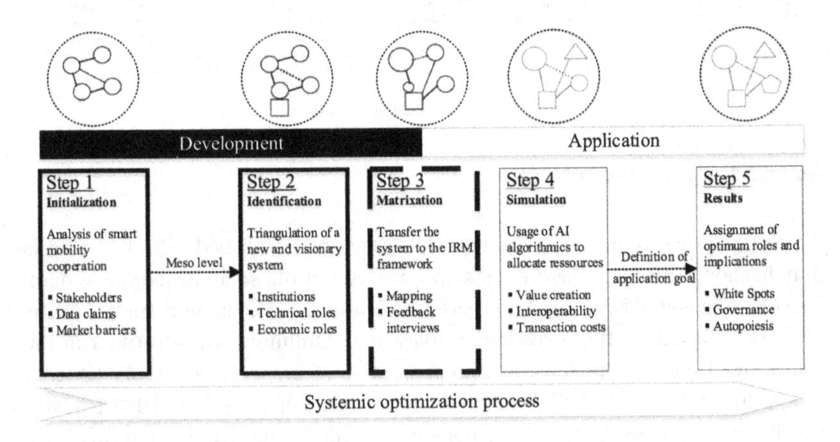

Figure 3.4 Application of the IRM to define a smart mobility model

The initialization serves to open up the subject area. Step 1 has already been vastly accomplished in Chapters 1 and 2. However, specific cooperation-related aspects require further elaboration. The nexus of cooperation is communication (Miller et al., 2002), which happens between actors in data systems. Therefore, the results of a stakeholder analysis of the digital mobility system are presented and future claims on data communication are formulated. This analysis yields an overview of possible market barriers in the existing regime.

In the identification phase, an IRM conceptualization of smart mobility is undertaken. The vital roles and institutions are derived from three sources: (a) prior applications of the IRM, (b) state-of-the-art data ecosystems, and (c) alternative approaches to understanding the concept of smart mobility. Not least from these studies, it becomes evident that it is sensible to converge the broad topic of smart mobility on a meso level first. From that level, more granular subsystems and use cases may be explored. In Step 3, the IRM framework is set up, and a temporal dimension is added. Finally, the experts from different disciplines discuss the results in feedback interviews. After validation, the significance of the model in the context of the MLP will be elicited.

3.2 Model Development

3.2.1 Initialization

The system of automobility is infrastructure- and above all vehicle-oriented. In contrast, smart systems are digital twins of such analog concepts and the processes occurring within them (Bhatti et al., 2021). Thus, nowadays, it is common to distinguish infrastructure and data ecosystems in substance. The question is how the world of mobility can become an authentic and integral space by interweaving digital and material systems. This is the point at which new types of interfaces and innovative communication formats arise. These projects use standardized access, adaptive services, and decentralized data exchange to create a hyperplane link between the two ecosystems. A flowing network of transport infrastructure and data system is established for smart mobility. This enables the renewal of business models in the sense of servitization. The value proposition is the shift from selling a mere product to a value-adding service. Ergo, businesses may no longer sell means of transport but a great variety of transport services orchestrated through a ubiquitous connection. Fernando et al. (2020) calculated a significant potential of servitization in mobility for efficiency and environment.

3.2.1.1 Stakeholder Analysis

Approaches such as the one presented above represent the starting point of IRM initialization. In order to supply an idea of the system under consideration, a stakeholder analysis was conducted. For this purpose, the efforts of three essays (Smith et al., 2018; Lyons et al., 2019; Qiuchen et al., 2021) were combined in a visualization. Next, the concept was consensually assessed by experienced technical and economic researchers in an online workshop. Thus, minor changes have been incorporated. The outcomes of this stakeholder analysis process are visualized in Figure 3.5. The two ecosystems are recognizable first of all. Within these spheres, the most important stakeholders are grouped by way of illustration. For clarity, a more refined granular classification has been omitted.

In the data system, existing data spaces (e.g., proprietary clouds or marketplaces) and data exchange systems (e.g., protocols and message types) were first identified for this purpose. Service providers include all those stakeholders who offer digital mobility services. The providers of trust (e.g., certification and standardization) retain a unique role, as this is the foundation of modern IT. Confidence and rules are the only way to ensure that data are secure and that social achievements such as civil rights and copyright are also reflected in the digital world. Finally, software providers support data systems or app development.

At the infrastructure level, such a support function is provided, for instance, by the providers of any kind of hardware (e.g., smartphones or spare parts). In general, the smart mobility infrastructure consists of transport structures, network structures, and energy structures. Vehicles move within this space. Thus, the classification of stakeholders is straightforward. Vehicle providers are dominantly OEMs, rental companies, or players in the second-hand market. Network players are network operators and network technology providers. Stakeholders from road or rail construction, operation and maintenance constitute transport structures.

Finally, the core of the smart mobility stakeholder space is formed by possible new groups as proposed by Smith et al. (2018), which also reflects Lyons et al. (2019)'s thoughts about a mobility intermediary. Consequently, smart mobility can solely function if two new instances are interposed between the transport provider (e.g., public transport) and the consumer. These entities are the integrator and the operator of smart mobility. Integrators offer multiple transportation providers to operators through technical integration, contracting, and financial clearing. Afterward, operators compose and offer services to end users from the

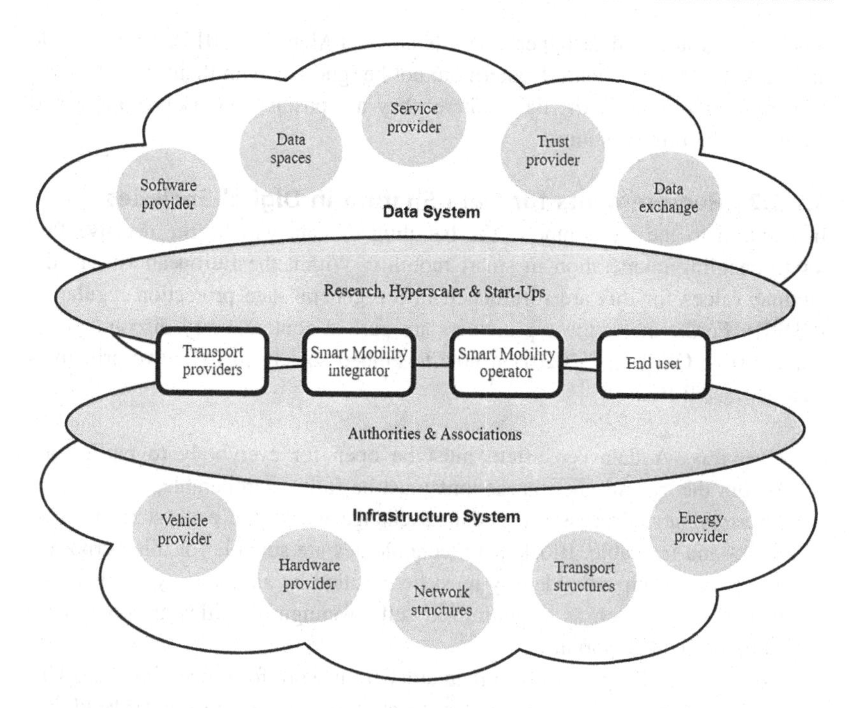

Figure 3.5 The stakeholder space of smart mobility

integrated data to offer a seamless smart mobility experience. Additionally, they provide transaction and informational services.

According to the present analysis, other stakeholders play a decisive role in the sphere of influence of smart mobility, although they are somewhat outside the data and infrastructure sectors. Examples of these stakeholders are public administration and interest groups, often organized in associations. Smart mobility is primarily a public responsibility (Docherty et al., 2018). In addition, the integration of mobility interests was identified as pivotal in succeeding in the context of this work. Therefore, these two stakeholders were placed directly next to the model core and more specifically below. Evaluators agreed that the relationship between administration and associations to the infrastructure is stronger than the data system. The last stakeholder group is just as important as the previous one and is situated at the model core. However, it has a more vital link to the data sphere. This is the place where research institutes or innovative startups that want to develop and implement new ideas traditionally congregate. The innovative group of data hyperscalers (Chui and Manyika, 2015), from Google and Tesla to Alibaba should furthermore not be ignored. With their market power and their know-how of the data sphere, they are pushing themselves more and more into the infrastructure.

3.2.1.2 Requirements for Data Sharing in Digital Societies

In addition to the environment, the handling of data will be the decisive factor in enabling cooperation in smart mobility. Within the European Union, the guiding values for this are formulated in the general data protection regulation (GDPR). Furthermore, these principles are part of contemporary discourse (e.g. Pohle, 2018; Qiu et al., 2020) on how to embed society in IoT. Five principles were extracted from it:

(1) *Openness:* A data ecosystem must be open for everybody to participate. Within the system, the transactions must be fully interoperable.
(2) *Transparency:* Processes and data exchange within the ecosystem must be track- and traceable. Blockchain technologies are suitable for this purpose.
(3) *Trust:* A participant's identity must be certified by an authority. In this context, concepts such as decentralized self-sovereign identities are considered to be of great importance.
(4) *Compliance:* There must be a regulatory framework for processing data. This includes, for example, the manifest location of servers and standardized data connectors.
(5) *Sovereignty:* The rights of the data must remain with the author unless otherwise agreed. Thus, data can exclusively be utilized for a defined purpose or a controllable duration of use.

3.2.1.3 Market and Cooperation Barriers

The available analyses show that there are currently numerous market entry barriers to smart mobility. Typical for oligopolies such as the automotive industry, these barriers are high and diverse. Moreover, they are likely to be co-responsible for the fact that many niche technologies have not yet been able to establish themselves. The IRM model accordingly intends to provide a solution basis for addressing the following barriers:

(1) *Structural:* The current automobile economy has achieved significant economies of scale, has built large supply networks and thus created interdependencies. These strengthen the regime and render it resistant to change.

Setup and development costs for new players and ideas are high, while OEMs themselves spend significant amounts on R&D. A fair ecosystem can break down structural barriers by being equally open to everyone. In order to create incentives for the participation of oligopolists, an added value for them must be specified. Such value may be found in the uptake of research joint-ventures, which are eligible tools for establishing emerging markets (Schulz et al., 2014). Furthermore, access must be free of cost, which requires public funding.

(2) *Strategic:* Automobile industrialists work to maintain their profits by deliberately buying out competing approaches and expanding their business model. Additionally, new firms are hindered from entering the market due to non-transparent background contracting. A role-based ecosystem may prohibit the establishment of such artificial barriers by design as market actors are replaceable as long as they fill out a designated role. Furthermore, smart contracts will foster transparency in business processes, making it easier for newcomers to understand and trust business mechanisms (Zheng et al., 2020).

(3) *Institutional*: Regulation can be a market barrier. Kosi and Bojnec (2013) found econometric evidence that institutional freedom[5] is even more beneficial to business entry rate than economic freedom.[6] Institutional freedoms found most beneficial for business entry were product freedom, property rights freedom and freedom from corruption. This suggests that a smart mobility ecosystem that accounts for data sharing requirements in the digital societies above will likely also reduce institutional barriers. One could continue to assume that self-organization and self-identification will, in the sense of new institutional economists such as Coase (1998), achieve this by lowering cooperation transaction costs.

3.2.2 Identification

In this section, I (1) review five prior applications of the IRM in the context of smart mobility, (2) derive roles and institutions from state-of-the-art cooperation architectures, and thus (3) formulate a unified model that integrates elements across the different concepts.

[5] Related to freedom from regulation.

[6] Related to the rate of net business formations.

3.2.2.1 Prior Applications

The IRM has a history of being applied in the environment of intelligent transport systems and dynamic mobility cooperation. Its aim is to enable the market introduction of co-operative driving systems by recognizing that technologically viable solutions are not necessarily economically successful (Schulz et al., 2019). Particularly noteworthy in this context are the collaborative German research projects CONVERGE, NoLimITS, and 5GNetMobil. In addition, the C-Mobile and 5GCroCo research initiatives are of interest at a European level.[7] These projects will be briefly outlined, and in particular, the institutional role models will be presented.

CONVERGE

The CONVERGE project was the first to test a decentralized service architecture for applications in the ITS environment (Wieker et al., 2014b). IRM played a key role in this, as it was the prerequisite for orchestrating the interests of participants and developing architecture in line with the requirements of a future operator model. Schneider (1995)'s traditional roles were employed as economic roles and adapted to a data context. They read as follows (originals in brackets):

- Business management
- Data preparation (Sales)
- Data gathering (Procurement)
- CONVERGE Services (Production)
- Human resources
- Financial management
- Controlling

In order to depict the ecosystem, the standard institutions of Hoddeson and Daitch (2002) were extended by the institutions of *state*, *enterprise*, and *university*. The technical roles were derived from the architecture and consisted of:

- System management
- System operation
- Service usage
- Policy framework
- Regulatory framework

[7] Please see https://fgvt.htwsaar.de/site/en/projects/ for further information on project partners, content and funding.

Presented roles are to be understood as meta roles, under which further necessary sub-roles (e.g., the certification authority as part of the system management or the C2X initialization body as part of the policy framework) could be aggregated. Within the IRM matrix, the four classic product life phases (Grether and Dean, 1952) were transferred to service life phases. The resultant model was empirically tested in expert workshops. Results suggested that CONVERGE was institutionally viable while realizing significant reductions in transaction costs (Schulz and Geis, 2016).

NoLimITS

The CONVERGE IRM approach was subsequently also transferred to the NoLimITS project, which demonstrated the function of a decentral architecture for electric vehicle communication. The aim was to provide alternative access to proprietary charging structures. The role modeling of Heinrich and Kauschke (2017) thus employed the following institutions: *state*, *municipalities*, *research*, *transport*, *service providers*, and *OEMs*. At the same time, the roles were further developed and regrouped according to application. As a result, the economic meta roles that emerged were:

- System leadership
- Quality management
- Marketing and sales
- Customer relations

The technical roles involved:

- System strategy
- System management
- System operation
- System provider

The IRM matrix was then developed using an online survey of project members representing the system network. The matching of roles and institutions resulted in an institutionally feasible market. Within this market, however, role conflicts can arise, for example, in system leadership. In addition, there is a latent danger that the network will be dominated by the institution of the OEMs, thus endangering the decentralized idea.

5GNetMobil
As part of the smart mobility project 5GNetMobil, an architecture for tactile automated driving was developed. The requirements for real-time systems of this type are high. In order to also offer an economic perspective, IRM was used once again. The aim was to determine whether there is sufficient willingness to renew new technical and economic roles in the market introduction phase. One particular interest was to balance interests between the telecommunications and automotive sectors, both in hardware and services. The economic roles combined the perspective of economics and business, suggesting that a tactile ecosystem requires:

- System leadership
- Service offer
- Knowledge and transfer
- Production and procurement
- Finance and support
- Operational management

The technical roles were again derived from the architecture and constituted similar elements to those of NoLimITS:

- System strategy
- System management
- System operation
- System user

The roles were allocated to the institutions by the consortium partners (each with their own and others' perspectives) in an online survey. For evaluation, a weighting vector was introduced that considered the industry's market power compared to SMEs. The market accordingly proved viable, even without the involvement of state institutions. Automotive and telecommunication industries were found to compete in terms of establishing and running platform solutions. Eventually, IRM evidenced that no single actor would be able to set up and operate the system on its own (Sanmateu, 2018).

Miscellaneous
Other projects have built on the ideas of the economic-institutional perspective of IRM. However, the model was used as a baseline theory rather than an applicable

tooling. Thus, the iKoPA project (Wieker, 2017) merely described the standard economic roles and additionally identified the following technical roles:

- Governance
- Backend
- Mobile station
- Communication network

The DECREE mobility data project (BMDV, 2021), on the other hand, was less parsimonious in describing new roles. Thus, 29 meta roles were identified for IRM processing. However, the use of a digital tool made these roles persistent. Publication of findings is pending.

At the European level, IRM has, inter alia, found its way into the ITS projects C-Mobile (Turetken, 2020) and 5GCroCo (Everingham, 2019). Accordingly, the architectures have been calibrated economically and technically. In the deployment of the IRM, however, the approach was adopted more in the sense of a stakeholder analysis. The following institutions were assigned to specific use cases in C-Mobile: citizens, businesses, service providers, consultancy, and government. In 5GCroCo, institutions were 5G industry, academia, road infrastructure operators, regulators, automotive industry, standard development organizations, and insurance companies.

In conclusion, it can be noted that a plethora of roles and institutions exist in previous IRM implementations. A smart mobility role model is likely to benefit from addressing these. However, inconsistencies can be discovered. For instance, the service provider is sometimes an institution, sometimes an economic, and sometimes a technical role. Therefore, a readjustment based on the existing applications seems to be warranted.

3.2.2.2 New Roles

Researchers outside the scope of IRM have also been thinking about roles in smart mobility. The levels of consideration are primarily of a technical nature. What these approaches have in common is that they are framed almost exclusively in terms of service. Transition is implied by the fact that mobility is becoming fully integrated, and technology is no longer a product but a means of service. Data are networked in a standardized way and become independent of assets. This means that business models no longer function in isolation but in cooperation. In this context, sustainability assessment is changing from a measurable to a controllable variable of an ecosystem. Flügge (2016) postulates the following seven IoS-roles:

- Service aggregator
- Service gateway
- Service hosting
- Service provider
- Service broker
- Service consumer
- Service channel maker

The European hypercloud project Gaia-X follows a similar philosophy. Gaia-X envisages a fully networked data infrastructure. This means that centralized and decentralized infrastructures will be networked into one. The aim is to create a trustworthy and user-friendly open-source network of cloud providers and their offerings. This bears maximal transparency for all institutions. Theoretically, anyone can become a part of the project (Gaia-X, 2022). The ecosystem is based on the three levels infrastructure, data, and federation. The decisive federation level consists of five central services that enable the framework for action, the fundamental law of cooperation, so to speak:

- Identity and trust
- Federated catalog
- Data sovereignty services
- Compliance
- Portal and integration

The economic dimension of service ecosystems, for example, has hardly been captured in research outside the scope of the IRM to date. However, a recent study by Xue et al. (2020) provides an innovative approach. The authors note that today's service systems are slowly converging with natural systems in their complexity and collaborative interactions. Therefore, they transfer the idea of entropy from ecology or biochemistry to economics. Entropy is a measure of the disorder in a system and the associated possibility of arrangement of the particles in a system. The value entropy model created in Xue et al. (2020)'s research argues that service ecosystems are highly dynamic value generation tools. The system inputs are customer value demands with negative entropy. The system output is of positive entropy. It consists of value creation for customers, reduction of operating costs, and uncertainty for system operators and managers. From this perspective, four economic roles may be recognized:

- System organization
- Service demand
- Value creation
- System management

3.2.2.3 Unification for Smart Mobility

Technical and economic roles

Inspired by prior applications and current developments, I theorized a set of meso and meta roles, which were supposed to absorb existing roles. Incorporating the meso level became imperative since data ecosystem and economy-of-things (EoT) approaches require an orchestrating entity yet must function organically and self-regulatory. Functional contradictions were to be eliminated by reviewing existing role definitions in the literature. The roles were also matched to the mobility context. Where deemed necessary, roles that had not yet been classified were subdivided into either economic or technical roles. The distinction was drawn based on the range of competencies required to fulfill the role (more technical or more economic). Hence, I arrived at four technical as well as four economic meso roles. These roles are based on the Gaia-X triad of data, infrastructure, and system levels.

The technical system core initially depends on the form of the organization. In addition, the participants in the decentralized network must agree on rules and governance maxims on the part of the economy. One echelon below this, system management takes place, which is realized technically by core services. Yet in turn, these require anchoring in management processes that are actually necessary for the economic environment, such as portfolio development, marketing, and customer care. The two-part system level on the technical and economic side can be inferred from the tables below (Table 3.1).

Table 3.1 Technical roles for smart mobility

Role	Description	Assigned references
System core	The allocation of fundamental organizational roles specifies the decentralized nature of the system network. The meso role defines the setup.	System organization (Xue et al., 2020); System strategy (Heinrich and Kauschke, 2017); Initialization Body (Fünfrocken et al., 2021)
System services	System participants benefit from a basic set of generic services that meet the state-of-the-art requirements for data treatment.	System management & operation (e.g. Sanmateu, 2018); CONVERGE Services (Fünfrocken et al., 2021); Federation services (Gaia-X, 2022)
Data tethering	This role contains the functional tasks involved in integrating the data ecosystem into the smart mobility system network.	Mobility platform operator (Heinrich and Kauschke, 2017); Service gateway (Flügge, 2016); Data connector (Gaia-X, 2022)
Infrastructure tethering	In order to facilitate intelligent mobility, relevant infrastructure systems must be sustainably connected to the network. The result is IoT.	Communication network (Wieker, 2017); Service hosting (Flügge, 2016), High-performance computing (Gaia-X, 2022)

In addition to the system level, the data level is significant. This is where generic services are developed into customer-relevant services and are offered on the decentralized market. Due to the high degree of heterogeneity, this mobility market must be connected in a technically stable manner and, on the economic side, serve all possible business models within smart mobility. On the infrastructure level, all entities must be smart and compatible with the system. Self-sovereign identities may be an integral part of such an undertaking (Rathee and Singh, 2022). Economically, business models based on the production of commodities (and their respective supply and value chains) must be accompanied by digital transformation (Table 3.2).

Table 3.2 Economic roles for smart mobility

Role	Description	Assigned references
System governance	The decentralized market is managed by agreeing on a set of rules and policies. These can be assigned and organized, e.g., by legal theme. If, for example, it is specified that decentralized digital identities must be used, the IoT becomes EoT (economy of things).	Business management (Fünfrocken et al., 2021); System leadership (Heinrich and Kauschke, 2017); Corporate governance (Wieker, 2017)
System value creation	Economic management of the system is required to meet the demand for services while value is created in terms of cost reduction and interoperability. Often this role is taken over by a standardization organization.	Value creation (Xue et al., 2020), Gaia-X association (Gaia-X, 2022); Sales, Marketing & Finance (Schneider, 1995); Quality management & customer relations (Heinrich and Kauschke, 2017)
Data economy	The monetization of data must be customer-oriented and preserve copyright, as it were. Both existing and new business models must thus be enabled in the system network if participation is to be incentivized. Decentral networks live on active participation.	Service demand (Xue et al., 2020); Service broker & channel maker (Flügge, 2016); Service offer (Sanmateu, 2018); Data owner (Gaia-X, 2022); Data gathering and preparation (Schulz and Geis, 2016);
Infrastructure economy	Existing value creation systems do not allow themselves to be organized well in platforms due to their heterogeneous nature and the latent danger of monopolization. Therefore, efforts must be made to ensure that existing classic product-based systems can be attractively mapped in the system network.	Production & procurement (Schneider, 1995); Support (Heinrich and Kauschke, 2017)

Institutions of smart mobility

Similar to consolidating roles, previously employed institutions and stakeholders were reviewed. In order to achieve harmonization, institutions were combined and adjusted to the required level of abstraction. As seen in Table 3.3, eight institutions are considered in the smart mobility ecosystem. On the market side, these are hyperscalers such as Google, Amazon or Alibaba. These operate their

Table 3.3 Institutions of smart mobility

Institution	Definition	Assigned references
Market		
Hyperscaler	Service-orientated internet businesses with strong international market power	Internet businesses (Sanmateu, 2018), Technology (Hoddeson and Daitch, 2002)
OEMs	Production-oriented businesses with strong international market power	Automotive industry (Sanmateu, 2018); OEM (Heinrich and Kauschke, 2017)
Telcos	Service-orientated communication businesses with international market power.	Telco (Sanmateu, 2018); 5G industry (Everingham, 2019)
Digitals	Service-orientated businesses promoting niche technologies with little market power.	SME (Sanmateu, 2018); Technology (Hoddeson and Daitch, 2002); Consultancy (Turetken, 2020)
Society	The assembly of citizens, norms, and culture demanding mobility.	Citizens (Turetken, 2020) , Culture (Geels, 2004)
Standards	Organizations that develop, curate, and manage standards.	R&D (Hoddeson and Daitch, 2002), Standard development (Everingham, 2019)
Academia	The community concerned with the pursuit of research, education, and scholarship.	R&D; Science; University (e.g. Schulz and Geis, 2016)
Transport	State businesses and organizations that operate transport and provide mobility.	Transport (Heinrich and Kauschke, 2017), Road infrastructure operator (Everingham, 2019)
Governance	The assembly of institutions that exercises executive, judicial, and legislative power.	Government/Municipalities (Heinrich and Kauschke, 2017); Regulators (Everingham, 2019), Policy (Geels, 2004)
State		

own ecosystems and could potentially be important or dangerous players in a decentralized network. Then follow OEMs, which are also immensely powerful on the market and typically comprise the oligopolies of vehicle manufacturers and IT equipment suppliers. Next, Telcos include telecommunication providers and mobile network operators. The institution of digitals has less market power and contains technology specialists, often smaller companies or start-ups.

At the center between market and state in Table 3.3 is society. Society's most crucial actor for smart mobility is the user, with their cultural characteristics and preferences. Structured slightly more in the direction of statehood, standardization organizations follow, which are becoming more and more important in the development of ecosystems. The same is true of the academic research and scientific community, which provides the foundations for new systems. Moreover, the administration, including government at all levels, must be treated as an institution. This is because it, too, can have not only a moderating-controlling function but also a participatory one in complex systems. Finally, one partly governmental, partly private entity remains to be considered for smart mobility: the current private or public mobility providers beyond individual ownership.

3.2.3 Matrixation

The IRM matrix is composed of a total of four interacting dimensions. First, on the y-axis, it consists of (1) a role-specific dimension in which the elicited eight meso roles and 36 functional meta roles that are decisive for ensuring the project's success are depicted. Second, institutions (2) are replicated on the x-axis according to a temporal dimension (3), which defines the period for which the IRM matrix is valid. In theory, four market phases are distinguished. However, since the scope of this dissertation is the development, emergence, and stable operation of smart mobility, the phase of decline and market saturation is omitted. Instead, an early conceptualization by Rogers (1962) considers three market phases, which essentially identifies the niche and mass-market phase in addition to development. Finally, the fourth dimension is mapped on the right side of the y-axis (4). It relates to operational role performance, i.e., matching roles and institutions based on the quality criteria of interest, ability, and trust. A possible evaluation via self-assessment and external assessment can provide insight into how the institutions perceive each other. Figure 3.6 includes all results. The associated meta roles originate from the assigned references.

	Meso	Meta	Development									Early market									Mass market								
Technical roles	System core	System federator																											
		System participant																											
		Third party																											
	System services	Access and integration system																											
		Data souvereignity system																											
		Service catalogue																											
		Identity and trust system																											
		Compliance system																											
	Data tethering	Data connector																											
		Data space manager																											
		Service prosumer																											
		Support servicing (e.g. AI, ML, BD)																											
		Mobility cloud platforming																											
	Infrastructure tethering	Hosting																											
		Communication networking																											
		Traffic control system																											
		High performance computing																											
		Edge system																											
Institutions			A	D	G	H	O	So	St	Te	Tr	A	D	G	H	O	So	St	Te	Tr	A	D	G	H	O	So	St	Te	Tr
Economic roles	System governance	Responsibility manager																											
		Standards manager																											
		Contract manager																											
		Policy manager																											
	System value creation	Portfolio developer																											
		Marketing manager																											
		Financial officer																											
		Participant relations caretaker																											
		Quality manager																											
	Data economy	Data aggregator																											
		Service broker																											
		Data service owner																											
		Monetarization provider																											
		Service channel maker																											
	Infrastructure economy	Decentral ressource manager																											
		Infrastructure producer																											
		Infrastructure operator																											
		Infrastructure support provider																											
		Mobility control center																											

Caption	
A	Academia
D	Digitals
G	Governance
H	Hyperscaler
O	OEMs
So	Society
St	Standards
Te	Telcos
Tr	Transport

Figure 3.6 IRM matrix for smart mobility

3.3 Feedback Interviews

In order to validate and enrich the IRM vision, expert interviews were the method of choice. This qualitative method is particularly suitable for collecting and evaluating non-standardized data in rather unknown research areas. Gläser and Laudel (2009) note that the procedure usually involves a small, non-representative sample to obtain knowledge saturation in a specific area. This entails carrying out standardized interviews with open questions until no more new knowledge can be gained. Such a threshold is often reached expeditiously if the selection of experts is effective.

3.3.1 Data Collection

Data were collected from April to May 2022. A total of eight experts were interviewed via web conferencing. The experts originated from academia (n = 5), business (n = 2) and administration (n = 1). A requirement was that they had previous experience with either smart mobility, service ecosystems, or IRM. All experts first received an easy image-based briefing on the research to sensitize them to the topic. Following this, they were asked a standardized set of questions. Such guided interviews are characterized by the fact that an interview guide pre-structures the conversation but allows for flexible adaptation to the interview situation (Lamnek and Krell, 2005). The interview guideline employed was rather short and consisted of two parts. In the first, three open-ended questions were formulated:

- Which roles must be filled for smart mobility to function technically?
- Which roles must be filled for smart mobility to prosper economically?
- Which institutions are decisive for smart mobility?

The second part was of a confirmatory nature. Experts were confronted with the developed IRM in Figure 3.6. The experts' creative reactions and comments were captured in the sense of a traditional brainstorming session (Al-Samarraie and Hurmuzan, 2018). Therefore, the interviewer asked:

- What do you think about the model? What do you like or dislike about it?

The experts' responses were subsequently recorded in written and paraphrased form by the interviewer, as recently suggested by Döringer (2021). On average, an interview lasted no longer than 35 minutes.

3.3.2 Data Evaluation

One common and intersubjectively verifiable text analysis method is the qualitative content analysis by Mayring (2015). As such, no individual interviews were conducted rather all interviews were assessed in an integrated manner. Following this methodology, the paraphrases of the first interview part were assigned deductively, i.e., according to prepared categories. In the context of the present research, these codes refer to *technical roles*, *economic roles*, and *institutions*. The second part, involving recording unprompted reactions, was coded inductively, i.e., spontaneously from the material. The categories that emerged were *positive*, *constructive*, and *other comments*. In the next step, data were submitted to a second researcher who performed the coding again. These two independent outcomes allow for reliability analysis of the findings in SPSS. Cohen's Kappa was 0.904, $p < 0.001$. According to Landis and Koch (1977), this implies an almost perfectly reliable assessment of the two evaluators.

3.3.3 Results

Notwithstanding this testing, a quantitative analysis of the coding table was performed to interpret the data numerically. The outcomes of this analysis are now presented according to the six code categories. Parallelly, some qualitative highlight results and citations are featured.

The aggregation of the interviews resulted in a total of 129 paraphrases, each of which was assigned to the six codes. In addition, the second level of coding was pulled in to granulate the paraphrases within the first coding lines. This will be presented in the upcoming paragraphs.

As in Figure 3.7, the interviewees mentioned 33 institutions. Furthermore, it is interesting to note that the experts see as many economic roles as technical roles, 25 respectively. Thus, the first part of the interview served its purpose well. Within the feedback part, the majority of participants were positive about the developed model, and no negative comment was captured. Nonetheless, there were also 16 constructive comments. These contained the next methodological steps and opportunities for improvement. In addition, the participants began

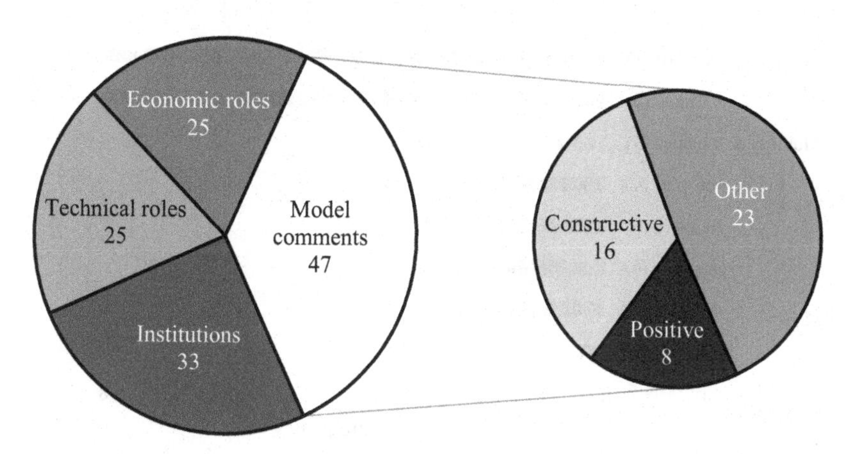

Figure 3.7 Quantitative composition of the coded paraphrases

engaging in conversation in the brainstorming part so that a further 23 free and unstructured comments could be documented. The focus was on both the model and its relevance, as well as personal opinions on the transition to smart mobility.

3.3.3.1 Institutions

The 33 cited institutions allowed for clustering them into eight categories. As can be observed in Figure 3.8, the experts' institutions are very close to the institutions developed in this work.[8] Only the *Mobility & logistics* category stands out. While it falls under *Transport* in the conceptualization of this work, the experts may see it as important to list it separately. Furthermore, the frequency of occurrence can be regarded as an initial indicator for assessing the importance of institutions. Accordingly, technology companies and the state appear to be the most critical parties responsible for advancing smart mobility.

The institutional concept of IRM includes current and future stakeholders. But by controlling results, we learn that the experts mainly refer to institutions that already exist today. At most, the presence of the hyperscalers, i.e., the large internet companies, represents an innovative element. Fittingly, experts state: '*Most of the current institutions of mobility will remain.*' (Vogt, 2022) and '*Larger IT companies naturally scent profits in such a data-driven environment.*' (Geissler, 2022).

[8] See the institutions of smart mobility in Table 3.3, Section 3.1.2.

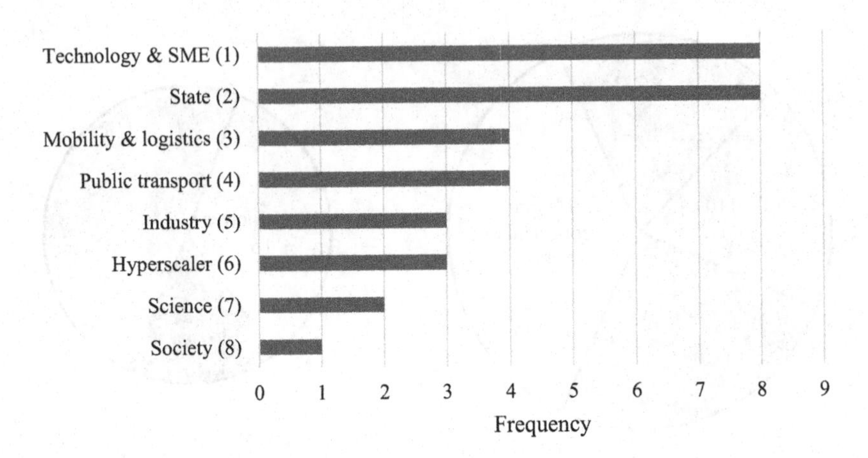

Figure 3.8 Ranked institutions by category and frequency

Hence, it is worth looking at the important categories. First, who is *the state*? Four of the experts generically speak of the government or the state per se. The other four see specific actors of the state as essential institutions, i.e., federal governments, municipalities, lobbyists, and regulatory authorities.

Second, who are the *technology* SMEs that smart mobility should be counting on? Again, in three cases, these SMEs are referred to rather generally. However, the experts offer two specifications. The first concerns the so-called hidden champions[9], who, in the view of the experts, should be able to deliver more than they manage to at the moment. The second refers to the SME data business branch that might expand once mobility becomes smart. Many specialized jobs should offer opportunities from data refinement and data trading to app provision.

3.3.3.2 Technical Roles

'Systems of means of transport must become interconnected.' (Adorff, 2022). This quote sums up the technical goal of smart mobility well. To achieve this, the experts mentioned 25 technical roles at different levels of abstractions. Thus, I categorized them toward data and infrastructure areas in the first place. After these results remained quite heterogeneous at the technical data layer, another tier was added, which distinguished between *data system*, *data services*, and *superior system*. Figure 3.9 displays the consequential distribution of technical roles.

[9] Hidden champions are medium-sized companies that have become world market leaders in niche segments.

In the area of *infrastructure systems*, communication networks are needed that provide links via mobile internet and digital infrastructures. Another prerequisite is overall connectivity in (stupid) hardware, which is mentioned as a technical role. In addition to networks, users and means of transport fall into the domain of manifest infrastructures, according to the interviewees.

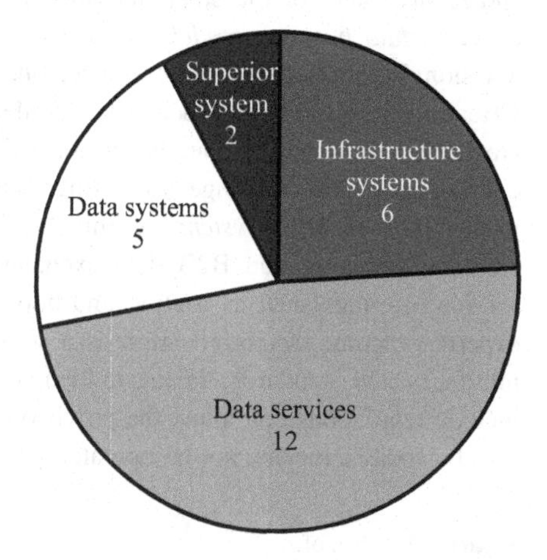

Figure 3.9 No. of technical roles by category

Five times, the respondents speak of *data systems* as the larger linking systems between reality and the digital world. In doing so, they are mainly referring to existing platforms and proprietary systems that collect data nowadays and generate information and services from it. However, the most commonly mentioned technical roles are related to specific *data services*. It may be that a higher-level system such as smart mobility can be better described by the sum of its parts. Appropriately enough, Wanke (2022) posits developing bottom-up: *'We must find and develop highly relevant use cases.'* Whatever the case, *data services*, for instance, include trust and identity management, booking and transaction processing, and interface provision. Furthermore, one expert claims that future key technologies such as AI, IoT, or distributed ledger technologies must be embedded in architectures to get the most out of smart mobility.

Eventually, two experts explicitly mention the need for a *superior system* connecting the infrastructure and the data world. Such a solution should connect platforms and incorporate a governance-standards for data exchange.

3.3.3.3 Economic Roles

Although Wieker (2022) ascertains in his interview that *'the economy of smart mobility is hard to predict'*, experts found 25 roles. Analogously to the technical roles, I was able to classify them into two categories: *ecosystem* and *data.* The *ecosystem* category deals with roles related to two sub-categories: the *marketing* and *management* of the smart mobility system itself. The economic *data* roles concern *data business models, data markets,* and *data governance.* From this, a vision becomes perceptible. Hence, one could imagine smart mobility as a virtual marketplace with a special set of rules within which various forms of value creation or transactions can be performed. In this context, Geilenberg (2022) cites an economic challenge suggesting: *'The own business case must be mirrored to the smart mobility system.'* Within the interviews, eight data business cases are featured, inter alia B2B data exchange, use-based payments or integrated service offerings such as routing and booking. In the field of *data markets,* the experts welcome decentralization as a solution approach for integrating diverse interfaces and standards. In the field of *data governance,* Vogt (2022) calls to include regulations that allow the provision of smart mobility, even if the regions can, by today's means, not be operated profitably (Figure 3.10).

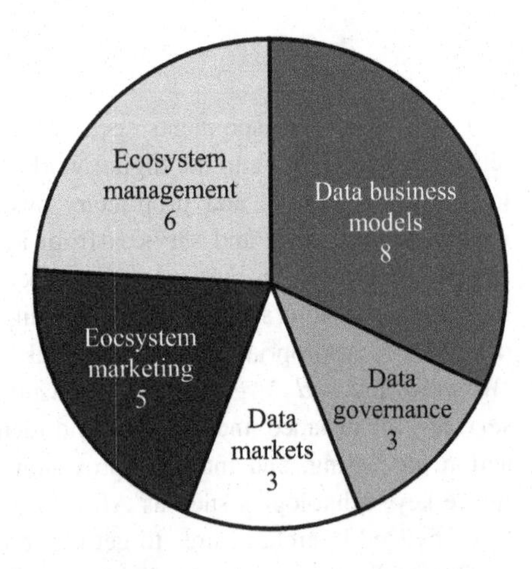

Figure 3.10 No. of economic roles by category

Concerning the ecosystem's marketing, experts bring up roles about collective funding, municipality consulting, and developing attractive incentives for system participation. At this point, it becomes obvious that the system itself requires business management. This encompasses *ecosystem management.* In his interview, Schulz (2022) raises a pertinent thought. According to him, a bundle of roles should deal exclusively with transaction costs. Thus, from an economic view, those system spots should be identified, at which the potential for lowering transaction costs is particularly large. After that, other functional roles could be used to evaluate and manage them. As a result, the system would unfold its optimal benefits.

3.3.3.4 Feedback

When visually confronted with the model[10], experts gave positive (n = 8) as well as constructive feedback (n = 16). No negative feedback was recorded. This does, to some extent, indicate that the model exerts its purpose of being meaningful for experts.

On the positive side, the interviewees especially valued the harmony and symmetry of the model, with its four key roles mirrored in the technical and economic parts. Engineer Vogt (2022) welcomes the equivalence: '*Economic roles eventually get a reference system.*' Furthermore, the layout and the introduction of meso roles were acknowledged. Wieker (2022) states: '*The approach of thinking like a computer operating system is very promising. Earlier approaches like CONVERGE had linked systems directly.*' Another positive response is that the conception is very useful for developing future architectures and policies (Geissler, 2022).

Constructive feedback generally concerned four issues: formal errors (n = 4), design improvements (n = 2), propositions of extensions (n = 4), and reflections on the next steps (n = 6). Formal comments were received, for example, noting that compliance and governance should be more clearly separated and that the institution of the *digitals* is still too vague. Concerning the design, Geilenberg (2022) suggested hiring a communication agency to illustrate the matrix more comprehensibly. Adorff (2022) proposed: '*The IRM matrix could be transposed, so that the technical-economic correspondences are better recognized.*' Furthermore, there were some suggestions to extend the model. Schulz (2022) and Wieker (2022), for example, claim that the roles of governance and transaction costs must be strengthened. Adorff (2022) continued proposing that the model might also contain a political or cultural dimension. Meanwhile, Geilenberg (2022) wondered whether mixed techno-economic roles might be useful.

Finally, experts put forward their outlook on which next steps might be undertaken with smart mobility IRM. Four experts would like to establish definitions for the functional meta roles and validate them via use cases. Dorlöchter (2022) recommends discussing whether the model features a central or a decentral internet approach as it contains both elements in its current form. This is yet another

[10] See Figure 3.6. *IRM matrix for smart mobility*, Section 3.2.3.

reason why he and Vogt (2022) consider it advisable to convert the model into an architecture that can be tested in practice.

3.3.3.5 Comments

During semi-standardized interviews, it is inevitable that issues beyond the scope of the guideline will arise. Fragments resulting from these free conversations were therefore collected and processed identically to the previous paraphrases.

After all, most comments were *recommendations for action* (n = 9). For instance, Geissler (2022) notes that a C-ITS directive is needed at the European level, but the EU member states have not yet been able to agree on one. Moreover, he suggests that the state *'may oblige to using certain systems'* to accelerate the development. Wieker (2022) adds that individual EU member states' legal systems still do not all allow for the same scope of smart mobility, which in turn represents a market barrier. Wanke (2022) comments on the overall development of overarching decentralized systems. In doing so, he notes that they have a communication problem with the public and that businesses and society do not understand the benefits. He therefore calls for appropriate information campaigns.

Concerning *challenges* for smart mobility, Vogt (2022) describes that the market creation of smart mobility requires balanced action. He explains a latent danger of overregulation that can result in non-competition and non-cooperation. This, in turn, would bring the system to a standstill right from the start. Nevertheless, state support remains vital. Hence, the greatest challenge for Vogt (2022) is finding a suitable instrument that promotes certain efforts on the one hand, whilst not discriminating against further market players on the other (Figure 3.11).

Figure 3.11 No. of comments by category

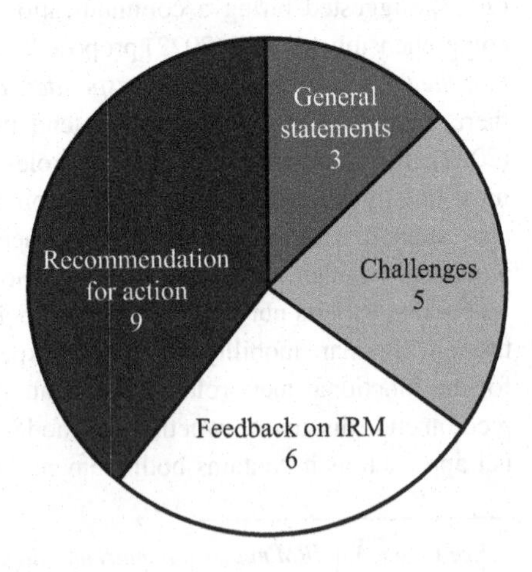

Finally, feedback on the IRM (n = 6) is provided by Geilenberg (2022) and Wieker (2022). They emphasize that the presented approach will reduce uncertainty and ensure marketability. They also wish to outline the next steps of how theorists and practitioners proceed with the model. Geilenberg (2022) stresses that the advantages of decentralism over classic operator models must be discussed.

3.4 Discussion

The findings are discussed in relation to targeted model validation, in relation to IRM theory and in relation to the overall transition to smart mobility.

3.4.1 Model Validation

Ultimately, the interviews confirm the IRM of smart mobility and realize some crucial advancements in roles, institutions, and design. Thus, I developed an alternative IRM, as illustrated in Figure 3.12. Some of the upgrades initiated by the experts have been implemented in this version. The next section will explain which ones and why.

First, I solved formal spelling and abbreviating issues related to *institutions*. Then I confronted experts' institutions with the institutions derived from literature. These matched fairly well. After I had initially combined *Mobility & logistics* and *Public transport* into *Transport*, solely experts declared the two as separate categories. In keeping the research efficient, I retained the original form. Some criticism was reported concerning the vagueness of *Digitals*. So, I refined the term to *Digital technology*.

Concerning *technical roles*, feedback was positive overall. Consequently, there was only minor need for amendment. I accordingly treated a formal support service issue that originally contained machine learning. However, the experts advised me that machine learning is a subtype of AI. Secondly, I introduced a new role to the system core. The role of *system designer* originates from interview observations. Hence, the question arose as to whether the IRM refers to a decentralized or centralized design. The answer is a decentralized design since that is why the meso level of the tripartite *system core* exists in the first place, by analogy with Gaia-X. Having said that, the model cannot help derive where this structure comes from. The *system designer*'s role is to define the structure of the system core (in this case, decentralized).

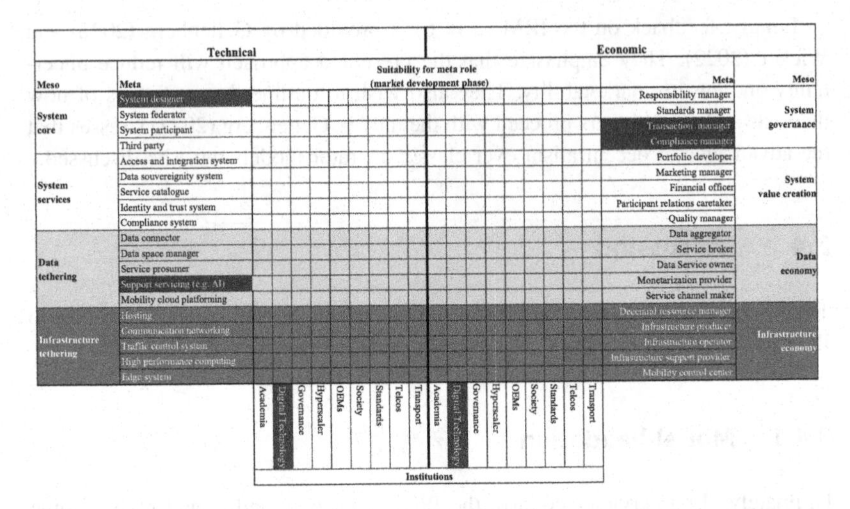

Figure 3.12 The improved IRM matrix for smart mobility (interview edits highlighted in black)

Economic roles caused the most controversy among interviewees. Here, the criticism centered around the issue of governance and compliance. On the one hand, the two should be better separated whilst on the other hand, the role of governance as a whole should be emphasized. In addition, Schulz (2022) elaborated on the vital role that transaction costs should play in the market development phase. So, I restructured the governance meta roles to account for all of this. The *transaction manager* and *compliance manager* were introduced as new roles. These rank one level above policies (compliance) and contracts (transactions)[11], so these two roles could be omitted.

The last step was to consider the commentators who expressed the wish for a better design and easier access, i.e., improved user experience. Consequently, I transposed parts of the matrix as suggested by Adorff (2022). Since the same number of roles existed economically and technically, this seemed appealing overall. In addition, I reduced the model complexity by focusing on one market phase only, namely the current one. These adaptations should help promote the model and its significance to a broader audience.

[11] See Figure 3.6. *IRM matrix for smart mobility*, Section 3.2.3.

3.4.2 Implications for IRM Theory

From a theoretical standpoint, three potential areas of development for IRM itself can be discussed through interview evaluation.[12] The first involves improving the model design, as discussed. In the future, the IRM matrix may become more application-oriented and flexible in scope. In this way, it is possible to dynamize roles according to market phases. As a result, some functions that are still important at the beginning of the technology introduction can lose their importance over time. Furthermore, flexible models can even potentially include completely new types of roles (e.g., cultural or social).

In any case, it will be necessary to continue developing the model. The experts in the present study agree that it is important to supplement the top-down approach of the smart mobility vision with a use case-based bottom-up convergence. Identifying the most promising use cases based on transaction costs is the key task for future research. Thus, the roles, which continue to be abstract today, can help solve concrete, practical problems of cooperation and reduce uncertainty. In theory, use case-based advances will likely lead to more sophisticated definitions of designated roles.

Thirdly, experts advise continuing with the application. This implies that the IRM process is not completed with matrix creation. As presented in Figure 3.3, Section 3.1.2, two more steps must follow. The aim is to collect as much valid empirical data as possible and feed it into the simulation, Step 4 of the IRM process. This can subsequently be used to determine whether the decentralized smart mobility market is viable. In addition, it can be helpful to pinpoint those roles that are not yet feasible today and that should consequently be addressed in the future through appropriate innovation policy measures.

3.4.3 Implications for Smart Mobility

Finally, if one employs the model as a visioning tool in MLP[13], possible influences on technology niches and the landscape come to light. The basis of this assumption is the recognition that IRM is leveraged through this approach, whereby architectures and policies can be developed as a result.

Communication architectures form the framework of modern ICT. In a digitizing world, their impact and required adaptability are constantly increasing.

---

[12] Please refer to Section 4.1.2 for a general discussion on IRM advances.

[13] See Figure 1.3. *Two views on smart mobility regime change within the MLP*, Section 1.3.3.

Communication networks in transport should ensure complete and bidirectional connectivity between intelligent tangible and intangible mobility resources, high reliability, and practicable interoperability (Marzal et al., 2018). Although developers are aware of these high requirements and technical solutions are available, many approaches fail to account for the economic reality. From the experts' point of view, it is, therefore, advisable to incorporate the economic and possibly social-cultural dimension of communication directly into developing new architectures. IRM can supply a basis for this. Henceforth, suppose a suitable smart mobility framework architecture were to emerge in a subsequent step. In that case, this could help enhance today's niche technologies in such a way that they could penetrate the regime and dock to its existing entities[14].

Furthermore, IRM can be exploited for another practical application: the development of funding and regulatory instruments. Once the IRM has defined the development goal and uncovered possible economic white spots, targeted measures and strategies can be developed from them (Schulz et al., 2021b). These can then be reflected at various governance levels. For example, results could be incorporated into the C-ITS directive at the European level to facilitate heterogeneous systems' convergence (Karkhanis et al., 2018). Nevertheless, the model could also be employed locally to evaluate the readiness for smart mobility in certain regions. After all, many experts think that smart mobility will initially develop within municipal transport associations (Wanke, 2022). Finally, policies and resulting projects may transform infrastructures and user requirements over time. Within the MLP, this would mean that the landscape and its pressures on the regime change.

[14] See Figure 3.3. *The five steps of IRM*, Section 3.1.2.

4 Conclusion

The starting point of this work was recognizing that the developments surrounding the smart mobility concept will strongly influence future transportation. As is being discussed by the transportation community, changes induced by ubiquitous connectivity or automation are very likely transformational. Within society, stakeholders and users have already begun facing an automobile regime that displays the first cracks. Uncertainty and ambiguity arise through the emergence of complex and highly volatile changes. Thus, the socio-economic dynamics of this new type of agility need to be managed (Millar et al., 2018). In this context, this thesis aims to establish a deeper understanding of the upcoming transition. For this purpose, the higher-level perspective of the MLP (Geels, 2004) was combined with the modeling of the UTAUT2 (Venkatesh et al., 2012) and the IRM (Schulz et al., 2019). Since this has as yet not been attempted, neither thematically nor in terms of content, the gaps in practical as well as theoretical research could be filled, and a new field of knowledge opened up. Three research questions were formulated:

- *Theoretical:* How can the transition to smart mobility be conceptualized?
- *Empirical:* What can the models say about the status and functionalities within the transition?
- *Practical:* How can the transition to smart mobility be successfully designed?

In order to answer these questions, relevant terms and definitions of smart mobility were introduced in Chapter 1. What's more, the overarching view of the MLP was presented and framing of the situation provided. This established a stable theoretical underpinning and scope for performing two different empirical studies.

L. Kauschke, *The Transition to Smart Mobility*, Mobilität – Innovation – Transformation, https://doi.org/10.1007/978-3-658-43001-6_4

The first study provided insights into the nucleus of all transformational efforts: the user. Hence, it revolved around technology adoption. Section 2.1, introduced the concept of individual user acceptance and highlighted different paths to its assessment. Within this process, definitions of acceptance and acceptability were sharpened. Furthermore, it was ascertained that acceptance is a dynamic measure. In order to facilitate such an understanding, a taxonomy based on the seminal work of Lyons et al. (2019) was developed. Subsequently, UTAUT2 was selected as the most comprehensive tool from numerous appealing modeling approaches. This step was followed by a comprehensive and unprecedented literature review on the acceptance of smart mobility. This systematization demonstrates that the acceptance of smart mobility could, from one angle, be similar to that of mobility technologies,, and similar to that of smart systems, from the other. A broad spectrum of factors will be necessary in order to be able to determine acceptance on a multidimensional basis (e.g., social, benefit-oriented, fun-oriented, habit-oriented). In Section 2.2, I thus derived a theoretically insightful and viable model. It extended UTAUT2 with the most relevant constructs of *perceived risk* and *collective efficacy*.

Additionally, it emphasized the contextual dimensions of *experience* and *level of system integration*. Section 2.3 empirically tested the model in three scenarios. Use cases included eBikes (N = 537), mobility-as-a-service (N = 531), and fully automated vehicles (N = 558). In addition, smart mobility as a whole (N = 517) was also examined using the model. Previously, data had been collected via an intelligent online survey and cleaned using different cleaning methods in SPSS. These data were analyzed with state-of-the-art SEM techniques in SmartPLS. Results suggest that acceptance of smart mobility primarily relies on five factors: *performance expectancy*, *facilitating conditions*, *social influence*, *habit*, and *hedonic motivation*. Furthermore, experience and system integration play a central role in niche technology adoption. Theoretical and practical implications were discussed in Section 2.4. As a consequence, future directions of research were put forward.

The second study was smaller and of qualitative nature. It applied the IRM to mobility transitions. This was to better understand the agility of institutions in a potential future smart mobility market. Thus, an institutional perspective was provided in Chapter 3. The chapter was structured as follows.: Section 3.1, discussed theory and methodology. This was followed by a literature review of earlier applications in the dynamic mobility environment. Section 3.2 proceeded with model development by running the first three out of five IRM phases. Roles and institutions were accordingly inferred from past projects, new theorems, and state-of-the-art architectures. As presented in Section 3.2 and 3.3, the present research discovered four key roles for technically operating a holistic and decentralized smart mobility system. As in computer operating systems, a hyper-level is superimposed on the data and infrastructure ecosystem, which implements policies in the overall system via core services. This approach is eventually transferable to the economics of mobility. Here, a decentralized management and marketing system is needed at the hyper-level. At the data and infrastructure level, existing and new business models as well as value creation systems must be enabled and stimulated through appropriate economic governance. The IRM matrix was subsequently discussed with eight experts from academia, business, and politics. The results of these interviews are discussed in Section 3.4. Overall, only minor changes had to be made to the IRM as a result of the interviews. However, these changes had a more significant impact on the model design.[1]

Finally, the present conclusion summarizes the thesis and highlights its different contributions. After Section 4.1.1, recaps the study´s overall contributions, Section 4.1.2 goes on to synthesize the essence of the study with the scholar of philosopher Jürgen Habermas. This yields a third perspective to change the regime of automobility, namely through public debate pressure. Lastly, Section 4.1.3 will provide an outlook on further research activities, i.a., in acceptance and cooperation research.

4.1 Contribution

In recent years, the digital revolution—with its sensors and big data, IoT, and AI—has also affected the transport sector. Thus, it has raised new hopes of mitigating the negative effects of transport on the environment by making mobility smarter. However, 'digital' is not a synonym for 'smart' (Fonzone et al., 2018). Smart means using data in a targeted way, safely, and integrated with society and politics. In this sense, data smarten mobility, particularly when transport operation is supported in real-time, enabling a flawless experience whilst providing a societal and economic benefit. In order to better understand this concept of smart mobility, I analyzed the perspectives of users and institutions. The theoretical, empirical, and practical research questions that were raised were answered by linking the MLP with the UTAUT and the IRM.[2]. The contributions of the thesis are accordingly comprehensive. Four areas of knowledge advancement were

[1] See Figure 3.12. *The improved IRM matrix for smart mobility*, Section 3.4.1.

[2] See Section 1.4.1 for the full list of research questions.

crystallized from the outcomes of the discussions in Parts B and C, which I will discuss below.

4.1.1 Acceptance Research

In order to investigate the acceptance of smart mobility, the existing theory was refined and strengthened. In this context, the first contribution of the present paper is a sharpening of the definition of acceptance as a progression of the thoughts of Adell et al. (2014). It is therefore important to distinguish between acceptance and acceptability, which, in turn, depends on the degree of anticipation and exposure. Building on this idea, I applied the SAE-level taxonomy approach to acceptance, following Lyon (2019)'s example. This demonstrates that acceptance is not a static parameter. Instead, acceptance may pass through various phases, with different acceptance factors significant for each phase. Further empirical validations are requested at this point.

The second major contribution of this thesis is a systemized and comprehensive literature review on smart mobility acceptance. This review covered a clear academic gap. It also underscored the need for cross-technological comparison, which was then undertaken empirically.

The third contribution to acceptance research is revealing smart mobility's direct and contextual acceptance factors. Here, the decisive roles of *habit* and *social influence* are particularly worth mentioning as possible vectors for fostering adoption. The different use cases, MaaS, FAV, and eBikes, revealed varying acceptance patterns. As such, MaaS is significantly more popular overall than the other technologies and appeals to a broad segment of the population, while it is predominantly men who are fascinated by FAVs. Ebikes, on the other hand, are a highly hedonic technology that requires better subjunctive norming throughout society. This work thus considerably contributes to understanding acceptance beyond pure cost-benefit relationships.

Finally, I argue that UTAUT2 leaves some room for revision and does not fit smart mobility perfectly, for which I suggest alternative modeling directions. Moreover, constructional shortcomings are addressed. Lastly, appropriate methods for model comparison are discussed as still needing to be improved to better test for statistically significant differences, i.e., in path coefficients.

4.1.2 Cooperations Research

The IRM represents a system dynamic approach to understanding cooperation and reducing complexity (Schulz and Franck, 2022). It can be applied to organizations along with exuberant systems. Its historic focus lies in the field of connected and collaborative driving. This work opens up this thematic arena by broadening the scope beyond the automotive spotlight to include smart mobility as a whole. In order to adequately represent this concept, meso roles were created that are superordinate to the 'normal' meta roles. This allows any smart mobility use case to be captured within the IRM. Beyond this contribution, further methodological advances for IRM research are targeted by the present elaborations. Five of them shall be mentioned:

- First, I clarified the IRM five-step proceeding by introducing development and an application phase.[3] This helps researchers navigate and build on each other.
- Second, the role model was validated through mixed exploratory-confirmatory interviews. This transferable methodological dichotomy yields more reliable results than a purely explorative proceeding.
- Third, model complexity was streamlined by removing the market decline phase, which appeared rather uninteresting for future innovations[4].
- Fourth, a symmetry between technical and economic roles was established. This aspect may fall under the superficial scheme of scientific beauty. However, as recent research suggests (Wassiliwizky and Menninghaus, 2021), scientific aesthetics are not to be neglected in terms of communication of results and intellectual access for laypeople.
- Fifth, the model was first employed as a visionary tool rather than a depiction of given architectures. This application function significantly expands the scope of the application.

4.1.3 Transition Research

Transition research analyzes structural changes in socio-economic systems issued from complex interactions in multiple domains (e.g., technology and civil society) and at multiple levels (e.g., national and international institutions). Halbe et al.

[3] See the five steps of IRM in Figure 3.3, Section 3.1.2.

[4] Within the improvement model, even more market phases were omitted (see Figure 3.12, Section 3.4.1).

(2015) continue to explain that in this context the model application's principal aim is understanding transition. In line with their suggestion to focus on certain aspects of transition, I conducted two model-to-model deep dives for smart mobility transition. Namely, in MLP, the domain of acceptance is addressed by UTAUT and that of strategic vision by IRM[5]. Whereas MLP has already been mixed with acceptance research (e.g. Schlüter and Weyer, 2019), the entanglement of roles and institutions is new. This connection thus adds an innovative component to future system analysis. In the next step, this procedure could map specific use cases or historic diffusion events. After all, providing this IRM-kind of techno-economic vision to transitions should help reduce institutional uncertainty and ambiguity.

4.1.4 Advancing Smart Mobility

Notwithstanding the academic value, this dissertation also delivers practical foundations for decision-makers and managers in politics and business. Collections of implications were presented in Sections 2.4.5 and 3.4. Four categories group the contributions: informational, design, marketing, and political value.

- *Informational value*: The conception of smart mobility communicates an integral understanding of the emerging digitization of the transport sector. This facilitates intelligent decisions. Appropriately, IRM provides vision and guidance for potentially insecure stakeholders and institutions. Hence, the research can help understand how decentralized hypernets will shape future mobility through standards, transparency, and interoperability. This expands the technical community with knowledge of socio-economic functionalities, which can thus be considered in service development from the very outset.
- *Design value*: At this point, the work offers another important contribution to developers and designers. The acceptance study provides numerous empirically supported design recommendations for smart mobility. The core of these recommendations is to place the user at the center of the design process. This means, for example, shaping smart mobility in such a way that it is fun to use, offers a high level of service reliability, and preserves the freedom of individual mobility.
- *Marketing value*: Attracting customers to alternative mobility systems in a stable system such as automobility has been a challenge for years. In this

[5] See the MLP of smart mobility in Figure 1.3, Section 1.3.3.

context, the present study accomplishes a psychology-based contribution that is intended to help gain a better understanding of customers' needs and design targeted strategies in order to address them. Marketing recommendations involve presenting smart mobility as an innovative and ecologically friendly alternative, promoting the gains in leisure time in automated vehicles or discovering public transport users as the new early adopters of MaaS technology.

- *Policy value:* This thesis promotes the need for smart governance for smart mobility. In addition to numerous political ideas for action, a proposal for a smart mobility roadmap is presented.[6] This is intended to stimulate public discussion. Furthermore, it is to broaden the view of policymakers on aspects beyond pure technology to include social and economic dimensions. As it is inevitable that different countries will exhibit different transition pathways, the typical recommendation is to employ a pluralistic policy approach and constantly monitor the progress.

4.2 Excursus: Habermas and Smart Mobility

If technology is a key to smart and sustainable transport, federal and local governance must meet the demands of change. Otherwise, there is a danger that society will receive the mobility it is given and not the one it wants or needs (Reardon, 2020). The question is not whether a transition is coming but how economies shape it. Therefore, the most significant contribution of this thesis is to point out that wise governance must be preceded by public debate. Articulating democratic will regarding the smart mobility transition has not yet taken place and thus politicians remain hesitant to act decisively. The public debate may well contain some controversy in this regard. For instance, Sovacool et al. (2019) found that children overwhelmingly agree on the future direction of car-based transport, while most elders do not. This raises the question of how discourse can be steered and moderated. To which findings might such a process lead? In this context, it is worth taking a look at the thoughts of philosopher and sociologist Jürgen Habermas. After all, he coined the concept of critical, public debate against the background of modernity, which for him was not yet complete unless society learned to communicate (Müller-Doohm, 2016).

[6] See the roadmap in Figure 2.34, Section 2.4.5.

Since the 1960s, his work has consequently revolved around the question of how a public debate oriented toward understanding facilitates the democratic constitutional state in the first place. The moral compass should be the ethics that arise from discourse. Norms may particularly emerge when we embrace the other and derive moral responsibility as an individual. Habermas develops such anticipation of counter-objections and the structuring of debates, among other things, in two of his most famous works: The Theory of Communicative Action (Habermas, 1981), and Facticity and Value (Habermas, 1992).

> *'There is no power for change greater than a community discovering what it cares about.'* (Wheatley, 2002)

Habermas' approach, aptly summarized by Margaret Wheatley, leads me to conclude that the strongly structured landscape of the MLP might be changed through a democratic effort. To my knowledge, however, Habermas himself did not concern himself much with mobility transitions. So, what can he nevertheless contribute to today's debates? Three avenues enrich the outlook on smart mobility:

4.2.1 The New Confusion

In recent decades, there has been a wide variety of contemporary diagnoses of society, all of which attest to far-reaching social change (Schwedes, 2021). Sociologists sometimes allude to the risk society, sometimes the experience society, and other times the network society. Each of these postmodern assessments describes correct aspects of current developmental dynamics, but none of these views have yet been conclusively accepted. Habermas already made the general diagnosis in the 1980 s when he spoke of a new confusion (Habermas, 1985). In it, he describes the danger that societies may break down in the face of the challenges of modernity if they do not develop new forms of cooperation and communication themselves.

4.2.2 The Public Debate as a Workshop on the Future

If we project this finding into today and relate it to the transition of mobility, most experts will agree that not much has changed. Although society is still intact, the magnitude of the challenge seems to be shaking its foundations. Habermas' proposed solution here would probably be the communicative reason and derived legislation. Humans are rational beings and live lives with open future horizons. In democratic societies, far-reaching policy measures such as smart mobility depend on broad legitimacy in the political arena. The more complex the matter, the more difficult it is to achieve the basic consensus required. In reality, our debate and discourse system is currently reaching its performance limits (Demetrious, 2019). In this respect, democratic digital participation and information systems could offer a great opportunity to grasp the plural world more effectively and efficiently. An inclusive transport policy may match mobility with citizens' personal and collective needs. This might help further think of techno-economic innovations as social innovations (Mehmood and Imran, 2021). Thus, liberal and solidary citizen participation in the sense of Habermas can help reduce public interest conflicts and accelerate transitions.

4.2.3 On Decisions

The debate demanded by Habermas, however, does not take place today either purely in public, purely in private, or in governmental dialogue. With the rise of social media, debate spaces are blurring. Constructive debates become even more difficult with these debate platforms, which are often emotionally charged and neither moderated nor curated. In this context, Habermas calls for strengthening parliaments and pragmatic political consultation. This means that more decision-making authority must once again be transferred from the executive to the legislative branch. Second, it means creating transparency in policy advice by having consultants come from the sphere of knowledge rather than the sphere of power (Drawer et al., 2021). As soon as such a condition is established, well-advised parliaments can enact reasonable laws. Challenges of our time, such as climate or mobility change, must frequently be met with solutions that require strategic endurance and anticipation of society's benefits. In this respect, wise collective decisions[7] may transform mobility. In the sense of Habermas, the currently perceived tension between the ideal and reality of structural change could be reduced accordingly (Habermas, 2022).

[7] According to game theorist Weimann (2013), these are the decisions that optimize the economic-ecological costs of society as a whole rather than the costs of an individual.

4.3 Outlook

In the end, I would like to offer some opportunities for further analysis and future implementation, which transcend beyond the individual perspectives of Chapters 2 and 3.

After all, this work has driven efforts to understand and address smart mobility transition not solely as a technological but also as a systemic and structural challenge. For this purpose, the MLP provided a layered socio-technical system in which the individual is embedded. Bringing in more tools such as IRM or UTAUT and reconnecting empirical findings to MLP´s abstract layers is a promising approach for future interdisciplinary analyses. At this point, electromobility offers itself as a potential basis for investigation since the question of which reconfiguration processes it triggers besides its obvious potential to substitute combustion vehicles has not yet been answered. Here, extending the IRM should also be considered to include additional roles (e.g., social or cultural). Further questions arise at the landscape level. These include: How can it be rendered tangible and possibly controllable, except through acceptance analyses? How can we learn from past shock events such as oil crises, wars, and pandemics in order to harness their effects for the benefit of smart mobility? In a recent Frontiers in Psychology call-for-papers, Bamberg et al. (2021) aptly reference the MLP and add: How can the individual elude its role in the context of such large-scale transformations?

From this work's point of view, it can be recommended to continue research on the individual acceptance of smart mobility. The central outlooks of the present acceptance study were the development of a more precise mobility-specific acceptance model and the continuous measurement of acceptance in the course of its diffusion. In this regard, Sovacool (2017)'s framework was discussed as a starting point for an elaborated modeling option. Overall, research should be conducted on content fit, construct validity, and variable parsimony. Even though the UTAUT provides sound results, mobility acceptance may be fundamentally different from information system schemes. A second spotlight can be cast on measurement in this context. Currently, studies are very often conducted with little to no technology exposure. Thus, it is mostly about estimated acceptability. With the introduction of the taxonomy perspective, this work has looked at acceptance as an inherently fluid process. This, in turn, depends, for example, on the experiential values of individuals or the level of system integration. From this perspective, I would like to encourage future research to work longitudinally. At best, when measuring acceptance of smart mobility over the course of time and changing landscapes, we can obtain truly stable and transferable conclusions.

An exciting thought would be to attempt to interweave the movements of the IRM and the UTAUT/TAM even better. Quite possibly, the degree of agile cooperation dynamics is something that people do indeed perceive and thus deliberately influence upon their adoption decision. One could consider developing and testing variables that refer to 'perceived transaction costs' or 'perceived interoperability'. Another way to intertwine both perspectives could be to introduce a weighting vector for the roles of the IRM. The weighting would have to be derived from the influence strengths of the acceptance factors. However, thought must be given to the extent to which there can or should be such a thing as role acceptance. Before all this, though, the next step must be empirically validating the IRM of smart mobility via use cases.[8]

In practice, implementing high-profile and value-creating use cases should be the foremost priority.[9] Purely theoretical considerations will always be limited due to insufficient knowledge of all the relevant factors. Despite this, the practice should still be thought of and developed in larger contexts and architectures. As proposed in the European Gaia-X or Catena-X standardization projects, a transparent, decentralized, and highly interoperable system yields hope. This hope is that actual secure, fair, and ubiquitous networking can be achieved between systems that today still remain isolated. Unfortunately, the public currently underestimates the game-changing potential of such solutions. Moreover, standardization processes are slow (Pfirsching, 2022). Henceforth, future research must focus on work to improving techno-scientific communication and collaboration itself.

Just then, when smart mobility gains public momentum, the transition can begin.

[8] See Section 3.4.2 for IRM implications from interview data.

[9] See Section 2.4.5 for recommendation for action.

References

Aarts, H., Dijksterhuis, A., 2000. Habits as knowledge structures: Automaticity in goal-directed behavior. Journal of Personality and Social Psychology 78, 53–63.

Abay, K.A., Mannering, F.L., 2016. An empirical analysis of risk-taking in car driving and other aspects of life. Accident; analysis and prevention 97, 57–68.

Adell, E., 2009. Driver experience and acceptance of driver support systems: A case of speed adaptation. Lund University, Lund.

Adell, E., Várhelyi, A., Nilsson, L., 2014. The Definition of Acceptance and Acceptability. In: Horberry, T., Regan, M.A., Stevens, A. (Eds.) Driver Acceptance of New Technology. Theory, Measurement and Optimization. CRC Press, Boca Raton, FL, pp. 11–22.

Adnan, N., Md Nordin, S., bin Bahruddin, M.A., Ali, M., 2018. How trust can drive forward the user acceptance to the technology? In-vehicle technology for autonomous vehicle. Transportation Research Part A: Policy and Practice 118, 819–836.

Adorff, C., 2022. Smart mobility role model interview, Web.

Ahlswede, A., 2020. e-bike sales in Germany up to 2019. Statista. https://de.statista.com/statistik/daten/studie/152721/umfrage/absatz-von-e-bikes-in-deutschland/. Accessed 29 May 2020.

Ahmed, W., Muhamad, S., Sentosa, I., Akter, H., Yafi, E., Ali, J., 2020. Predicting IoT Service Adoption towards Smart Mobility in Malaysia: SEM-Neural Hybrid Pilot Study. International Journal of Advanced Computer Science and Applications 11, 524–531.

Ainslie, G., Haslam, N., 1992. Hyperbolic discounting. In G. Loewenstein & J. Elster (Eds.) Choice over time, Russell Sage Foundation, pp. 57–92.

Ajzen, I., 1985. From Intentions to Actions: A Theory of Planned Behavior. In: Kuhl, J., Beckmann, J. (Eds.) Action control. From cognition to behavior, vol. 34, 1st ed. Springer, Berlin, pp. 11–39.

Ajzen, I., 1991. The theory of planned behavior. Organizational Behavior and Human Decision Processes 50, 179–211.

Ajzen, I., 2002. Perceived Behavioral Control, Self-Efficacy, Locus of Control, and the Theory of Planned Behavior 1. Journal of Applied Social Psychology 32, 665–683.

Ajzen, I., Fishbein, M., 1980. Understanding attitudes and predicting social behavior. Prentice-Hall, Englewood Cliffs, New Jersey, 278 pp.

Ajzen, I., Fishbein, M., 2000. Attitudes and the Attitude-Behavior Relation: Reasoned and Automatic Processes. European Review of Social Psychology 11, 1–33.

L. Kauschke, *The Transition to Smart Mobility*, Mobilität – Innovation – Transformation, https://doi.org/10.1007/978-3-658-43001-6

Ajzen, I., Madden, T.J., 1986. Prediction of goal-directed behavior: Attitudes, intentions, and perceived behavioral control. Journal of Experimental Social Psychology 22, 453–474.

Alalwan, A.A., Dwivedi, Y.K., Rana, N.P., Algharabat, R., 2018. Examining factors influencing Jordanian customers' intentions and adoption of internet banking: Extending UTAUT2 with risk. Journal of Retailing and Consumer Services 40, 125–138.

Alessandrini, A., Alfonsi, R., Site, P.D., Stam, D., 2014. Users' Preferences towards Automated Road Public Transport: Results from European Surveys. Transportation Research Procedia 3, 139–144.

Alexandre, B., Reynaud, E., Osiurak, F., Navarro, J., 2018. Acceptance and acceptability criteria: a literature review. Cognition, Technology & Work 20, 165–177.

AllBright, 2018. The power of monoculture, Stockholm, Berlin. https://static1.squarespace.com/static/5c7e8528f4755a0bedc3f8f1/t/5d78d6831d5e0e514fe34628/1568200336024/Allbright+Bericht_September+2018_klein.pdf. Accessed 21 December 2021.

AllBright, 2021. Growth without women: Stock exchange newcomers are old friends. AllBright Foundation, Stockholm, Berlin. https://static1.squarespace.com/static/5c7e8528f4755a0bedc3f8f1/t/60ca5e501233fb3164140c1d/1623875167416/Allbright-Bericht-Juni-2021_Bo%CC%88rsenneulinge.pdf. Accessed 21 December 2021.

Allen, P.J., Bennett, K., 2008. SPSS for the health & behavioural sciences. Thomson, Australia, 284 pp.

Almaiah, M.A., Alamri, M.M., Al-Rahmi, W., 2019. Applying the UTAUT Model to Explain the Students' Acceptance of Mobile Learning System in Higher Education. IEEE Access 7, 174673–174686.

Al-Ruithe, M., Benkhelifa, E., Hameed, K., 2019. A systematic literature review of data governance and cloud data governance. Personal Ubiquitous Computers 23, 839–859.

Al-Samarraie, H., Hurmuzan, S., 2018. A review of brainstorming techniques in higher education. Thinking Skills and Creativity 27, 78–91.

Altmann, M., Graesel, C., 1998. The acceptance of hydrogen technologies. Dissertation, Technical University of Munich.

Angelis, M. de, Puchades, V.M., Fraboni, F., Pietrantoni, L., Prati, G., 2017. Negative attitudes towards cyclists influence the acceptance of an in-vehicle cyclist detection system. Transportation Research Part F: Traffic Psychology and Behaviour 49, 244–256.

Astrachan, C.B., Patel, V.K., Wanzenried, G., 2014. A comparative study of CB-SEM and PLS-SEM for theory development in family firm research. Journal of Family Business Strategy 5, 116–128.

Aswani, R., Ilavarasan, P.V., Kar, A.K., Vijayan, S., 2018. Adoption of public WiFi using UTAUT2: An exploration in an emerging economy. Procedia Computer Science 132, 297–306.

Axsen, J., Sovacool, B.K., 2019. The roles of users in electric, shared and automated mobility transitions. Transportation Research Part D: Transport and Environment 71, 1–21.

Backhaus, K., Erichson, B., Plinke, W., Weiber, R., 2018. Multivariate Analysemethoden: Eine anwendungsorientierte Einführung, 15th ed. Springer Gabler, Berlin, Heidelberg, 625 pp.

Backstrom, C.H., Hursh-César, G., 1981. Survey research, 2nd ed. Wiley, New York, 436 pp.

Bagozzi, R., 2007. The Legacy of the Technology Acceptance Model and a Proposal for a Paradigm Shift. Journal of the Association for Information Systems 8, 244–254.

Bagozzi, R.P., 1981. Attitudes, intentions, and behavior: A test of some key hypotheses. Journal of Personality and Social Psychology 41, 607–627.

Bagozzi, R.P., Baumgartner, H., 1994. The Evaluation of Structural Equation Models and Hypothesis Testing. In: Bagozzi, R.P. (Ed.) Principles of Marketing Research. Blackwell Business, Cambridge (Massachusetts), pp. 386–422.

Bagozzi, R.P., Yi, Y., 1988. On the evaluation of structural equation models. Journal of the Academy of Marketing Science 16, 74–94.

Bagozzi, R.P., Yi, Y., Phillips, L.W., 1991. Assessing Construct Validity in Organizational Research. Administrative Science Quarterly 36, 421.

Bamberg, S., 2013. Applying the stage model of self-regulated behavioral change in a car use reduction intervention. Journal of Environmental Psychology 33, 68–75.

Bamberg, S., Fischer, D., Geiger, S.M., 2021. Editorial: The Role of the Individual in the Great Transformation Toward Sustainability. Frontiers in Psychology 12, 710897.

Bamberg, S., Fujii, S., Friman, M., Gärling, T., 2011. Behaviour theory and soft transport policy measures. Transport Policy 18, 228–235.

Bandura, A., 1977. Self-efficacy: Toward a unifying theory of behavioral change. Advances in Behaviour Research and Therapy 1, 139–161.

Bandura, A., 1986a. Social foundations of thought and action. Englewood Cliffs, New Jersey, 1986.

Bandura, A., 1986b. The explanatory and predictive scope of self-efficacy theory. Journal of social and clinical psychology 4, 359–373.

Bansal, P., Kockelman, K.M., Singh, A., 2016. Assessing public opinions of and interest in new vehicle technologies: An Austin perspective. Transportation Research Part C: Emerging Technologies 67, 1–14.

Baptista, G., Oliveira, T., 2015. Understanding mobile banking: The unified theory of acceptance and use of technology combined with cultural moderators. Computers in Human Behavior 50, 418–430.

Baptista, G., Oliveira, T., 2019. Gamification and serious games: A literature meta-analysis and integrative model. Computers in Human Behavior 92, 306–315.

Barnes, S.B., 2006. A privacy paradox: Social networking in the United States. First Monday 11.

Barr, S., 2018. Personal mobility and climate change. Wiley Interdisciplinary Reviews: Climate Change 9, 542.

Barth, B., Flaig, B.B., Schäuble, N., Tautscher, M. (Eds.), 2018. Praxis der Sinus-Milieus®: Gegenwart und Zukunft eines modernen Gesellschafts- und Zielgruppenmodells. Springer Fachmedien, Wiesbaden.

Barth, M., Jugert, P., Fritsche, I., 2016. Still underdetected – Social norms and collective efficacy predict the acceptance of electric vehicles in Germany. Transportation Research Part F: Traffic Psychology and Behaviour 37, 64–77.

Barth, M.J., Wu, G., Boriboonsomsin, K., 2015. Intelligent Transportation Systems and Greenhouse Gas Reductions. Curr Sustainable Renewable Energy Rep 2, 90–97.

Baudier, P., Ammi, C., Deboeuf-Rouchon, M., 2018. Smart home: Highly-educated students' acceptance. Technological Forecasting and Social Change, 119355.

Bauer, R.A., 1960. Consumer behavior as risk taking. Chicago, Illinois.

Becker, J.-M., Ismail, I.R., 2016. Accounting for sampling weights in PLS path modeling: Simulations and empirical examples. European Management Journal 34, 606–617.

Becker, J.-M., Rai, A., Ringle, C.M., Völckner, F., 2013. Discovering Unobserved Heterogeneity in Structural Equation Models to Avert Validity Threats. Management Information Systems Quarterly 37, 665–694.

Bem, S.L., 1981. Gender schema theory: A cognitive account of sex typing. Psychological review 88, 354–364.

Benbasat, I., Barki, H., 2007. Quo vadis TAM? Journal of the Association for Information Systems 8, 211–218.

Ben-Elia, E., Avineri, E., 2015. Response to Travel Information: A Behavioural Review. Transport Reviews 35, 352–377.

Benevolo, C., Dameri, R.P., D'Auria, B., 2016. Smart Mobility in Smart City. In: Torre, T., Braccini, A.M., Spinelli, R. (Eds.) Empowering Organizations, vol. 11. Springer International Publishing, Cham, pp. 13–28.

Berliner, R.M., Hardman, S., Tal, G., 2019. Uncovering early adopter's perceptions and purchase intentions of automated vehicles: Insights from early adopters of electric vehicles in California. Transportation Research Part F: Traffic Psychology and Behaviour 60, 712–722.

Bhatti, G., Mohan, H., Singh, R.R., 2021. Towards the future of smart electric vehicles: Digital twin technology. Renewable and Sustainable Energy Reviews 141, 110801.

Bigerna, S., Bollino, C.A., Micheli, S., 2016. Socio-economic acceptability for smart grid development – a comprehensive review. Journal of Cleaner Production 131, 399–409.

Bigerna, S., Bollino, C.A., Micheli, S., Polinori, P., 2017. A new unified approach to evaluate economic acceptance towards main green technologies using the meta-analysis. Journal of Cleaner Production 167, 1251–1262.

Blanca Hernández, J. Jiménez, M. J. Martín, 2009. Adoption vs acceptance of e-commerce: two different decisions. European journal of marketing 43, 1232–1245.

BMDV, 2021. Development of a tool for data management of mobility providers – DECREE. https://www.bmvi.de/SharedDocs/DE/Artikel/DG/mfund-projekte/decree.html. Accessed 6 April 2022.

Bögel, P.M., Upham, P., 2018. Role of psychology in sociotechnical transitions studies: Review in relation to consumption and technology acceptance. Environmental Innovation and Societal Transitions 28, 122–136.

Bogner, A., Littig, B., Menz, W., 2014. Interviews mit Experten: eine praxisorientierte Einführung. Springer-Verlag.

Bollen, K.A., Long, J.S. (Eds.), 1993. Testing structural equation models. SAGE Publications, Inc, Los Angeles.

Bollen, K.A., Ting, K.F., 2000. A tetrad test for causal indicators. Psychological methods 5, 3–22.

Bouwman, H., Ló, C., Nicolá, p., s, N.A., Castillo, F.J.M., van Hattum, P., 2012. Consumer lifestyles: alternative adoption patterns for advanced mobile services. International Journal of Mobile Communications 10, 169.

Braumoeller, B.F., 2004. Hypothesis Testing and Multiplicative Interaction Terms. International Organization 58, 807–820.

Breckler, S.J., 1984. Empirical validation of affect, behavior, and cognition as distinct components of attitude. Journal of Personality and Social Psychology 47, 1191–1205.

Brown, Venkatesh, 2005. Model of Adoption of Technology in Households: A Baseline Model Test and Extension Incorporating Household Life Cycle. MIS Quarterly 29, 399.

Buckley, L., Kaye, S.-A., Pradhan, A.K., 2018. Psychosocial factors associated with intended use of automated vehicles: A simulated driving study. Accident, analysis and prevention 115, 202–208.

Bühner, M., 2011. Einführung in die Test- und Fragebogenkonstruktion. Pearson Deutschland GmbH, München.

Bultel, Y., Aurousseau, M., Ozil, M., Perrin, L., 2007. Risk Analysis on a Fuel Cell in Electric Vehicle Using the MADS/MOSAR Methodology. Process Safety and Environmental Protection 85, 241–250.

Bundestag, 2022. Annual Economic Report 2022: First reading. https://www.bundestag.de/dokumente/textarchiv/2022/kw04-de-jahreswirtschaftsbericht-877112. Accessed 9 February 2022.

Burghard, U., Dütschke, E., 2018. Who wants shared mobility? Lessons from early adopters and mainstream drivers on electric carsharing in Germany. Transportation Research Part D: Transport and Environment, 96–109.

Burnett, G., Diels, C., 2014. Driver acceptance of in-vehicle information, assistance and automated systems: An overview. Driver Acceptance of New Technology: Theory, Measurement and Optimization, 137–151.

Burnett, J.S., Mitzner, T.L., Charness, N., Rogers, W.A., 2011. Understanding Predictors of Computer Communication Technology Use by Older Adults. Proceedings of the Human Factors and Ergonomics Society ... Annual Meeting. Human Factors and Ergonomics Society. Annual Meeting 55, 172–176.

Casey, T., Wilson-Evered, E., 2012. Predicting uptake of technology innovations in online family dispute resolution services: An application and extension of the UTAUT. Computers in Human Behavior 28, 2034–2045.

Castañeda, J.A., Muñoz-Leiva, F., Luque, T., 2007. Web Acceptance Model (WAM): Moderating effects of user experience. Information & Management 44, 384–396.

Catenacci, M., Verdolini, E., Bosetti, V., Fiorese, G., 2013. Going electric: Expert survey on the future of battery technologies for electric vehicles. Energy Policy 61, 403–413.

CCV, 2021. [Case] Mobility as a Service (MaaS) in Public Transport. https://www.ccv.eu/en/2021/case-mobility-as-a-service-maas-in-public-transport/#:~:text=MaaS%20in%20the%20city%20of%20Helsinki&text=The%20main%20platform%20is%20called,pay%2Das%20you%20go%20model. Accessed 5 February 2022.

Chan, D., 2009. So why ask me? Are self-report data really that bad? Statistical and methodological myths and urban legends: Doctrine, verity and fable in the organizational and social sciences, 309–336.

Chaney, R.A., Hall, P.C., Crowder, A.R., Crookston, B.T., West, J.H., 2019. Mountain biker attitudes and perceptions of eMTBs (electric-mountain bikes). Sport Sci Health 15, 577–583.

Chang, S.-J., van Witteloostuijn, A., Eden, L., 2010. From the Editors: Common method variance in international business research. Journal of International Business Studies 41, 178–184.

Chao, C.-M., 2019. Factors Determining the Behavioral Intention to Use Mobile Learning: An Application and Extension of the UTAUT Model. Frontiers in Psychology.

Chen, C.-F., Chao, W.-H., 2011. Habitual or reasoned? Using the theory of planned behavior, technology acceptance model, and habit to examine switching intentions toward public transit. Transportation Research Part F: Traffic Psychology and Behaviour 14, 128–137.

Chen, H.S., Chen, C.Y., 2013. A Study on Willingness to Pay of Hydrogen Energy and Fuel Cell Technologies. Applied Mechanics and Materials, 448–453.

Chen, K., Chan, A., 2011. A review of technology acceptance by older adults. Gerontechnology 10, 1–12.

Chen, L., Gillenson, M.L., Sherrell, D.L., 2002. Enticing online consumers: an extended technology acceptance perspective. Information & Management 39, 705–719.

Chen, L.S.-L., Kuan, C.J., Lee, Y.-H., Huang, H.-L., 2011. Applicability of the UTAUT model in playing online game through mobile phones: Moderating effects of user experience, First International Technology Management Conference, IEEE, https://doi.org/10.1109/ITMC.2011.5996035.

Chen, S.-Y., 2016. Using the sustainable modified TAM and TPB to analyze the effects of perceived green value on loyalty to a public bike system. Transportation Research Part A: Policy and Practice 88, 58–72.

Chin, W.W., 1998. Issues and Opinion on Structural Equation Modeling. MIS Quarterly, 1–10.

Chismar, W.G., Wiley-Patton, S., 2003. Does the extended technology acceptance model apply to physicians, in: 36th Annual Hawaii International Conference on System Sciences. Proceedings of the 36th Annual Hawaii International Conference on System Sciences, 2003. Proceedings of the, Big Island, HI, USA. 2003. IEEE, 8 pp.

Chui, M., Manyika, J., 2015. Competition at the digital edge: 'Hyperscale' businesses. https://www.mckinsey.com/industries/technology-media-and-telecommunications/our-insights/competition-at-the-digital-edge-hyperscale-businesses, Accesssed 20 March 2022.

Chung, J.E., Park, N., Wang, H., Fulk, J., McLaughlin, M., 2010. Age differences in perceptions of online community participation among non-users: An extension of the Technology Acceptance Model. Computers in Human Behavior 26, 1674–1684.

Chuttur, M., 2009. Overview of the technology acceptance model: Origins, developments and future directions. Working Papers on Information Systems, 9–37.

CIVITAS, 2015. Smart choices for cities: Gender equality and mobility: mind the gap! https://civitas.eu/sites/default/files/civ_pol-an2_m_web.pdf. Accessed 7 January 2022.

Clear, J., 2018. Atomic habits: Tiny changes, remarkable results; an easy & proven way to build good habits & break bad ones. Avery, New York, New York, 306 pp.

Coase, R., 1998. The new institutional economics. The American economic review 88, 72–74.

Cocca, S., Fabry, C., Stryja, C., 2015. Electric mobility services: results from expert surveys. Fraunhofer Institute for Industrial Engineering, Stuttgart.

Cocron, P., Bühler, F., Neumann, I., Franke, T., Krems, J.F., Schwalm, M., Keinath, A., 2011. Methods of evaluating electric vehicles from a user's perspective–the MINI E field trial in Berlin. IET Intelligent Transport Systems 5, 127–133.

Cohen, J., 1988. Statistical power analysis for the behavioral sciences, 2nd ed. Erlbaum, Hillsdale, New Jersey, 567 pp.

Cohen, J., 1992. A power primer. Psychological bulletin 112, 155 pp.

Compeau, D., Higgins, C.A., Huff, S., 1999. Social Cognitive Theory and Individual Reactions to Computing Technology: A Longitudinal Study. MIS Quarterly 23, 145 pp.

Conner, M., Armitage, C.J., 1998. Extending the Theory of Planned Behavior: A Review and Avenues for Further Research. Journal of Applied Social Pyschology 28, 1429–1464.

Cook, R.D., Weisberg, S., 1980. Characterizations of an empirical influence function for detecting influential cases in regression. Technometrics 22, 495–508.

Cookson, G., Pishue, B., 2017. INRIX Connected and Autonomous Vehicle Consumer Survey.

Cronbach, L.J., 1951. Coefficient alpha and the internal structure of tests. Psychometrika 16, 297–334.

Crowne, D.P., Marlowe, D., 1960. A new scale of social desirability independent of psychopathology. Journal of Consulting Psychology 24, 349.

Csikszentmihalyi, M., 1988. The flow experience and its significance for human psychology. 05213428.

Cunningham, M.L., Regan, M.A., Horberry, T., Weeratunga, K., Dixit, V., 2019. Public opinion about automated vehicles in Australia: Results from a large-scale national survey. Transportation Research Part A: Policy and Practice 129, 1–18.

Davis, F.D., 1989. Perceived Usefulness, Perceived Ease of Use, and User Acceptance of Information Technology. MIS Quarterly 13, 319pp.

Davis, F.D., Bagozzi, R.P., Warshaw, P.R., 1992. Extrinsic and Intrinsic Motivation to Use Computers in the Workplace1. Journal of Applied Social Psychology 22, 1111–1132.

DeLone, W.H., McLean, E.R., 1992. Information Systems Success: The Quest for the Dependent Variable. Information systems research 3, 60–95.

Demetrious, K., 2019. Energy wars: Global public relations and public debate in the 21st century. Public Relations Inquiry 8, 7–22.

Devall, B., 1991. Deep ecology and radical environmentalism. Society & Natural Resources 4, 247–258.

Diamantopoulos, 2011. Incorporating Formative Measures into Covariance-Based Structural Equation Models. MIS Quarterly 35, 335.

Diamantopoulos, A., 2005. The COARSE procedure for scale development in marketing: a comment. International Journal of Research in Marketing 22, 1–9.

Dijkstra, T.K., Henseler, J., 2015a. Consistent and asymptotically normal PLS estimators for linear structural equations. Computational Statistics & Data Analysis 81, 10–23.

Dijkstra, T.K., Henseler, J., 2015b. Consistent partial least squares path modeling. Management Information Systems Quarterly 39, 297–316.

Dillon, A., Morris, M., 1996. User Acceptance of Information Technology: Theories and Models. Annual Review of Information Science and Technology 31, 7–15.

Dishaw, M.T., Strong, D.M., 1999. Extending the technology acceptance model with task–technology fit constructs. Information & Management 36, 9–21.

Docherty, I., Marsden, G., Anable, J., 2018. The governance of smart mobility. Transportation Research Part A: Policy and Practice 115, 114–125.

Dodds, W.B., Monroe, K.B., Grewal, D., 1991. Effects of Price, Brand, and Store Information on Buyers' Product Evaluations. Journal of Marketing Research 28, 307.

Donald, I.J., Cooper, S.R., Conchie, S.M., 2014. An extended theory of planned behaviour model of the psychological factors affecting commuters' transport mode use. Journal of Environmental Psychology 40, 39–48.

Donaldson, S.I., Grant-Vallone, E.J., 2002. Understanding self-report bias in organizational behavior research. Journal of business and Psychology 17, 245–260.

Döringer, S., 2021. 'The problem-centred expert interview'. Combining qualitative interviewing approaches for investigating implicit expert knowledge. International Journal of Social Research Methodology 24, 265–278.

Dorlöchter, L., 2022. Smart mobility role model interview, Web.

Dowling, G.R., Staelin, R., 1994. A model of perceived risk and intended risk-handling activity. Journal of consumer research 21, 119–134.

Drawer, C., Bruckhoff, D., Schmidt, M., 2021. Das wohlberatene Parlament. Orte und Prozesse der Politikberatung des Deutschen BT, 1st ed. GRIN Verlag, München, 28 pp.

Du, W., Yang, J., Powis, B., Zheng, X., Ozanne-Smith, J., Bilston, L., Wu, M., 2013. Understanding on-road practices of electric bike riders: an observational study in a developed city of China. Accident; analysis and prevention 59, 319–326.

Dudenhöffer, K., 2013. Why electric vehicles failed. Journal of Management Control 24, 95–124.

Duhigg, C., 2012. The power of habit: Why we do what we do in life and business. Random House, New York, NY, 371 pp.

Dwivedi, Y.K., Rana, N.P., Chen, H., Williams, M.D., 2011. A Meta-analysis of the Unified Theory of Acceptance and Use of Technology (UTAUT). In: Nüttgens, M., Gadatsch, A., Kautz, K., Schirmer, I., Blinn, N. (Eds.) Governance and Sustainability in Information Systems. Managing the Transfer and Diffusion of IT, vol. 366. Springer Berlin Heidelberg, Berlin, Heidelberg, pp. 155–170.

Dwivedi, Y.K., Rana, N.P., Jeyaraj, A., Clement, M., Williams, M.D., 2019. Re-examining the Unified Theory of Acceptance and Use of Technology (UTAUT): Towards a Revised Theoretical Model. Information System Frontiers 21, 719–734.

Ebert, T.A.E., Raithel, S., 2011. Leitfaden zur Messung von Konstrukten. In: Schwaiger, M., Meyer, A. (Eds.) Theorien und Methoden der Betriebswirtschaft. Handbuch für Wissenschaftler und Studierende. Franz Vahlen, München, pp. 503–532.

Edgington, E.S., Onghena, P., 2007. Randomization tests, 4th ed. Chapman & Hall/CRC, Boca Raton, Fla., 345 pp.

Eisinga, R., Grotenhuis, M.t., Pelzer, B., 2013. The reliability of a two-item scale: Pearson, Cronbach, or Spearman-Brown? International journal of public health 58, 637–642.

Eklund, P., Dou, E., and Gretzel, U., 2016. Location privacy acceptance: attitudes to transport-based location-aware mobile applications on a university campus. ACIS 2016 Proceedings.

Elder-Vass, D., 2019. No price without value: towards a theory of value and price. Cambridge Journal of Economics, 1485–1498.

Emsenhuber, E.-M., 2012. Determinants of the Acceptance of Electric Vehicles: An empirical analysis. master thesis, Aarhus.

Enders, C.K., 2003. Performing Multivariate Group Comparisons Following a Statistically Significant MANOVA. Measurement and Evaluation in Counseling and Development 36, 40–56.

Escobar-Rodríguez, T., Carvajal-Trujillo, E., 2014. Online purchasing tickets for low cost carriers: An application of the unified theory of acceptance and use of technology (UTAUT) model. Tourism Management 43, 70–88.

European Commission, 2020. European Climate Law: 0036(COD).

European Parliament, 2019. CO_2 emissions from cars: facts and figures. https://www.europarl.europa.eu/news/en/headlines/society/20190313STO31218/co2-emissions-from-cars-facts-and-figures-infographics. Accessed 10 February 2022.

Everingham, C., 2019. Deliverable D5.1: Description of 5GCroCo Business Potentials. https://5gcroco.eu/images/templates/rsvario/images/5GCroCo_D5_1.pdf. Accessed 12 February 2022.

Everitt, B., Skrondal, A., 2010. The Cambridge dictionary of statistics, 4th ed. Cambridge Univ. Press, Cambridge, 468 pp.

Farag, S., Lyons, G., 2012. To use or not to use? An empirical study of pre-trip public transport information for business and leisure trips and comparison with car travel. Transport Policy 20, 82–92.

Fassott, G., 2006. Operationalisierung latenter variablen in Strukturgleichungsmodellen: eine Standortbestimmung. Schmalenbachs Zeitschrift für betriebswirtschaftliche Forschung 58, 67–88.

Fazel, L., 2013. Acceptance of electric mobility. Dissertation, Technical University of Chemnitz.

Featherman, M.S., Pavlou, P.A., 2003. Predicting e-services adoption: a perceived risk facets perspective. International Journal of Human-Computer Studies 59, 451–474.

Ferguson, C.J., 2016. An effect size primer: A guide for clinicians and researchers. In: Kazdin, A.E. (Ed.) Methodological issues and strategies in clinical research (4th ed.). American Psychological Association, Washington, pp. 301–310.

Fernando, C., Soo, V.K., Doolan, M., 2020. Life Cycle Assessment for Servitization: A Case Study on Current Mobility Services. Procedia Manufacturing 43, 72–79.

Field, A., 2011. Discovering statistics using SPSS: (and sex and drugs and rock ‘n’ roll), 3rd ed. Sage, Los Angeles, Calif., 821 pp.

Fishbein, M., Ajzen, I., 1975. Belief, attitude, intention, and behavior: An introduction to theory and research, 1st ed. Addison-Wesley, Reading, Mass., 578 pp.

Fishbein, M., Ajzen, I., 2005. Theory-based behavior change interventions: comments on Hobbis and Sutton. Journal of health psychology 10, 27–31; 37–43.

Fleury, S., Tom, A., Jamet, E., Colas-Maheux, E., 2017. What drives corporate carsharing acceptance? A French case study. Transportation Research Part F: Traffic Psychology and Behaviour 45, 218–227.

Flink, J.J., 1975. The car culture, 1st ed. MIT Press, Cambridge (Massachusetts), 260 pp.

Flügge, B., 2016. Smart Mobility. Springer, Wiesbaden.

Fonzone, A., Saleh, W., Rye, T., 2018. Smart urban mobility – Escaping the technological Sirens. Transportation Research Part A: Policy and Practice 115, 1–3.

Fornell, C., Larcker, D.F., 1981. Evaluating Structural Equation Models with Unobservable Variables and Measurement Error. Journal of Marketing Research 18, 39pp.

Friis, F., 2020. An alternative explanation of the persistent low EV-uptake: The need for interventions in current norms of mobility demand. Journal of Transport Geography 83, 102635.

Fritsche, I., Barth, M., Jugert, P., Masson, T., Reese, G., 2018. A social identity model of pro-environmental action (SIMPEA). Psychological review 125, 245–269.

Fünfrocken, M., Vogt, J., Wieker, H., 2021. The CONVERGE project – A systems network for ITS.

Gaia-X, 2022. Gaia-X European Association for Data and Cloud AISBL: Factsheet. https://gaia-x.eu/what-gaia-x/factsheet. Accessed 7 February 2022.

Galanis, A., Papanikolaou, A., Eliou, N., 2014. Bikeability Audit in Urban Road Environment. International Journal of Operations Research and Information Systems 5, 21–39.

Geels, F.W., 2004. From sectoral systems of innovation to socio-technical systems. Research Policy 33, 897–920.

Geels, F.W., 2012. A socio-technical analysis of low-carbon transitions: introducing the multi-level perspective into transport studies. Journal of Transport Geography 24, 471–482.

Geels, F.W., 2019. Socio-technical transitions to sustainability: a review of criticisms and elaborations of the Multi-Level Perspective. Current Opinion in Environmental Sustainability 39, 187–201.

Geels, F.W., Schot, J., 2007. Typology of sociotechnical transition pathways. Research Policy 36, 399–417.

Gefen, D., Karahanna, E., Straub, D.W., 2003. Inexperience and experience with online stores: The importance of tam and trust. IEEE Transactions on Engineering Management 50, 307–321.

Gefen, D., Straub, D., 2000. The Relative Importance of Perceived Ease of Use in IS Adoption: A Study of E-Commerce Adoption. Journal of the Association for Information Systems 1, 1–30.

Gefen, D., Straub, D.W., 1997. Gender Differences in the Perception and Use of E-Mail: An Extension to the Technology Acceptance Model. MIS Quarterly 21, 389.

Geilenberg, V., 2022. Smart mobility role model interview, Web.

Geis, I., Kauschke L., Schulz, W., 2016. Improving electric mobility with ITS. ITS European Congress 2016.

Geis, I., Schulz, W., 2017. Incentivizing modal change – exploring the effect of multimodal information and ticketing systems for medium and long distances in Europe. Journal of the Transportation Research Board.

Geisser, S., 1974. A predictive approach to the random effect model. Biometrika 61, 101–107.

Geissler, T., 2022. Smart mobility role model interview, Web.

Geotab, 2018. What is smart mobility? https://www.geotab.com/blog/what-is-smart-mobility/. Accessed 17 February 2022.

Giddens, A., 1984. The constitution of society: Outline of the theory of structuration. University of California Press.

Giesecke, R., Surakka, T., Hakonen, M., 2016. Conceptualizing Mobility as a Service, in: 2016 Eleventh International Conference on Ecological Vehicles and Renewable Energies (EVER). 2016 Eleventh International Conference on Ecological Vehicles and Renewable Energies (EVER), pp. 1–11.

Gimpel, H., Graf, V., Graf-Drasch, V., 2020. A comprehensive model for individuals' acceptance of smart energy technology – A meta-analysis. Energy Policy 138, 111196.

Girod, B., Mayer, S., Nägele, F., 2017. Economic versus belief-based models: Shedding light on the adoption of novel green technologies. Energy Policy 101, 415–426.

Gkiotsalitis, K., Cats, O., 2021. Public transport planning adaption under the COVID-19 pandemic crisis: literature review of research needs and directions. Transport Reviews 41, 374–392.

Gläser, J., Laudel, G., 2009. Experteninterviews und qualitative Inhaltsanalyse: als Instrumente rekonstruierender Untersuchungen. Springer-Verlag.

Globisch, J., Schneider, U., Dütschke, E., 2013. Acceptance of electric vehicles by commercial users in the pilot regions in Germany. eceee Summer Study Proceedings, pp. 973–983.

Goldsmith, R.E., Kim, D., Flynn, L.R., Kim, W.-M., 2005. Price sensitivity and innovativeness for fashion among Korean consumers. The Journal of social psychology 145, 501–508.

Goodhue, D.L., Thompson, R.L., 1995. Task-Technology Fit and Individual Performance. MIS Quarterly 19, 213.

Google Trends, 2022. Smart mobility worldwide 2011–2021. https://trends.google.de/trends/explore?date=2011-01-10%202021-01-10&geo=DE&q=smart%20mobility. Accessed 8 February 2022.

Götz, O., Liehr-Gobbers, K., Krafft, M., 2010. Evaluation of structural equation models using the partial least squares (PLS) approach. In: Esposito Vinzi, V. (Ed.) Handbook of Partial Least Squares. Concepts, Methods and Applications. Springer-Verlag Berlin Heidelberg, Berlin, Heidelberg, pp. 691–711.

Greenwood, P.E., Nikulin, M.S., 1996. A guide to chi-squared testing. Wiley, New York, NY, 280 pp.

Grether, E.T., Dean, J., 1952. Managerial Economics. The American economic review 42, 452–455.

Gunasinghe, A., Hamid, J.A., Khatibi, A., Azam, S.F., 2020. The adequacy of UTAUT-3 in interpreting academician's adoption to e-Learning in higher education environments. Interactive Technology and Smart Education 17, 86–106.

Gupta, B., Dasgupta, S., Gupta, A., 2008. Adoption of ICT in a government organization in a developing country: An empirical study. The Journal of Strategic Information Systems 17, 140–154.

Gustafsod, P.E., 1998. Gender Differences in Risk Perception: Theoretical and Methodological Perspectives. Risk Analysis 18, 805–811.

Ha, I., Yoon, Y., Choi, M., 2007. Determinants of adoption of mobile games under mobile broadband wireless access environment. Information & Management 44, 276–286.

Habermas, J., 1981. Theorie des kommunikativen Handelns. Suhrkamp Frankfurt.

Habermas, J., 1985. Die neue Unübersichtlichkeit. Merkur 39, 1–14.

Habermas, J., 1992. Faktizität und Geltung. Suhrkamp Frankfurt am Main.

Habermas, J., 2022. Ein neuer Strukturwandel der Öffentlichkeit und die deliberative Politik, 1st ed. Suhrkamp, Berlin, 100 pp.

Habib, A., Alsmadi, D., Prybutok, V.R., 2019. Factors that determine residents' acceptance of smart city technologies. Behaviour & Information Technology 11, 1–14.

Haboucha, C.J., Ishaq, R., Shiftan, Y., 2017. User preferences regarding autonomous vehicles. Transportation Research Part C: Emerging Technologies 78, 37–49.

Hair, J.F., 2014. Multivariate data analysis, 7th ed. Pearson, Harlow, 734 pp.

Hair, J.F., Howard, M.C., Nitzl, C., 2020. Assessing measurement model quality in PLS-SEM using confirmatory composite analysis. Journal of Business Research 109, 101–110.

Hair, J.F., Hult, G.T.M., Ringle, C.M., Sarstedt, M., 2017. A primer on partial least squares structural equation modeling (PLS-SEM). Sage, Los Angeles, London, New Delhi, Singapore, Washington DC, Melbourne, 363 pp.

Hair, J.F., Ringle, C.M., Sarstedt, M., 2011. PLS-SEM: Indeed a Silver Bullet. Journal of Marketing Theory and Practice 19, 139–152.

Hair, J.F., Sarstedt, M., Ringle, C.M., Gudergan, S., 2018. Advanced issues in partial least squares structural equation modeling. Sage, Los Angeles, London, New Delhi, Singapore, Washington DC, Melbourne, 254 pp.

Halbe, J., Reusser, D.E., Holtz, G., Haasnoot, M., Stosius, A., Avenhaus, W., Kwakkel, J.H., 2015. Lessons for model use in transition research: A survey and comparison with other research areas. Environmental Innovation and Societal Transitions 15, 194–210.

Hamari, J., Koivisto, J., 2015. Why do people use gamification services? International Journal of Information Management 35, 419–431.

Hardin, G., 1968. The tragedy of the commons. Science 162, 1243–1248.

Hartl, B., Sabitzer, T., Hofmann, E., Penz, E., 2018. 'Sustainability is a nice bonus' the role of sustainability in carsharing from a consumer perspective. Journal of Cleaner Production 202, 88–100.

He, J., Freeman, L., 2010. Are Men More Technology-Oriented Than Women? The Role of Gender on the Development of General Computer Self-Efficacy of College Students. Journal of Information Systems Education 21, 203–212.

Heacock, P. (Ed.), 2009. Cambridge academic content dictionary: With more than 2.000 content-area words; from algebra to zoology; CD-ROM dictionary and thesaurus in one, 1st ed. Cambridge Univ. Press, Cambridge, 1137 pp.

Heinrich, L., Kauschke, L., 2017. Deliverable D4 – evaluation and recommendation: New economic developments for ITS in electric mobility.

Heiskanen, E., Hodson, M., Mourik, R.M., Raven, R., Feenstra, C.F., Alcantud, A., Brohmann, B., Daniels, A., Di Fiore, M., Farkas, B., 2008. Factors influencing the societal acceptance of new energy technologies: meta-analysis of recent European projects. Work Package 2.

Henseler, J., 2018. Partial least squares path modeling: Quo vadis? Quality & quantity 52, 1–8.

Henseler, J., Hubona, G., Ray, P.A., 2016a. Using PLS path modeling in new technology research: updated guidelines. Industrial Management & Data Systems 116, 2–20.

Henseler, J., Ringle, C.M., Sarstedt, M., 2015. A new criterion for assessing discriminant validity in variance-based structural equation modeling. Journal of the Academy of Marketing Science 43, 115–135.

Henseler, J., Ringle, C.M., Sarstedt, M., 2016b. Testing measurement invariance of composites using partial least squares. International Marketing Review 33, 405–431.

Henseler, J., Ringle, C.M., Sinkovics, R.R., 2009. The use of partial least squares path modeling in international marketing. In: Sinkovics, R.R. (Ed.) New challenges to international marketing, vol. 20. Emerald, London, pp. 277–319.

Henseler, J., Sarstedt, M., 2013. Goodness-of-fit indices for partial least squares path modeling. Computer Statistics 28, 565–580.

Herb, T., 2013. Organizational architecture of cooperative systems. Straßenverkehrstechnik.

Herrero, Á., San Martín, H., Garcia-De los Salmones, M.d.M., 2017. Explaining the adoption of social networks sites for sharing user-generated content: A revision of the UTAUT2. Computers in Human Behavior 71, 209–217.

Herzenberg, S., Alic, J., Wial, H., 2018. New rules for a new economy: Employment and Opportunity in Post-Industrial America, Cornell University Press, New York.

Hildebrandt, L., Temme, D., 2006. Probleme der Validierung mit Strukturgleichungsmodellen: SFB 649 Discussion Paper. Collaborative Research Center 649 – Economic Risk.

Hillemacher, L., Hufendiek, K., Bertsch, V., Wiechmann, H., Gratenau, J., Jochem, P., Fichtner, W., 2013. Ein Rollenmodell zur Einbindung der Endkunden in eine smarte Energiewelt. Zeitung der Energiewirtschaft 37, 195–210.

Ho, C.Q., Hensher, D.A., Mulley, C., Wong, Y.Z., 2018. Potential uptake and willingness-to-pay for Mobility as a Service (MaaS): A stated choice study. Transportation Research Part A: Policy and Practice 117, 302–318.

Hoddeson, L., Daitch, V., 2002. True genius: The life and science of John Bardeen; the only winner of two Nobel Prizes in physics. Joseph Henry Press, Washington, DC, 467 pp.

Hoffmann, S., Weyer, J., Longen, J., 2017. Discontinuation of the automobility regime? An integrated approach to multi-level governance. Transportation Research Part A: Policy and Practice 103, 391–408.

Hoffmeyer-Zlotnik, J.H.P., Warner, U., 2014. Soziodemographische Standards. In: Handbuch Methoden der empirischen Sozialforschung, Wiesbaden. Springer, pp. 733–743.

Hohenberger, C., Spörrle, M., Welpe, I.M., 2016. How and why do men and women differ in their willingness to use automated cars? The influence of emotions across different age groups. Transportation Research Part A: Policy and Practice 94, 374–385.

Holden, E., Banister, D., Gössling, S., Gilpin, G., Linnerud, K., 2020. Grand Narratives for sustainable mobility: A conceptual review. Energy Research & Social Science 65, 101454.

Homburg, C., Giering, A., 1998. Konzeptualisierung und Operationalisierung komplexer Konstrukte : Ein Leitfaden für die Marketingforschung. In: Hildebrandt, L., Homburg, C. (Eds.) Die Kausalanalyse. Ein Instrument der empirischen betriebswirtschaftlichen Forschung. Schäffer-Poeschel Verlag, Stuttgart, pp. 111–146.

Hornbæk, K., Hertzum, M., 2017. Technology Acceptance and User Experience. ACM Trans. Computers in Human Interactions 24, 1–30.

Hotchkiss, L., 1976. A Technique for Comparing Path Models Between Subgroups. Sociological Methods & Research 5, 53–76.

Howell, D.C., 2010. Statistical methods for psychology, 7th ed. Wadsworth Cengage Learning, Belmont, Calif., 768 pp.

Hoyle, R.H., 1995. Structural equation modeling: Concepts, issues, and applications. Sage, London.

Hu, L., Bentler, P.M., 1999. Cutoff criteria for fit indexes in covariance structure analysis: Conventional criteria versus new alternatives. Structural Equation Modeling: A Multidisciplinary Journal 6, 1–55.

Hu, P.J.-H., Clark, T.H., Ma, W.W., 2003. Examining technology acceptance by school teachers: a longitudinal study. Information & Management 41, 227–241.

Huijts, N., E.J.E. Molin, B. van Wee, 2014. Hydrogen fuel station acceptance: A structural equation model based on the technology acceptance framework. Journal of Environmental Psychology 38, 153–166.

Huijts, N., Groot, J. de, Molin, E., van Wee, B., 2013. Intention to act towards a local hydrogen refueling facility: Moral considerations versus self-interest. Transportation Research Part A: Policy and Practice 48, 63–74.

Huijts, N., Molin, E., Steg, L., 2012. Psychological factors influencing sustainable energy technology acceptance: A review-based comprehensive framework. Renewable and Sustainable Energy Reviews 16, 525–531.

Hung, N.B., Lim, O., 2020. A review of history, development, design and research of electric bicycles. Applied Energy 260, 114323.

Hussain, A., Mkpoguijogu, E., 2017. Usability metrics and methods for public transportation applications: A systematic review. Journal of Engineering Science and Technology, 98–105.

Hwang, J., Lee, J.-S., Kim, H., 2019. Perceived innovativeness of drone food delivery services and its impacts on attitude and behavioral intentions: The moderating role of gender and age. International Journal of Hospitality Management 81, 94–103.

Igbaria, M., Guimaraes, T., Davis, G.B., 1995. Testing the Determinants of Microcomputer Usage via a Structural Equation Model. Journal of Management Information Systems 11, 87–114.

Im Il, Kim, Y., Han, H.-J., 2008. The effects of perceived risk and technology type on users' acceptance of technologies. Information & Management 45, 1–9.

Irani, T., 2000. Prior Experience, Perceived Usefulness and the Web: Factors Influencing Agricultural Audiences' Adoption of Internet Communication Tools. Journal of Applied Communications 84.

ISO, 1998. Ergonomic requirements for office work with visual display terminals (VDTs): Part 11: Guidance on usability, 1st ed. ISO 13.180 Ergonomics; 35.180 IT terminal and other peripheral equipment. https://www.iso.org/standard/16883.html. Accessed 16 March 2020.

ITS International, 2019. Singapore aims to set MaaS benchmark. https://www.itsinternational.com/its17/feature/singapore-aims-set-maas-benchmark. Accessed 5 March 2022.

Jackson, C.M., Chow, S., Leitch, R.A., 1997. Toward an Understanding of the Behavioral Intention to Use an Information System. Decision Sciences 28, 357–389.

Jacoby, J., Kaplan, L.B., 1972. The Components of Perceived Risk. Proceedings of the Third Annual Conference of the Association for Consumer Research, 382–393.

Jarvenpaa, S.L., Todd, P.A., 1996. Consumer Reactions to Electronic Shopping on the World Wide Web. International Journal of Electronic Commerce 1, 59–88.

Jeekel, H., 2017. Social Sustainability and Smart Mobility: Exploring the relationship. Transportation Research Procedia 25, 4296–4310.

Jia, L., Dianne, H., Shiwei, S., 2014. The Effect of Technology Usage Habits on Consumers Intention to Continue Use Mobile Payments. Technology Usage Habits and Behavioral Intention. Twentieth Americas Conference on Information Systems.

Jing, P., Xu, G., Chen, Y., Shi, Y., Zhan, F., 2020. The Determinants behind the Acceptance of Autonomous Vehicles: A Systematic Review. Sustainability 12, 1719.

Jittrapirom, P., Caiati, V., Feneri, A.-M., Ebrahimigharehbaghi, S., González, M.J.A., Narayan, J., 2017. Mobility as a Service: A Critical Review of Definitions, Assessments of Schemes, and Key Challenges. UP 2, 13.

Jittrapirom, P., Marchau, V., van der Heijden, R., Meurs, H., 2018. Future implementation of mobility as a service (MaaS): Results of an international Delphi study. Travel Behaviour and Society, 281–294.

Jochem, P., Babrowski, S., Fichtner, W., 2015. Assessing CO 2 emissions of electric vehicles in Germany in 2030. Transportation Research Part A: Policy and Practice 78, 68–83.

Jones, T., Harms, L., Heinen, E., 2016. Motives, perceptions and experiences of electric bicycle owners and implications for health, wellbeing and mobility. Journal of Transport Geography 53, 41–49.

Jugert, P., Greenaway, K.H., Barth, M., Büchner, R., Eisentraut, S., Fritsche, I., 2016. Collective efficacy increases pro-environmental intentions through increasing self-efficacy. Journal of Environmental Psychology 48, 12–23.

Kagermann, H., 2017. Die Mobilitätswende: Die Zukunft der Mobilität ist elektrisch, vernetzt und automatisiert. In: Hildebrandt, A., Landhäußer, W. (Eds.) CSR und Digitalisierung: Der digitale Wandel als Chance und Herausforderung für Wirtschaft und Gesellschaft. Springer Berlin Heidelberg, Berlin, Heidelberg, pp. 357–371.

Kaiser, H.F., 1974. An index of factorial simplicity. Psychometrika 39, 31–36.

Kalton, G., Schuman, H., 1982. The Effect of the Question on Survey Responses: A Review. Journal of the Royal Statistical Society. Series A (General) 145, 42.

Kanger, L., Geels, F.W., Sovacool, B., Schot, J., 2019. Technological diffusion as a process of societal embedding: Lessons from historical automobile transitions for future electric mobility. Transportation Research Part D: Transport and Environment 71, 47–66.

Kapser, S., Abdelrahman, M., 2020. Acceptance of autonomous delivery vehicles for last-mile delivery in Germany – Extending UTAUT2 with risk perceptions. Transportation Research Part C: Emerging Technologies 111, 210–225.

Karahanna, E., Detmar W. Straub, Norman L. Chervany, 1999. Information technology adoption across time: A cross-sectional comparison of pre-adoption and post-adoption beliefs. MIS Quarterly: Management Information Systems 23, 183–213.

Karkhanis, P., van den Mark, B., Saurab, R. (Eds.), 2018. Defining the C-ITS reference architecture. IEEE, 148–151.

Karlsson, I., Mukhtar-Landgren, D., Smith, G., Koglin, T., Kronsell, A., Lund, E., Sarasini, S., Sochor, J., 2020. Development and implementation of Mobility-as-a-Service – A qualitative study of barriers and enabling factors. Transportation Research Part A: Policy and Practice 131, 283–295.

Kauschke, L., 2020. Deliverable D2 – competence region smart mobility: empiric results. KoSMoS project report, Saarbruecken. kosmos-project.eu.

Kauschke, L., Adorff, C., Kany, S., 2021. Testing hydrogen – Results of an acceptance study: Deliverable D2.2 EMPOWER project. University of Applied Sciences Saarland, Saarbrücken.

Kauschke, L., Maringer, S., 2019. Requirement analysis and new perspectives: Mobility, human and smart mobility. University of Applied Sciences Saarbrücken, Saarland. https://kosmos-project.eu/anforderungsanalyse-mobilitaet-mensch-und-smart-mobility-heute-und-in-zukunft/. Accessed 17 September 2020.

Kauschke, L., Schulz, W., 2017. Upgrading electric mobility with Intelligent Transport Systems: A technology acceptance perspective. Transportation Research Procedia.

Kaye, S.-A., Somoray, K., Rodwell, D., Lewis, I., 2021. Users' acceptance of private automated vehicles: A systematic review and meta-analysis. Journal of safety research 79, 352–367.

Kenyon, S., Lyons, G., 2003. The value of integrated multimodal traveler information and its potential contribution to modal change. Transportation Research Part F: Traffic Psychology and Behaviour 6, 1–21.

Khalilzadeh, J., Tasci, A.D., 2017. Large sample size, significance level, and the effect size: Solutions to perils of using big data for academic research. Tourism Management 62, 89–96.

Khechine, H., Lakhal, S., Ndjambou, P., 2016. A meta-analysis of the UTAUT model: Eleven years later. Canadian Journal of Administrative Sciences / Revue Canadienne des Sciences de l'Administration 33, 138–152.

Khechine, H., Lakhal, S., Pascot, D., Bytha, A., 2014. UTAUT model for blended learning: The role of gender and age in the intention to use webinars. Interdisciplinary Journal of E-Learning and Learning Objects 10, 33–52.

Kim, S.H., 2008. Moderating effects of Job Relevance and Experience on mobile wireless technology acceptance: Adoption of a smartphone by individuals. Information & Management 45, 387–393.

Kim, S.S., Malhotra, N.K., 2005. A Longitudinal Model of Continued Is Use: An Integrative View of Four Mechanisms Underlying Postadoption Phenomena. Management science 51, 741–755.

Kim, S.S., Malhotra, N.K., Narasimhan, S., 2005. Research note—two competing perspectives on automatic use: A theoretical and empirical comparison. Information systems research 16, 418–432.

King, W.R., He, J., 2006. A meta-analysis of the technology acceptance model. Information & Management 43, 740–755.

Kinnear, T.C., Taylor, J.R., Ahmed, S.A., 1974. Ecologically concerned consumers: who are they? Ecologically concerned consumers can be identified. Journal of Marketing 38, 20–24.

Kline, R.B., 2015. Principles and practice of structural equation modeling. Guilford publications, New York.

Klöckner, C.A., 2014. The dynamics of purchasing an electric vehicle – A prospective longitudinal study of the decision-making process. Transportation Research Part F: Traffic Psychology and Behaviour 24, 103–116.

Kmenta, J., 1990. Elements of econometrics, 2nd ed. Macmillan, New York, 786 pp.

Kock, N., 2015. Common method bias in PLS-SEM: A full collinearity assessment approach. International Journal of e-Collaboration (ijec) 11, 1–10.

Kock, N., Hadaya, P., 2018. Minimum sample size estimation in PLS-SEM: The inverse square root and gamma-exponential methods. Information Systems Journal 28, 227–261.

Kollmann, T., 2013. Acceptance of innovative user goods and systems: Consequences for the introduction of telecommunication- and multimediasystems. Springer, Wiesbaden.

König, M., Neumayr, L., 2017. Users' resistance towards radical innovations: The case of the self-driving car. Transportation Research Part F: Traffic Psychology and Behaviour 44, 42–52.

Kormos, C., Gifford, R., 2014. The validity of self-report measures of pro-environmental behavior: A meta-analytic review. Journal of Environmental Psychology 40, 359–371.

Kosi, T., Bojnec, Š., 2013. Institutional barriers to business entry in advanced economics. Journal of Business Economics and Management 14, 317–329.

Koufaris, M., 2002. Applying the Technology Acceptance Model and Flow Theory to Online Consumer Behavior. Information systems research 13, 205–223.

Kraljic, A., Pestek, A., 2016. An application of UTAUT2 model in exploring the impact of quality of technology on mobile Internet. Economic Review: Journal of Economics and Business 14, 66–76.

Kulas, J.T., Stachowski, A.A., 2009. Middle category endorsement in odd-numbered Likert response scales: Associated item characteristics, cognitive demands, and preferred meanings. Journal of Research in Personality 43, 489–493.

Kumar, R.R., Alok, K., 2020. Adoption of electric vehicle: A literature review and prospects for sustainability. Journal of Cleaner Production 253, 119911.

Kunze, F., Boehm, S., Bruch, H., 2013. Age, resistance to change, and job performance. Journal of Managerial Psych 28, 741–760.

Kyriakidis, M., Happee, R., Winter, J. de, 2015. Public opinion on automated driving: Results of an international questionnaire among 5000 respondents. Transportation Research Part F: Traffic Psychology and Behaviour 32, 127–140.

Lamnek, S., Krell, C., 2005. Qualitative Sozialforschung. Psychologie Verlags Union München.

Landis, J.R., Koch, G.G., 1977. An application of hierarchical kappa-type statistics in the assessment of majority agreement among multiple observers. Biometrics, 363–374.

Larue, G.S., Rakotonirainy, A., Haworth, N.L., Darvell, M., 2015. Assessing driver acceptance of Intelligent Transport Systems in the context of railway level crossings. Transportation Research Part F: Traffic Psychology and Behaviour 30, 1–13.

Law, J., 2009. Actor network theory and material semiotics. The new Blackwell companion to social theory 3, 141–158.

Le Bris, J., 2015. Die individuelle Mobilitätspraxis und Mobilitätskarrieren von Pedelec-Besitzern. Adoption und Appropriation von Elektrofahrrädern., Tübingen, 600 pp.

Lee, E., Park, N.-K., Han, J.H., 2013. Gender Difference in Environmental Attitude and Behaviors in Adoption of Energy-Efficient Lighting at Home. Journal of Sustainable Development 6.

Lee, S.Y., 2014. Examining the factors that influence early adopters' smartphone adoption: The case of college students. Telematics and Informatics 31, 308–318.

Lee, Y., Larsen, K.R., 2003. The Technology Acceptance Model: Past, Present, and Future. Communications of the Association for Information Systems 12.

Legris, P., Ingham, J., Collerette, P., 2003. Why do people use information technology? A critical review of the technology acceptance model. Information & Management 40, 191–204.

Lenz, B., Kolarova, V., Stark, K., 2019. Gender Issues in the Digitalized 'Smart' Mobility World – Conceptualization and Empirical Findings Applying a Mixed Methods Approach. In: Krömker, H. (Ed.) HCI in Mobility, Transport, and Automotive Systems, vol. 11596. Springer International Publishing, Cham, pp. 378–392.

Lian, J.-W., Yen, D.C., 2014. Online shopping drivers and barriers for older adults: Age and gender differences. Computers in Human Behavior 37, 133–143.

Liébana-Cabanillas, F., Sánchez-Fernández, J., Muñoz-Leiva, F., 2014. The moderating effect of experience in the adoption of mobile payment tools in Virtual Social Networks: The m-Payment Acceptance Model in Virtual Social Networks (MPAM-VSN). International Journal of Information Management 34, 151–166.

Likert, R., 1932. A technique for the measurement of attitudes. Archives of psychology.

Liljamo, T., Liimatainen, H., Pöllänen, M., 2018. Attitudes and concerns on automated vehicles. Transportation Research Part F: Traffic Psychology and Behaviour 59, 24–44.

Limayem, M., Hirt, S.G., Cheung, C.M.K., 2007. How Habit Limits the Predictive Power of Intention: The Case of Information Systems Continuance. MIS Quarterly, 705–737.

Lin, B., Tan, R., 2017. Estimation of the environmental values of electric vehicles in Chinese cities. Energy Policy 104, 221–229.

Liu, P., Yang, R., Xu, Z., 2019a. Public Acceptance of Fully Automated Driving: Effects of Social Trust and Risk/Benefit Perceptions. Risk Analysis 39, 326–341.

Liu, Z., Tan, H., Kuang, X., Hao, H., Zhao, F., 2019b. The Negative Impact of Vehicular Intelligence on Energy Consumption. Journal of Advanced Transportation 2019, 1–11.

Loo, R., 2002. A caveat on using single-item versus multiple-item scales. Journal of Managerial Psych 17, 68–75.

Lowry, P.B., Gaskin, J., 2014. Partial Least Squares (PLS) Structural Equation Modeling (SEM) for Building and Testing Behavioral Causal Theory: When to Choose It and How to Use It. IEEE Trans. Profess. Commun. 57, 123–146.

Lucke, D., 1995. Akzeptanz: Legitimität in der 'Abstimmungsgesellschaft'. VS Verlag für Sozialwissenschaften, Wiesbaden, 452 pp.

Luhmann, N., 1995. Social systems. Stanford University Press.

Luhmann, N., 1997. Die Gesellschaft der Gesellschaft, 1st ed. Suhrkamp, Frankfurt am Main, 594 pp.

Lutz, C., Hoffmann, C.P., Bucher, E., Fieseler, C., 2018. The role of privacy concerns in the sharing economy. Information, Communication & Society 21, 1472–1492.

Lyons, G., 2011. Technology fix versus behaviour change. Mobilities: new perspectives on transport and society, 159–177.

Lyons, G., 2018. Getting smart about urban mobility – Aligning the paradigms of smart and sustainable. Transportation Research Part A: Policy and Practice 115, 4–14.

Lyons, G., Hammond, P., Mackay, K., 2019. The importance of user perspective in the evolution of MaaS. Transportation Research Part A: Policy and Practice 121, 22–36.

Ma, Q., Chan, A.H.S., Chen, K., 2016. Personal and other factors affecting acceptance of smartphone technology by older Chinese adults. Applied ergonomics 54, 62–71.

MaaS Global, 2017. Written evidence submitted by MaaS Global Oy (MAS0026). Submission to the UK Parliament's Transport Committee inquiry into Mobility as a Service.

Macedo, I.M., 2017. Predicting the acceptance and use of information and communication technology by older adults: An empirical examination of the revised UTAUT2. Computers in Human Behavior 75, 935–948.

Madigan, R., Louw, T., Dziennus, M., Graindorge, T., Ortega, E., Graindorge, M., Merat, N., 2016. Acceptance of Automated Road Transport Systems (ARTS): An Adaptation of the UTAUT Model. Transportation Research Procedia 14, 2217–2226.

Madigan, R., Louw, T., Wilbrink, M., Schieben, A., Merat, N., 2017. What influences the decision to use automated public transport? Using UTAUT to understand public acceptance of automated road transport systems. Transportation Research Part F: Traffic Psychology and Behaviour 50, 55–64.

Mahlke, S., Thüring, M., 2007. Studying antecedents of emotional experiences in interactive contexts, in: Proceedings of the SIGCHI Conference on Human Factors in Computing Systems – CHI'07. the SIGCHI Conference, San Jose, California, USA. 28.04.2007 – 03.05.2007. ACM Press, New York, New York, USA, pp. 915–918.

Maibach, M., Schreyer, C., Sutter, D., van Essen, H.P., Boon, B.H., Smokers, R., Schroten, A., Doll, C., Pawlowska, B., Bak, M., 2008. Handbook on estimation of external costs in the transport sector. 336 pp.

Malhotra, Y., Galletta, D.F., 1999. Extending the technology acceptance model to account for social influence: Theoretical bases and empirical validation. Proceedings of the 32nd Annual Hawaii International Conference on Systems Sciences 32.

Mallat, N., Rossi, M., Tuunainen, V.K., Öörni, A., 2009. The impact of use context on mobile ser-vices acceptance: The case of mobile ticketing. Information & Management 46, 190–195.

Manders, M., Klaassen, K., 2018. Unpacking the Smart Mobility Concept in the Dutch Context Based on a Text Mining Approach. Sustainability, 11, 6583.

Manfreda, A., Ljubi, K., Groznik, A., 2019. Autonomous vehicles in the smart city era: An empirical study of adoption factors important for millennials. International Journal of Information Management, 102050.

Marchau, V.A.W.J., 2000. Technology assessment of automated vehicle guidance: Prospects for automated driving implementation. Delft University Press, Delft, 209 pp.

Markus, K.A., Borsboom, D., 2013. Frontiers of test validity theory: Measurement, causation, and meaning. Routledge, New York.

Martins, C., Oliveira, T., Popovič, A., 2014. Understanding the Internet banking adoption: A unified theory of acceptance and use of technology and perceived risk application. International Journal of Information Management 34, 1–13.

Marzal, S., Salas, R., González-Medina, R., Garcerá, G., Figueres, E., 2018. Current challenges and future trends in the field of communication architectures for microgrids. Renewable and Sustainable Energy Reviews 82, 3610–3622.

Matheus, R., Janssen, M., Janowski, T., 2021. Design principles for creating digital transparency in government. Government Information Quarterly 38, 101550.

Maurer, M., Gerdes, J.C., Lenz, B., Winner, H., 2015. Autonomes Fahren. Springer Berlin Heidelberg, Berlin, Heidelberg, 726 pp.

Mayer, A., 2019. Motivations and barriers to electric bike use in the U.S.: views from online forum participants. International Journal of Urban Sustainable Development, 1–9.

Mayring, P., 2015. Qualitative content analysis: Theoretical background and procedures. In: Approaches to qualitative research in mathematics education. Springer, pp. 365–380.

Mazur, C., Contestabile, M., Offer, G.J., Brandon, N.P., 2015. Assessing and comparing German and UK transition policies for electric mobility. Environmental Innovation and Societal Transitions 14, 84–100.

McAndrews, Z., Richardson, J., Stopa, L., 2018. Psychometric properties of acceptance measures: A systematic review. Journal of Contextual Behavioral Science, 261–277.

McKelvie, S.J., Standing, L., St. Jean, D., Law, J., 1993. Gender differences in recognition memory for faces and cars: Evidence for the interest hypothesis. Bulletin of the psychological society 31, 447–448.

McQueen, M., MacArthur, J., Cherry, C., 2019. The E-Bike Potential: Estimating the Effect of E-Bikes on Person Miles Travelled and Greenhouse Gas Emissions, Transportation Research and Education Center (TREC), https://doi.org/10.15760/trec.242.

Mehmood, A., Imran, M., 2021. Digital social innovation and civic participation: Toward responsible and inclusive transport planning. European Planning Studies 29, 1870–1885.

Mentzer, J.T., Flint, D.J., 1997. Validity in logistics research. Journal of business logistics 18, 199.

Messick, S., 1960. Dimensions of social desirability. Journal of Consulting Psychology 24, 279.

Meuser, M., Nagel, U., 2009. Das Experteninterview — konzeptionelle Grundlagen und methodische Anlage. In: Pickel, S., Pickel, G., Lauth, H.-J., Jahn, D. (Eds.) Methoden der vergleichenden Politik- und Sozialwissenschaft: Neue Entwicklungen und Anwendungen. VS Verlag für Sozialwissenschaften, Wiesbaden, pp. 465–479.

Millar, C.C., Groth, O., Mahon, J.F., 2018. Management innovation in a VUCA world: Challenges and recommendations. California management review 61, 5–14.

Miller, A.L., 2012. Investigating Social Desirability Bias in Student Self-Report Surveys. Educational Research Quarterly 36 (1), 30–47.

Miller, J., Ulrich, R., 2019. The quest for an optimal alpha. PLOS ONE 14 (1), e0208631.

Miller, J.H., Butts, C.T., Rode, D., 2002. Communication and cooperation. Journal of Economic Behavior & Organization 47, 179–195.

Mitchell, V.-W., 1999. Consumer perceived risk: conceptualizations and models. European Journal of Marketing 33, 163–195.

Mitzner, T.L., Boron, J.B., Fausset, C.B., Adams, A.E., Charness, N., Czaja, S.J., Dijkstra, K., Fisk, A.D., Rogers, W.A., Sharit, J., 2010. Older Adults Talk Technology: Technology Usage and Attitudes. Computers in Human Behavior 26, 1710–1721.

Mohamed, M., Higgins, C.D., Ferguson, M., Réquia, W.J., 2018. The influence of vehicle body type in shaping behavioural intention to acquire electric vehicles: A multi-group structural equation approach. Transportation Research Part A: Policy and Practice 116, 54–72.

Molin, E.J., Brookhuis, K.A., 2007. Modeling acceptability of the intelligent speed adapter. Transportation Research Part F: Traffic Psychology and Behaviour 10, 99–108.

Møller, M., Haustein, S., Bohlbro, M.S., 2018. Adolescents' associations between travel behaviour and environmental impact: A qualitative study based on the Norm-Activation Model. Travel Behaviour and Society 11, 69–77.

Montoya, A.K., 2019. Moderation analysis in two-instance repeated measures designs: Probing methods and multiple moderator models. Behavior research methods 51, 61–82.

Moore, G.C., Benbasat, I., 1991. Development of an instrument to measure the perceptions of adopting an information technology innovation. Information systems research 2, 192–222.

Morgan-Thomas, A., Veloutsou, C., 2013. Beyond technology acceptance: Brand relationships and online brand experience. Journal of Business Research 66, 21–27.

Morosan, C., DeFranco, A., 2016. It's about time: Revisiting UTAUT2 to examine consumers' intentions to use NFC mobile payments in hotels. International Journal of Hospitality Management 53, 17–29.

Morris, E., 2009. Sex and the SUV: Men, Women, and Travel Behavior. https://freakonomics.com/2009/11/25/sex-and-the-suv-men-women-and-travel-behavior/?c_page=3. Accessed 9 December 2021.

Morris, M.G., Venkatesh, V., Ackerman, P.L., 2005. Gender and Age Differences in Employee Decisions About New Technology: An Extension to the Theory of Planned Behavior. IEEE Transactions on Engineering Management 52, 69–84.

Mourato, S., Saynor, B., Hart, D., 2004. Greening London's black cabs: a study of driver's preferences for fuel cell taxis. Energy Policy 32, 685–695.

Mueller-Seitz, G., Seiter, M., Wenz, P., 2016. What is smart city? Springer, Berlin.

Mukhtar-Landgren, D., Paulsson, A., 2020. Governing smart mobility: policy instrumentation, technological utopianism, and the administrative quest for knowledge. Administrative Theory & Praxis, 1–19.

Müller-Doohm, S., 2016. Habermas: A biography. John Wiley & Sons, New Jersey.

Mulley, C., Kronsell, A., 2018. Workshop 7 report: The 'uberisation' of public transport and mobility as a service (MaaS): Implications for future mainstream public transport. Research in Transportation Economics 69, 568–572.

Mumford, J.G., 2006. Improving risk communication: Strategies for public acceptance of new technology involving high impact low frequency risk. University of Surrey, UK.

Murman, D.L., 2015. The Impact of Age on Cognition. Seminars in hearing 36, 111–121.

National People's Congress, 2014. National plan for tackling climate change 2014–2020.

Nelson, R., 1994. The Co-evolution of Technology, Industrial Structure, and Supporting Institutions. Industrial Corporation Change 3, 47–63.

Neumann, I., Cocron, P., Franke, T., Krems, J.F., 2010. Electric vehicles as a solution for green driving in the future? A field study examining the user acceptance of electric vehicles. European Conference on Human Interface Design for Intelligent Transport Systems, Berlin.

Niehaves, B., Plattfaut, R., 2014. Internet adoption by the elderly: employing IS technology acceptance theories for understanding the age-related digital divide. European Journal of Information Systems 23, 708–726.

Nielsen, J., 1994. Usability engineering. Morgan Kaufmann, Amsterdam.

Nietsch, M., Schott, G., 2021. The Legal Framework for Ridesharing Businesses and the Case of Uber in Germany. In: Global Perspectives on Legal Challenges Posed by Ridesharing Companies. Springer, pp. 163–183.

Nikitas, A., 2018. Understanding bike-sharing acceptability and expected usage patterns in the context of a small city novel to the concept: A story of 'Greek Drama'. Transportation Research Part F: Traffic Psychology and Behaviour 56, 306–321.

Nikitas, A., Wallgren, P., Rexfelt, O., 2016. The paradox of public acceptance of bike sharing in Gothenburg. Proceedings of the Institution of Civil Engineers – Engineering Sustainability 169, 101–113.

Nikolaeva, A., Adey, P., Cresswell, T., Lee, J.Y., Nóvoa, A., Temenos, C., 2019. Commoning mobility: Towards a new politics of mobility transitions. Transportation in British geography 44, 346–360.

Nilsson, L., 1996. Safety effects of adaptive cruise controls in critical traffic situations. Swedish National Road and Transport Research Institute, Yokohama.

Nitzl, C., 2010. Eine anwenderorientierte Einführung in Partial Least Square (PLS)-Methode. SSRN Electronic Journal, 72.

Nobis, C., 2019. Mobility in Germany – analyses for cyclists and pedestrians 70.905, Berlin. https://nationaler-radverkehrsplan.de/de/node/21214. Accessed 29 May 2020.

Nordhoff, S., Kyriakidis, M., van Arem, B., Happee, R., 2019. A multi-level model on automated vehicle acceptance (MAVA): a review-based study. Theoretical Issues in Ergonomics Science 20, 682–710.

Nordhoff, S., Louw, T., Innamaa, S., Lehtonen, E., Beuster, A., Torrao, G., Bjorvatn, A., Kessel, T., Malin, F., Happee, R., Merat, N., 2020. Using the UTAUT2 model to explain public acceptance of conditionally automated (L3) cars: A questionnaire study among 9,118 car drivers from eight European countries. Transportation Research Part F: Traffic Psychology and Behaviour 74, 280–297.

Nordhoff, S., Malmsten, V., van Arem, B., Liu, P., Happee, R., 2021. A structural equation modeling approach for the acceptance of driverless automated shuttles based on constructs from the Unified Theory of Acceptance and Use of Technology and the Diffusion of Innovation Theory. Transportation Research Part F: Traffic Psychology and Behaviour 78, 58–73.

Nordhoff, S., Winter, J. de, Kyriakidis, M., van Arem, B., Happee, R., 2018a. Acceptance of Driverless Vehicles: Results from a Large Cross-National Questionnaire Study. Journal of Advanced Transportation 2018, 1–22.

Nordhoff, S., Winter, J. de, Madigan, R., Merat, N., van Arem, B., Happee, R., 2018b. User acceptance of automated shuttles in Berlin-Schöneberg: A questionnaire study. Transportation Research Part F: Traffic Psychology and Behaviour 58, 843–854.

Nordlund, A., Jansson, J., Westin, K., 2018. Acceptability of electric vehicle aimed measures: Effects of norm activation, perceived justice and effectiveness. Transportation Research Part A: Policy and Practice 117, 205–213.

Nunnally, J.C., 1978. An Overview of Psychological Measurement. In: Wolman, B.B. (Ed.) Clinical Diagnosis of Mental Disorders: A Handbook. Springer US, Boston, MA, pp. 97–146.

Nyblom, Å., 2014. Making plans or 'just thinking about the trip'? Understanding people's travel planning in practice. Journal of Transport Geography 35, 30–39.

Nye, D.E., 2001. Consuming power: A social history of American energies, 1st ed. MIT Press, Cambridge, Mass., 331 pp.

O'Garra, T., Mourato, S., Pearson, P., 2005. Analysing awareness and acceptability of hydrogen vehicles: A London case study. International Journal of Hydrogen Energy 30, 649–659.

Ono, K., Tsunemi, K., 2017. Identification of public acceptance factors with risk perception scales on hydrogen fueling stations in Japan. International Journal of Hydrogen Energy 42, 10697–10707.

Ooi, K.-B., Tan, G.W.-H., 2016. Mobile technology acceptance model: An investigation using mobile users to explore smartphone credit card. Expert Systems with Applications 59, 33–46.

ORAD, 2014. Taxonomy and Definitions for Terms Related to On-Road Motor Vehicle Automated Driving Systems, 201806th ed. SAE International, 400 Commonwealth Drive, Warrendale, PA, United States, https://doi.org/10.4271/J3016_201401.

Pakusch, C., Bossauer, P., 2017. User Acceptance of Fully Autonomous Public Transport, in: Proceedings of the 14th International Joint Conference on e-Business and Telecommunications. 14th International Conference on e-Business, Madrid, Spain. 24.07.2017 – 26.07.2017. SCITEPRESS – Science and Technology Publications, pp. 52–60.

Palau-Saumell, R., Forgas-Coll, S., Sánchez-García, J., Robres, E., 2019. User Acceptance of Mobile Apps for Restaurants: An Expanded and Extended UTAUT-2. Sustainability, 12–20.

Panagiotopoulos, I., Dimitrakopoulos, G., 2018. An empirical investigation on consumers' intentions towards autonomous driving. Transportation Research Part C: Emerging Technologies 95, 773–784.

Park, C., Kim, H., Yong, T., 2017a. Dynamic characteristics of smart grid technology acceptance. Energy Procedia 128, 187–193.

Park, E., 2020. User acceptance of smart wearable devices: An expectation-confirmation model approach. Telematics and Informatics 47, 101318.

Park, E., Cho, Y., Han, J., Kwon, S.J., 2017b. Comprehensive Approaches to User Acceptance of Internet of Things in a Smart Home Environment. IEEE Internet of Things Journal 4, 2342–2350.

Park, E., Jooyoung, L., Yongwoo, C., 2018. Understanding the Emergence and Social Acceptance of Electric Vehicles as Next-Generation Models for the Automobile Industry. Sustainability 10, 662.

Park, J., Ha, S., 2014. Understanding Consumer Recycling Behavior: Combining the Theory of Planned Behavior and the Norm Activation Model. Family and Consumer Sciences Research Journal 42, 278–291.

Park, Y., Chen, J.V., 2007. Acceptance and adoption of the innovative use of smartphone. Industrial Management & Data Systems 107, 1349–1365.

Parkhurst, G., Seedhouse, A., 2019. Will the 'smart mobility' revolution matter? In: Docherty, I., Shaw, J. (Eds.) Transport matters. Policy Press, Bristol, Chicago, IL, pp. 349–380.

Paternoga, S., Pieper, N., Woisetschläger, D.M., Beuscher, G., Wachalski, T., 2013. Acceptance of electric vehicles – Hopeless venture or big chance?, Wolfsburg.

Pavlou, P.A., 2003. Consumer Acceptance of Electronic Commerce: Integrating Trust and Risk with the Technology Acceptance Model. International Journal of Electronic Commerce 7, 101–134.

Payre, W., Cestac, J., Delhomme, P., 2014. Intention to use a fully automated car: Attitudes and a-priori acceptability. Transportation Research Part F: Traffic Psychology and Behaviour 27, 252–263.

Payre, W., Diels, C., 2020. I want to brake free: The effect of connected vehicle features on driver behaviour, usability and acceptance. Applied ergonomics 82, 102932.

Pepper, S.C., 1926. Emergence. The Journal of Philosophy 23, 241.

Perneger, T.V., Courvoisier, D.S., Hudelson, P.M., Gayet-Ageron, A., 2015. Sample size for pretests of questionnaires. Quality of Life Research 24, 147–151.

Pfeiffer, J., Entress-Fuersteneck, M. von, Urbach, N., Buchwald, A., 2016. Quantify-me: consumer acceptance of wearable self-tracking devices, 1–16.

Pfirsching, V., 2022. Interview: Gaia-X – Stuck in the concept stage? https://genesis-aka.net/language/german/2022/04/22/interview-gaia-x-im-konzeptstadium-stecken-geblieben/. Accessed 20 May 2022.

Pianelli, C., Saad, F., Abric, J.-C., 2007. Social representations and acceptability of LAVIA (French ISA system), Bejing, China.

Piao, J., McDonald, M., Hounsell, N., Graindorge, M., Graindorge, T., Malhene, N., 2016. Public Views towards Implementation of Automated Vehicles in Urban Areas. Transportation Research Procedia 14, 2168–2177.

Plazier, P.A., Weitkamp, G., van den Berg, Agnes E., 2017. The potential for e-biking among the younger population: A study of Dutch students. Travel Behaviour and Society 8, 37–45.

Podsakoff, N.P., 2003. Common method biases in behavioral research: a critical review of the literature and recommended remedies. Journal of Applied Psychology 885, 10.1037.

Pohle, J., 2018. Datenschutz und Technikgestaltung, Dissertation, HU Berlin, DOI: 10.18452/19136.

Polydoropoulou, A., Pagoni, I., Tsirimpa, A., 2018. Ready for Mobility as a Service? Insights from stakeholders and end-users. Travel Behaviour and Society 21, 295–306.

Potoglou, D., Kanaroglou, P.S., 2007. Household demand and willingness to pay for clean vehicles. Transportation Research Part D: Transport and Environment 12, 264–274.

Prati, G., Marín Puchades, V., Angelis, M. de, Pietrantoni, L., Fraboni, F., Decarli, N., Guerra, A., Dardari, D., 2018. Evaluation of user behavior and acceptance of an on-bike system. Transportation Research Part F: Traffic Psychology and Behaviour 58, 145–155.

Pribyl, O., Blokpoel, R., Matowicki, M., 2020. Addressing EU climate targets: Reducing CO_2 emissions using cooperative and automated vehicles. Transportation Research Part D: Transport and Environment 86, 102437.

Punel, A., Stathopoulos, A., 2017. Modeling the acceptability of crowdsourced goods deliveries: Role of context and experience effects. Transportation Research Part E: Logistics and Transportation Review 105, 18–38.

Qiu, J., Tian, Z., Du, C., Zuo, Q., Su, S., Fang, B., 2020. A survey on access control in the age of internet of things. IEEE Internet of Things Journal 7, 4682–4696.

Qiuchen, W., Jannicke, H.B., Sebastiaan, M., 2021. The complexity of stakeholder influence on MaaS: A study on multi-stakeholder perspectives in Shenzhen self-driving mini-bus case. Research in Transportation Economics, 101070.

Rachels, J.R., Rockinson-Szapkiw, A.J., 2018. The effects of a mobile gamification app on elementary students' Spanish achievement and self-efficacy. Computer Assisted Language Learning 31, 72–89.

Rahm, E., Do, H.H., 2000. Data cleaning: Problems and current approaches. IEEE Data Engineering Bulletin 23, 3–13.

Raman, A., Don, Y., 2013. Preservice teachers' acceptance of learning management software: An application of the UTAUT2 model. International Education Studies 6, 157–164.

Ramírez-Correa, P., Rondán-Cataluña, F.J., Arenas-Gaitán, J., Martín-Velicia, F., 2019. Analyzing the acceptation of online games in mobile devices: An application of UTAUT2. Journal of Retailing and Consumer Services 50, 85–93.

Ramli, N.A., Latan, H., Nartea, G.V., 2018. Why Should PLS-SEM Be Used Rather Than Regression? Evidence from the Capital Structure Perspective. In: Avkiran, N.K., Ringle,

C.M. (Eds.) Partial least squares structural equation modeling. Recent advances in banking and finance, vol. 267, 1st ed. Springer, Cham, Switzerland, pp. 171–209.

Rathee, T., Singh, P., 2022. A Self-Sovereign Identity Management System Using Blockchain. In: Cyber Security and Digital Forensics. Springer, pp. 371–379.

Reardon, L., 2020. Smart mobility as a catalyst for policy change towards low carbon mobility? In: Shaping smart mobility futures: Governance and policy instruments in times of sustainability transitions. Emerald Publishing Limited.

Regan, D.T., Fazio, R., 1977. On the consistency between attitudes and behavior: Look to the method of attitude formation. Journal of Experimental Social Psychology 13, 28–45.

Reid, C., 2017. Bike Boom: The unexpected resurgence of cycling. Island Press, Washington, 270 pp.

Ricci, M., Bellaby, P., Flynn, R., 2008. What do we know about public perceptions and acceptance of hydrogen? A critical review and new case study evidence. International Journal of Hydrogen Energy 33, 5868–5880.

Richards, J.A., Johnson, M.P., 2014. A Case for Theoretical Integration: Combining Constructs from the Theory of Planned Behavior and the Extended Parallel Process Model to Predict Exercise Intentions. SAGE Open 4.

Richardson, H.A., Simmering, M.J., Sturman, M.C., 2009. A Tale of Three Perspectives. Organizational Research Methods 12, 762–800.

Rigdon, E.E., 2016. Choosing PLS path modeling as analytical method in European management research: A realist perspective. European Management Journal 34, 598–605.

Ringle, Sarstedt, Straub, 2012. Editor's Comments: A Critical Look at the Use of PLS-SEM in 'MIS Quarterly'. MIS Quarterly 36, 1–12.

Ringle, C., Wende, S., Becker, J.-M., 2015. SmartPLS 3.

Ringle, C.M., Sarstedt, M., 2016. Gain more insight from your PLS-SEM results. Industrial Management & Data Systems 116, 1865–1886.

Roberts, J.A., 1995. Profiling Levels of Socially Responsible Consumer Behavior: A Cluster Analytic Approach and Its Implications for Marketing. Journal of Marketing Theory and Practice 3, 97–117.

Rödel, C., Stadler, S., Meschtscherjakov, A., Tscheligi, M., 2014. Towards Autonomous Cars, in: Proceedings. AutomotiveUI 2014 : 6th International Conference on Automotive User Interfaces and Interactive Vehicular Applications : September 17–19, 2014, Seattle, Washington, USA. the 6th International Conference, Seattle, WA, USA. 9/17/2014 – 9/19/2014. Association for Computing Machinery, New York, New York, pp. 1–8.

Rogers, E.M., 1962. Diffusion of innovations. Free Press, New York.

Rogers, E.M., 1995. Diffusion of innovations, 4th ed. Free Press, New York.

Rotmans, J., Kemp, R., van Asselt, M., 2001. More evolution than revolution: transition management in public policy. Foresight 3, 15–31.

Rousseau, D.M., Sitkin, S.B., Burt, R.S., Camerer, C., 1998. Not so different after all: A cross-discipline view of trust. Academy of Management Review 23, 393–404.

Roy, S.K., Balaji, M.S., Quazi, A., Quaddus, M., 2018. Predictors of customer acceptance of and resistance to smart technologies in the retail sector. Journal of Retailing and Consumer Services 42, 147–160.

Ru, X., Wang, S., Chen, Q., Yan, S., 2018. Exploring the interaction effects of norms and attitudes on green travel intention: An empirical study in eastern China. Journal of Cleaner Production 197, 1317–1327.

Saad, F., 2006. Some critical issues when studying behavioural adaptations to new driver support systems. Cognition, Technology & Work 8, 175–181.

Sammer, G., Meth, D., Gruber, C.J., 2008. Electric vehicles: the user perspective. e & i Electronics and Information Technology, pp. 393–400.

San Martín, H., Herrero, Á., 2012. Influence of the user's psychological factors on the online purchase intention in rural tourism: Integrating innovativeness to the UTAUT framework. Tourism Management 33, 341–350.

Sanmateu, M.-A., 2018. 5GNetMobil Deliverable D1.2: Analysis of value chain and market players. Project report.

Sarstedt, M., Mooi, E., 2014. A concise guide to market research: The process, data, and methods using IBM SPSS Statistics, 2nd ed. Springer Berlin Heidelberg, Berlin, Heidelberg, 347 pp.

Schade, J., Schlag, B., 2003. Acceptability of urban transport pricing strategies. Transportation Research Part F: Traffic Psychology and Behaviour 6, 45–61.

Schaefer, P.K., Schmidt, K., Knese, D., 2014. Acceptance of Electric Vehicles and New Mobility Behavior: The Example of Rhine-Main Region. In: Hülsmann, M., Fornahl, D. (Eds.) Evolutionary paths towards the mobility patterns of the future. Springer, Heidelberg, pp. 319–334.

Schäfer, M., Keppler, D., 2013. Models in technology-orientated acceptance research: Overview and reflections. discussion paper.

Schepers, J., Wetzels, M., 2007. A meta-analysis of the technology acceptance model: Investigating subjective norm and moderation effects. Information & Management 44, 90–103.

Schikofsky, J., Dannewald, T., Kowald, M., 2020. Exploring motivational mechanisms behind the intention to adopt mobility as a service (MaaS): Insights from Germany. Transportation Research Part A: Policy and Practice 131, 296–312.

Schlüter, J., Weyer, J., 2019. Carsharing as a means to raise acceptance of electric vehicles: An empirical study on regime change in automobility. Transportation Research Part F: Traffic Psychology and Behaviour 60, 185–201.

Schmidt, A., Donsbach, W., 2016. Acceptance factors of hydrogen and their use by relevant stakeholders and the media. International Journal of Hydrogen Energy 41, 4509–4520.

Schneider, D., 1995. Allgemeine Betriebswirtschaftslehre. de Gruyter, Berlin.

Schneidewind, U., 2018. Transformative Literacy. Gesellschaftliche Veränderungsprozesse verstehen und gestalten. Gaia: Okologische Perspektiven in Natur-, Geistes- und Wirtschaftswissenschaften 22, 82–86.

Schuitema, G., Steg, L., van Kruining, M., 2011. When Are Transport Pricing Policies Fair and Acceptable? Social Justice Research 24, 66–84.

Schulte, I., 2004. Issues affecting the acceptance of hydrogen fuel. International Journal of Hydrogen Energy 29, 677–685.

Schulz, W., Geis, I., 2016. Making Public-Private Cooperation – Empirical Evidence from an Institutional Economic Role Model on the Example of Cooperative Transport Systems.

Schulz, W., Joisten, N., Arnegger, B., 2019. Development of the Institutional Role Models as a Contribution to the Implementation of Co-Operative Transport Systems. SSRN Electronic Journal. https://doi.org/10.2139/ssrn.3421107.

Schulz, W.H., 2005. Application of System Dynamics to Empirical Industrial Organization – The Effects of the New Toll System. Jahrbuch für Wirtschaftswissenschaften / Review of Economics 56, 205–227.

Schulz, W.H., 2022. Smart mobility role model interview, Web.
Schulz, W.H., Franck, O., 2022. The Institutional Role Model: A System-Dynamic Approach to Reduce Complexity. International journal for sustainable development 17, 351–361.
Schulz, W.H., Franck, O., Smolka, S., 2021a. Die Theorie der institutionellen Rollenmodelle – der Restrukturierungsansatz für Unternehmen zur Bewältigung der COVID-19 Krise. In: Schulz, W.H., Joisten, N., Edye, C.F. (Eds.) Mobilität nach COVID-19. Springer Fachmedien, Wiesbaden, pp. 1–32.
Schulz, W.H., Franck, O., Smolka, S., 2021b. Die Theorie der institutionellen Rollenmodelle als Grundlagentheorie für Transformationsprozesse in Organisationen. In: Bodemann, M., Fellner, W., Just, V. (Eds.) Zukunftsfähigkeit durch Innovation. Gabler, pp. 77–100.
Schulz, W.H., Franck, O., Smolka, S., Geilenberg, V., 2021c. Nachhaltigkeit und Ressourceneffizienz bei unternehmensübergreifenden Kooperationen: Die Theorie der Institutionellen Rollenmodelle als Grundlage für Best Practices. In: Wellbrock, W., Ludin, D. (Eds.) Nachhaltiger Konsum. Springer Fachmedien Wiesbaden, Wiesbaden, pp. 349–361.
Schulz, W.H., Wieker, H., Kichniawy, J., 2014. Research Joint Ventures as a European Policy Instrument Beneath Directives and Action Plans: Transitions, Interlocking and Permeability of Political, Technological and Economical Requirements. SSRN Electronic Journal. https://doi.org/10.2139/ssrn.2606414.
Schumacker, R.E., Lomax, R.G., 2010. A beginner's guide to structural equation modeling, 3rd ed. Routledge, New York, 510 pp.
Schwanen, T., 2015. Beyond instrument: smartphone app and sustainable mobility. European Journal of Transport and Infrastructure Research 15(4). https://doi.org/10.18757/ejtir.2015.15.4.3104.
Schwartz, S.H., 1977. Normative Influences on Altruism. In: Berkowitz, L. (Ed.) Advances in Experimental Social Psychology, vol. 10. Academic Press, pp. 221–279.
Schwedes, O., 2021. „Objekt der Begierde'. In: Schwedes, O., Keichel, M. (Eds.) Das Elektroauto. Springer Fachmedien Wiesbaden, Wiesbaden, pp. 49–76.
Seele, P., Lock, I., 2017. The game-changing potential of digitalization for sustainability: possibilities, perils, and pathways. Sustainability Science 12, 183–185.
Sener, I.N., Zmud, J., Williams, T., 2019. Measures of baseline intent to use automated vehicles: A case study of Texas cities. Transportation Research Part F: Traffic Psychology and Behaviour 62, 66–77.
Sepasgozar, S.M., Hawken, S., Sargolzaei, S., Foroozanfa, M., 2019. Implementing citizen centric technology in developing smart cities: A model for predicting the acceptance of urban technologies. Technological Forecasting and Social Change 142, 105–116.
Shackel, B., 1981. The concept of usability. In: Bennet, J.L., Case, D., Sandelin, J., smith, M. (Eds.) Visual display terminals: usability issues and health concerns. Prentice Hall, New York, pp. 45–88.
Sharma, Yetton, Crawford, 2009. Estimating the Effect of Common Method Variance: The Method—Method Pair Technique with an Illustration from TAM Research. MIS Quarterly 33, 473.
Shaw, J., Docherty, I., 2014. The transport debate. Policy Press, Bristol, 236 pp.
Shaw, N., Sergueeva, K., 2019. The non-monetary benefits of mobile commerce: Extending UTAUT2 with perceived value. International Journal of Information Management 45, 44–55.

Shuhaiber, A., Mashal, I., 2019. Understanding users' acceptance of smart homes. Technology in Society 58, 101110.
Simsekoglu, Ö., Nayum, A., 2019. Predictors of intention to buy a battery electric vehicle among conventional car drivers. Transportation Research Part F: Traffic Psychology and Behaviour 60, 1–10.
Singh, Y.J., 2019. Is smart mobility also gender-smart? Journal of Gender Studies 29, 832–846.
Smith, G., Sochor, J., Karlsson, I.M., 2018. Mobility as a Service: Development scenarios and implications for public transport. Research in Transportation Economics 69, 592–599.
Soong, R., 2000. Early Adopters of Technological Innovations. TGI Mobile Systems. http://www.zonalatina.com/Zldata99.htm. Accessed 9 February 2021.
Souza, A.A. de, Sanches, S.P., Ferreira, M.A., 2014. Influence of Attitudes with Respect to Cycling on the Perception of Existing Barriers for Using this Mode of Transport for Commuting. Procedia – Social and Behavioral Sciences 162, 111–120.
Sovacool, B.K., 2017. Experts, theories, and electric mobility transitions: Toward an integrated conceptual framework for the adoption of electric vehicles. Energy Research & Social Science 27, 78–95.
Sovacool, B.K., 2020. Acceptance model based on the integrated conceptual framework.
Sovacool, B.K., Kester, J., Heida, V., 2019. Cars and kids: Childhood perceptions of electric vehicles and sustainable transport in Denmark and the Netherlands. Technological Forecasting and Social Change 144, 182–192.
Sprei, F., 2018. Disrupting mobility. Energy Research & Social Science 37, 238–242.
Stone, M., 1974. Cross-Validatory Choice and Assessment of Statistical Predictions. Journal of the Royal Statistical Society. Series B (Methodological) 36, 111–147.
Stone, R.N., Winter, F.W., 1987. Risk: Is it still uncertainty times consequences. Proceedings of the American Marketing Association, 261–265.
Stradling, S., 2006. Moving around: some aspects of the psychology of transport. Foresight Intelligent Infrastructure Systems Project.
Straub, D., Gefen, D., 2004. Validation Guidelines for IS Positivist Research. Communications of the Association for Information Systems 13.
Straub, D., Limayem, M., Karahanna-Evaristo, E., 1995. Measuring system usage: Implications for IS theory testing. Management science 41, 1328–1342.
Streukens, S., Leroi-Werelds, S., 2016. Bootstrapping and PLS-SEM: A step-by-step guide to get more out of your bootstrap results. European Management Journal 34, 618–632.
Stryker, N., 1989. The Shaping of Social Organization: Social Rule System Theory with Applications, by Tom R. Burns and Helena Flam, Sage Publications, London, 1987, 432 pp. Syst. Res. 6, 174.
Subramanian, G.H., 1994. A Replication of Perceived Usefulness and Perceived Ease of Use Measurement. Decision Sciences 25, 863–874.
Sue, V.M., Ritter, L.A., 2012. Conducting online surveys. Sage, Los Angeles.
Suki, N.M., Suki, N.M., 2017. Determining students' behavioural intention to use animation and storytelling applying the UTAUT model: The moderating roles of gender and experience level. The International Journal of Management Education 15, 528–538.
Sullivan, G.M., Feinn, R., 2012. Using Effect Size-or Why the P Value Is Not Enough. Journal of graduate medical education 4, 279–282.

Sulz, S., 2018. Cooperation in mobility: A critical analysis. Masterthesis, Friedrichshafen, 255 pp.

Šumak, B., Heričko, M., Pušnik, M., 2011. A meta-analysis of e-learning technology acceptance: The role of user types and e-learning technology types. Computers in Human Behavior 27, 2067–2077.

Šumak, B., Šorgo, A., 2016. The acceptance and use of interactive whiteboards among teachers: Differences in UTAUT determinants between pre- and post-adopters. Computers in Human Behavior 64, 602–620.

Susanto, T., Goodwin, R., 2010. Factors influencing citizen adoption of SMS-based eGovernment services. Electronic journal of e-government, 16.

Taherdoost, H., 2018. A review of technology acceptance and adoption models and theories. Procedia Manufacturing 22, 960–967.

Talmar, M., Walrave, B., Podoynitsyna, K.S., Holmström, J., Romme, A.G.L., 2020. Mapping, analyzing and designing innovation ecosystems: The Ecosystem Pie Model. Long Range Planning 53, 101850.

Tamilmani, K., Rana, N.P., Dwivedi, Y., Sahu, G.P., Roderick, S., 2018a. Exploring the Role of 'Price Value' for Understanding Consumer Adoption of Technology: A Review and Meta-analysis of UTAUT2 based Empirical Studies. Pacific Asia Conference on Information Systems.

Tamilmani, K., Rana, N.P., Dwivedi, Y.K., 2017. A Systematic Review of Citations of UTAUT2 Article and Its Usage Trends. In: Kar, A.K., Ilavarasan, P.V., Gupta, M., Dwivedi, Y.K., Mäntymäki, M., Janssen, M., Simintiras, A., Al-Sharhan, S. (Eds.) Digital Nations – Smart Cities, Innovation, and Sustainability, vol. 10595. Springer International Publishing, Cham, pp. 38–49.

Tamilmani, K., Rana, N.P., Dwivedi, Y.K., 2018b. Mobile Application Adoption Predictors: Systematic Review of UTAUT2 Studies Using Weight Analysis, in: Challenges and Opportunities in the Digital Era, Cham. 2018. Springer International Publishing, pp. 1–12.

Tamilmani, K., Rana, N.P., Dwivedi, Y.K., 2019a. Use of 'Habit' Is not a Habit in Understanding Individual Technology Adoption: A Review of UTAUT2 Based Empirical Studies, in: Smart Working, Living and Organizing, Cham. 2019. Springer International Publishing, pp. 277–294.

Tamilmani, K., Rana, N.P., Dwivedi, Y.K., 2020. Consumer Acceptance and Use of Information Technology: A Meta-Analytic Evaluation of UTAUT2. Information Systems Frontiers, 987–1005.

Tamilmani, K., Rana, N.P., Prakasam, N., Dwivedi, Y.K., 2019b. The battle of Brain vs. Heart: A literature review and meta-analysis of 'hedonic motivation' use in UTAUT2. International Journal of Information Management 46, 222–235.

Tamilmani, K., Rana, N.P., Wamba, S.F., Dwivedi, R., 2021. The extended Unified Theory of Acceptance and Use of Technology (UTAUT2): A systematic literature review and theory evaluation. International Journal of Information Management 57, 102269.

Tamor, M.A., Gearhart, C., Soto, C., 2013. A statistical approach to estimating acceptance of electric vehicles and electrification of personal transportation. Transportation Research Part C: Emerging Technologies 26, 125–134.

Tarigan, A.K., Bayer, S.B., Langhelle, O., Thesen, G., 2012. Estimating determinants of public acceptance of hydrogen vehicles and refueling stations in greater Stavanger. International Journal of Hydrogen Energy 37, 6063–6073.

Tarka, P., 2018. An overview of structural equation modeling: its beginnings, historical development, usefulness and controversies in the social sciences. Quality & quantity 52, 313–354.

Taylor, S., Todd, P., 1995a. Assessing IT Usage: The Role of Prior Experience. MIS Quarterly 19, 561.

Taylor, S., Todd, P.A., 1995b. Understanding information technology usage: A test of competing models. Information systems research 6, 144–176.

Tenenhaus, M., Amato, S., Esposito Vinzi, V., 2004. A global goodness-of-fit index for PLS structural equation modeling, 739 pp.

Teo, T., 2009. Is there an attitude problem? Reconsidering the role of attitude in the TAM. British Journal of Educational Technology 40, 1139–1141.

Teo, T., Fan, X., Du, J., 2015. Technology acceptance among pre-service teachers: Does gender matter? Australasian Journal of Educational Technology 31, 253–251.

Teo, T., Noyes, J., 2011. An assessment of the influence of perceived enjoyment and attitude on the intention to use technology among pre-service teachers: A structural equation modeling approach. Computers & Education 57, 1645–1653.

2019. The European Green Deal: COM(2019) 640 final. In: Communication from the commission.

The White House, 2016. FACT SHEET: President Obama's 21st Century Clean Transportation System.

Thompson, R.L., Higgins, C.A., Howell, J.M., 1991. Personal Computing: Toward a Conceptual Model of Utilization. MIS Quarterly 15, 125.

Toft, M., Schuitema, G., Thøgersen, J., 2014. Responsible technology acceptance: Model development and application to consumer acceptance of Smart Grid technology. Applied Energy 134, 392–400.

Triandis, H.C., 1977. Interpersonal behavior. Brooks/Cole Pub. Co.

Tsou, H.-T., Chen, J.-S., Chou, Y., Chen, T.-W., 2019. Sharing Economy Service Experience and Its Effects on Behavioral Intention. Sustainability 11, 5050.

Turetken, O., 2020. Deliverable D4.5 C-Mobile project: Final business models.

Tversky, A., Kahneman, D., 1974. Judgment under Uncertainty: Heuristics and Biases. Science 185, 1124–1131.

Uber, 2017. Written evidence submitted by Uber (MAS0015). Submission to the UK Parliament's Transport Committee inquiry into Mobility as a Service.

Ullman, J.B., Bentler, P.M., 2006. Structural Equation Modeling. In: Weiner, I.B. (Ed.) Handbook of psychology. Wiley, Hoboken, New Jersey.

Urban Audit, 2013. Modal Split in European Cities. http://www.urbanaudit.org/, Accessed 22 October 2021.

Urry, J., 2004. The 'System' of Automobility. Theory, Culture & Society 21, 25–39.

Urry, J., 2008. Climate change, travel and complex futures 1. The British journal of sociology 59, 261–279.

van Bree, B., Verbong, G., Kramer, G.J., 2010. A multi-level perspective on the introduction of hydrogen and battery-electric vehicles. Technological Forecasting and Social Change 77, 529–540.

van Cauwenberg, J., Bourdeaudhuij, I. de, Clarys, P., Geus, B. de, Deforche, B., 2019. E-bikes among older adults: benefits, disadvantages, usage and crash characteristics. Transportation 46, 2151–2172.

van den Heuvel, C., Kao, P.-J., Matyas, M., 2020. Factors driving and hindering business model innovations for mobility sector start-ups. Research in Transportation Business & Management 37, 100568.

van der Heijden, H., 2004. User Acceptance of Hedonic Information Systems. MIS Quarterly 28.

van der Laan, J.D., Heino, A., Waard, D. de, 1997. A simple procedure for the assessment of acceptance of advanced transport telematics. Transportation Research Part C: Emerging Technologies 5, 1–10.

van Noorden, R., Maher, B., Nuzzo, R., 2014. The top 100 papers. Nature 514, 550–553.

van Raaij, E.M., Schepers, J.J., 2008. The acceptance and use of a virtual learning environment in China. Computers & Education 50, 838–852.

Venkatesh, V., 2000. Determinants of Perceived Ease of Use: Integrating Control, Intrinsic Motivation, and Emotion into the Technology Acceptance Model. Information systems research 11, 342–365.

Venkatesh, V., Bala, H., 2008. Technology Acceptance Model 3 and a Research Agenda on Interventions. Decision Sciences 39, 273–315.

Venkatesh, V., Brown, S.A., Maruping, L.M., Bala, H., 2008. Predicting different conceptualizations of system use: the competing roles of behavioral intention, facilitating conditions, and behavioral expectation. MIS Quarterly, 483–502.

Venkatesh, V., Davis, F.D., 2000. A Theoretical Extension of the Technology Acceptance Model: Four Longitudinal Field Studies. Management science 46, 186–204.

Venkatesh, V., Morris, M.G., 2000. Why Do not Men Ever Stop to Ask for Directions? Gender, Social Influence, and Their Role in Technology Acceptance and Usage Behavior. MIS Quarterly 24, 115.

Venkatesh, V., Morris, M.G., Davis, G.B., Davis, F.D., 2003. User Acceptance of Information Technology: Toward a Unified View. MIS Quarterly 27, 425–478.

Venkatesh, V., Thong, J., Xu, X., 2016. Unified Theory of Acceptance and Use of Technology: A Synthesis and the Road Ahead. Journal of the Association for Information Systems 17, 328–376.

Venkatesh, V., Thong, J.Y.L., Xu, X., 2012. Consumer acceptance and use of information technology: extending the unified theory of acceptance and use of technology. MIS Quarterly, 157–178.

Verhaeghen, P., Salthouse, T.A., 1997. Meta-analyses of age–cognition relations in adulthood: Estimates of linear and nonlinear age effects and structural models. Psychological Bulletin 122, 231.

Verplanken, B., Aarts, H., van Knippenberg, A., Moonen, A., 1998. Habit versus planned behaviour: a field experiment. The British journal of social psychology 37, 111–128.

Vlassenroot, S., Brookhuis, K., Marchau, V., Witlox, F., 2010. Towards defining a unified concept for the acceptability of Intelligent Transport Systems (ITS): A conceptual analysis based on the case of Intelligent Speed Adaptation (ISA). Transportation Research Part F: Traffic Psychology and Behaviour 13, 164–178.

Vlassenroot, S., Molin, E., Kavadias, D., 2011. What drives acceptability of Intelligent Speed Assistance? (ISA). European Journal of Transport and Infrastructure Research 2011, 256–273.

Vogt, J., 2022. Smart mobility role model interview, Web.

Waard, D. de, van der Hulst, M., Hoedemaeker, M., Brookhuis, K.A., 1999. Driver Behavior in an Emergency Situation in the Automated Highway System. Transportation Human Factors 1, 67–82.

Walter, S., Ulli-Beer, S., Wokaun, A., 2012. Assessing customer preferences for hydrogen-powered street sweepers: A choice experiment. International Journal of Hydrogen Energy 37, 12003–12014.

Wang, N., Tang, L., Pan, H., 2018a. Analysis of public acceptance of electric vehicles: An empirical study in Shanghai. Technological Forecasting and Social Change 126, 284–291.

Wang, Y., Douglas, M.A., Hazen, B.T., Dresner, M., 2018b. Be green and clearly be seen: How consumer values and attitudes affect adoption of bicycle sharing. Transportation Research Part F: Traffic Psychology and Behaviour 58, 730–742.

Wang, Y., Gu, J., Wang, S., Wang, J., 2019. Understanding consumers' willingness to use ride-sharing services: The roles of perceived value and perceived risk. Transportation Research, Part C: Emerging Technologies, 504–519.

Wang, Y., Wang, S., Wang, J., Wei, J., Wang, C., 2018c. An empirical study of consumers' intention to use ride-sharing services: using an extended technology acceptance model. Transportation 47, 397–415.

Wang, Y.-S., Wu, M.-C., Wang, H.-Y., 2009. Investigating the determinants and age and gender differences in the acceptance of mobile learning. British Journal of Educational Technology 40, 92–118.

Wanke, T., 2022. Smart mobility role model interview, Web.

Warshaw, P.R., 1980. A new model for predicting behavioral intentions: An alternative to Fishbein. Journal of Marketing Research 17, 153–172.

Wassiliwizky, E., Menninghaus, W., 2021. Why and how should cognitive science care about aesthetics? Trends in Cognitive Sciences 25, 437–449.

Weichhart, P., 2003. Gesellschaftlicher Metabolismus und Action Settings. Die Verknüpfung von Sach- und Sozialstrukturen im alltagsweltlichen Handeln.

Weimann, J., 2013. Umweltökonomik: eine theorieorientierte Einführung. Springer, Berlin.

Weston, R., Gore, P.A., 2006. A Brief Guide to Structural Equation Modeling. The Counseling Psychologist 34, 719–751.

Wetzels, M., Odekerken-Schroder, G., van Oppen, C., 2009. Using PLS Path Modeling for Assessing Hierarchical Construct Models: Guidelines and Empirical Illustration. MIS Quarterly 33, 177–195.

Wheatley, M., 2002. Turning to one another: Simple conversations to restore hope to the future. The Journal for Quality and Participation 25, 8.

Wheaton, B., Muthen, B., Alwin, D.F., Summers, G.F., 1977. Assessing Reliability and Stability in Panel Models. Sociological Methodology 8, 84.

Whittle, C., Whitmarsh, L., Haggar, P., Morgan, P., Parkhurst, G., 2019. User decision-making in transitions to electrified, autonomous, shared or reduced mobility. Transportation Research Part D: Transport and Environment 71, 302–319.

Wieker, H., 2016. Connected and automated driving, October 2016, Venice.

Wieker, H., 2017. iKoPA Deliverable D1v1: Requirements Analysis and System Architecture. htw saar.
Wieker, H., 2022. Smart mobility role model interview, Web.
Wieker, H., Eckhart, W., Vogt, J., Fünfrocken, M., 2014a. CONVERGE – ITS design like Internet, Helsinki, Finnland SIS 09 – Towards deployment of cooperative systems.
Wieker, H., Schulz, W., Geis, I., Maringer, S., Fünfrocken, M., 2014b. Institutional Role Models in ITS – CONVERGE as an example of a new approach for business models, Presented at ITS World Congress, Detroit.
Wietschel, M., 2011. Socio-political issues of electromobility. Fraunhofer Institute for System Innovation, Karlsruhe.
Williams, M., Rana, N., Dwivedi, Y., Lal, B., 2011. Is UTAUT really used or just cited for the sake of it? A systematic review of citations of UTAUT's originating article. European Conference on Information Systems.
Willits, F., Theodori, G., Luloff, A., 2016. Another Look at Likert Scales. Journal of Rural Social Sciences 31.
Wintermann, O., 2021. Digital change and sustainability. https://makronom.de/arbeitswelt-wie-der-digitale-wandel-zu-mehr-nachhaltigkeit-fuehren-kann-39169. Accessed 16 February 2022.
Wold, H., 1980. Model construction and evaluation when theoretical knowledge is scarce: Theory and application of partial least squares. In: Evaluation of econometric models. Elsevier, pp. 47–74.
Wold, H., 1982. Soft Modeling: The Basic Design and Some Extensions. In: Joreskog, K.G., Wold, H. (Eds.) Systems under Indirect Observations, vol. 2. North-Holland, Amsterdam, pp. 1–54.
Wolff, S., Madlener, R., 2019. Driven by change: Commercial drivers' acceptance and efficiency perceptions of light-duty electric vehicle usage in Germany. Transportation Research Part C: Emerging Technologies 105, 262–282.
Wolsink, M., 2018. Social acceptance revisited: gaps, questionable trends, and an auspicious perspective. Energy Research & Social Science 46, 287–295.
Woodcock, A., Romer Christensen, H., Levin, L., 2020. TInnGO: Challenging Gender Inequality in Smart Mobility. PIS 66, 1–5.
Wu, J., Liao, H., Wang, J.-W., Chen, T., 2019. The role of environmental concern in the public acceptance of autonomous electric vehicles: A survey from China. Transportation Research Part F: Traffic Psychology and Behaviour 60, 37–46.
Wu, J.-H., Wang, S.-C., 2005. What drives mobile commerce? Information & Management 42, 719–729.
Xu, Z., Zhang, K., Min, H., Wang, Z., Zhao, X., Liu, P., 2018. What drives people to accept automated vehicles? Findings from a field experiment. Transportation Research Part C: Emerging Technologies 95, 320–334.
Xue, X., Chen, Z., Wang, S., Feng, Z., Duan, Y., Zhou, Z., 2020. Value Entropy: A Systematic Evaluation Model of Service Ecosystem Evolution. IEEE transactions on service computing 1, doi: https://doi.org/10.1109/TSC.2020.3016660.
Yang, H., Yoo, Y., 2004. It's all about attitude: revisiting the technology acceptance model. Decision Support Systems 38, 19–31.
Ye, J., Zheng, J., Yi, F., 2020. A study on users' willingness to accept mobility as a service based on UTAUT model. Technological Forecasting and Social Change 157, 120066.

Yen, B.T., Mulley, C., Burke, M., 2018. Gamification in transport interventions: Another way to improve travel behavioural change. Cities 85, 140–149.

Yen, Y.-S., Wu, F.-S., 2016. Predicting the adoption of mobile financial services: The impacts of perceived mobility and personal habit. Computers in Human Behavior 65, 31–42.

Yetano Roche, M., Mourato, S., Fischedick, M., Pietzner, K., Viebahn, P., 2010. Public attitudes towards and demand for hydrogen and fuel cell vehicles: A review of the evidence and methodological implications. Energy Policy 38, 5301–5310.

Ylitalo, J., 2009. Controlling for Common Method Variance with Partial Least Squares Path modeling: A Monte Carlo Study. undefined.

Zeithaml, V.A., 1988. Consumer Perceptions of Price, Quality, and Value: A Means-End Model and Synthesis of Evidence. Journal of Marketing 52, 22pp.

Zhang, T., Tao, D., Qu, X., Zhang, X., Lin, R., Zhang, W., 2019. The roles of initial trust and perceived risk in public's acceptance of automated vehicles. Transportation Research Part C: Emerging Technologies 98, 207–220.

Zheng, Z., Xie, S., Dai, H.-N., Chen, W., Chen, X., Weng, J., Imran, M., 2020. An overview on smart contracts: Challenges, advances and platforms. Future Generation Computer Systems 105, 475–491.

Zhou, F., Zheng, Z., Whitehead, J., Washington, S., Perrons, R.K., Page, L., 2020. Preference heterogeneity in mode choice for car-sharing and shared automated vehicles. Transportation Research Part A: Policy and Practice 132, 633–650.

Zhou, J., Rau, P.-L.P., Salvendy, G., 2014. Older adults' use of smart phones: an investigation of the factors influencing the acceptance of new functions. Behaviour & Information Technology 33, 552–560.

Zimmer, R., Zschiesche, M., Hölzinger, N., 2009. The role of trust and familiarity in risk communication, Independent institute for environmental concern, Berlin.

Zoellick, J.C., Kuhlmey, A., Schenk, L., Schindel, D., Blüher, S., 2019. Amused, accepted, and used? Attitudes and emotions towards automated vehicles, their relationships, and predictive value for usage intention. Transportation Research Part F: Traffic Psychology and Behaviour 65, 68–78.

Printed in the United States
by Baker & Taylor Publisher Services